THE SOCIAL PSYCHOLOGY OF SCIENCE

THE CONDUCT OF SCIENCE SERIES

Steve Fuller, Ph.D., Editor
DEPARTMENT OF COMMUNICATION
UNIVERSITY OF PITTSBURGH

The Social Psychology of Science
WILLIAM R. SHADISH AND STEVE FULLER

Philosophy of Science and Its Discontents, Second Edition
STEVE FULLER

The Scientific Attitude, Second Edition
FREDERICK GRINNELL

Politics and Technology
JOHN STREET

THE SOCIAL PSYCHOLOGY OF SCIENCE

EDITED BY

William R. Shadish
Steve Fuller

THE GUILFORD PRESS
New York London

A Division of Guilford Publications, Inc.
72 Spring Street, New York, NY 10012

Printed in the United States of America

This book is printed on acid-free paper.

Last digit is print number: 9 8 7 6 5 4 3 2 1

Library of Congress Cataloging-in-Publication Data

The Social psychology of science / edited by
William R. Shadish and Steve Fuller
p. cm. — (Conduct of science series)
Includes bibliographical references and index.
ISBN 0-89862-021-X
1. Science—Philosophy. 2. Science—Methodology.
3. Science—Social aspects. 4. Science—Psychological aspects.
I. Shadish, William R. II. Fuller, Steve, 1959– . III. Series.
Q175.S6393 1994
501—dc20 93-18825
CIP

Life is but one continual course of instruction. The hand of the parent writes on the heart of the child the first faint characters which time deepens into strength so that nothing can efface them. —R. HILL

With deep love and gratitude,
to my parents,

William R. Shadish, Sr.
and Maryjane A. Cartmell

W. R. S.

To my ever charitable wife,

Sujatha Raman

S. F.

Contributors

Teresa M. Amabile, Ph.D., Department of Psychology, Brandeis University, Waltham, Massachusetts

Donald T. Campbell, Ph.D., Department of Social Relations, Lehigh University, Bethlehem, Pennsylvania

Thomas D. Cook, Ph.D., Departments of Psychology and Sociology, Northwestern University, Evanston, Illinois

Steve Fuller, Ph.D., Department of Communication, University of Pittsburgh, Pittsburgh, Pennsylvania

Michael E. Gorman, Ph.D., Department of Technology, Culture, and Communications, School of Engineering and Applied Science, University of Virginia, Charlottesville, Virginia

Arie W. Kruglanski, Ph.D., Department of Psychology, University of Maryland, College Park, Maryland

William Lawless, Ph.D., Department of Mathematics, Paine College, Augusta, Georgia

Norman Miller, Ph.D., Department of Psychology, University of Southern California, Los Angeles, California

Vicki E. Pollock, Ph.D., Department of Psychology, University of Southern California, Los Angeles, California

Robert Rosenthal, Ph.D., Department of Psychology, Harvard University, Cambridge, Massachusetts

Robert Rosenwein, Ph.D., Department of Sociology and Anthropology, Lehigh University, Bethlehem, Pennsylvania

William R. Shadish, Ph.D., Department of Psychology, Memphis State University, Memphis, Tennessee

Joan E. Sieber, Ph.D., Department of Psychology, California State University, Hayward, California

Stephen Turner, Ph.D., Department of Philosophy, University of South Florida, Tampa, Florida

Ryan D. Tweney, Ph.D., Department of Psychology, Bowling Green University, Bowling Green, Ohio

Ron Westrum, Ph.D., Department of Sociology, Eastern Michigan University, Ypsilanti, Michigan

John M. Wilkes, Ph.D., Social Science and Policy Studies, Worcester Polytechnic Institute, Worcester, Massachusetts

Preface

This book is an errant offspring of work started by the psychology of science group at Memphis State University beginning in about 1985. That group produced an edited book (Gholson, Shadish, Neimeyer, & Houts, 1989) that overviewed the psychology of science, discussing such matters as its justifications, its accomplishments, its role in metascience, and its agenda. But our treatment of the social psychology of science in that book was inadequate in several respects. Only two chapters were devoted to the social psychology of science. One of those (Shadish's on science evaluation) was not obviously social psychological in content, and neither chapter was by a social psychologist trained primarily in psychology. One aim of the present book, therefore, is to remedy that lacuna. Toward that end we have brought together contributors dominated by social psychologists writing on mainstream topics from psychology's view of that field. Their contributions to this book cover topics that range from social cognition to interpersonal processes, and begin to outline a conceptual and research agenda for the social psychology of science.

This book is secondarily the offspring of a desire to explore whether some versions of the psychology of science might be at least partly compatible with the social views of science so prevalent in science studies today. Steve Fuller, a philosopher of science strongly associated with science studies, has been active in this regard. After hearing about the Memphis group's work, he invited us to present papers at the 1987 Sociology of the Sciences Yearbook conference (Fuller, DeMey, Shinn, & Woolgar, 1989). His own work in "social epistemology" (Fuller 1988, 1992, 1993a, 1993b) is largely devoted to revealing and removing the artificial barriers that stand in the way of a joint psychology–sociology of science enterprise.

Part of the goal of this book is to remedy a simple misunderstanding—many members of the science studies community view the psychology of science as isomorphic with the cognitive psychology of science. This is understandable given all the attention that cognitive science has received in science studies in recent years. But it is also unfortunate: While few psychologists would want to eliminate cognition as a topic of study, equally few would want to limit psychological contributions solely to cognition. Further exacerbating the problem, cognitive contributions are sometimes held up as exemplars of why psychology of science must necessarily oppose sociology of science. Indeed, such tensions exist between the cognitive and the social, but they are nowhere near as extreme as the unfortunate tensions that seem to exist between psychology and sociology of science. Hence the present book tries to find some rickety but still passable bridges over the gap between the more individually focused aspects of psychology such as cognition and the more socially oriented aspects of science studies generally. These are bridges that must be cautiously traversed from time to time by those of us who believe that science is profoundly shaped by both social and psychological aspects. Sometimes the bridge will collapse, and sometimes not, but it's hard to tell without walking it. Fortunately, because it is just a thought experiment, it won't kill us to try.

However, we should remember that the social psychology of science is itself just one subfield in a larger psychology of science whose shape is still emerging. Since our 1989 book, for example, the members of the psychology of science group at Memphis State have largely gone their separate ways, exploring developments in such diverse areas as general psychology of science (Shadish & Neimeyer, 1989), cognitive psychology of science (Gholson & Houts, 1989; Houts & Gholson, 1989), behavioral psychology of science (Houts & Haddock, 1992), scientometric analyses of psychology (Neimeyer, Baker, & Neimeyer, 1990), history of psychology (Neimeyer, 1990), and science evaluation (Shadish, 1989). These diverse directions, along with the present book, reflect the diversity of psychology itself. We should not expect the psychology of science to be just one thing; the social psychology of science is just one small part of the picture.

A special thanks go from the editors to the contributors to this volume, many of whom went out of their way to present their research so as to appeal to a broader science studies audience. The editors wish to thank Seymour Weingarten and Peter Wissoker of The Guilford Press for taking on this extensive project and being so patient about its development. Also, thanks go to Fuller's research assistant, Jim Collier, who did much of the word processing and some of the inevitable editorial patch-up work. Shadish wishes to thank the Center for Applied Psychological Research for financial support that contributed to our ability to produce this book. The Center is funded by a grant from the State of Tennessee Centers of Excellence program. Thanks also go to Subira Ajanaku, Mary Church, Steve DePaola, and Paul Wilson for providing him with logistical help. Finally, Shadish would like to thank his dear wife, Betty Duke Shadish—time goes by quickly, but there's nobody I'd rather watch it go by with than you.

William R. Shadish
Memphis State University

Steve Fuller
Virginia Polytechnic Institute

References

Fuller, S. (1988). *Social epistemology.* Bloomington: Indiana University Press.

Fuller, S. (1992). Epistemology radically naturalized. In R. N. Giere (Ed.), *Cognitive models of science: Minnesota studies in the philosophy of science* (Vol. 15, pp. 427–459). Minneapolis: University of Minnesota Press.

Fuller, S. (1993a). *Philosophy of science and its discontents* (2nd ed.). New York: Guilford Press.

Fuller, S. (1993b). *Philosophy, rhetoric, and the end of knowledge: The coming of science and technology studies.* Madison: University of Wisconsin Press.

Fuller, S., De Mey, M., Shinn, T., & Woolgar, S. (Eds.). (1989). *The cognitive turn: Sociological and psychological perspectives on science.* Dordrecht, The Netherlands: Kluwer.

Gholson, B., & Houts, A. C. (1989). Toward a cognitive psychology of science. *Social Epistemology, 3,* 107–127.

Gholson, B., Shadish, W. R., Neimeyer, R. A., & Houts, A. C. (Eds.). (1989). *Psychology of science: Contributions to metascience.* Cambridge, England: Cambridge University Press.

Houts, A. C., & Gholson, B. (1989). Brownian notions: One historicist philosopher's resistance to psychology of science via three truisms and ecological validity. *Social Epistemology, 3,* 139–146.

Houts, A. C., & Haddock, C. K. (1992). Answers to philosophical and sociological uses of psychologism in science studies: A behavioral psychology of science. In R. N. Giere (Ed.), *Cognitive models of science: Minnesota studies in the philosophy of science* (Vol. 15, pp. 367–399). Minneapolis: University of Minnesota Press.

Neimeyer, R. A. (1990). George A. Kelly: In Memoriam. *History of Psychology, 22,* 3–14.

Neimeyer, R. A., Baker, K. D., & Neimeyer, G. J. (1990). The current status of personal construct theory: Some scientometric data. In R. Neimeyer & G.

Neimeyer (Eds.), *Advances in personal construct theory* (Vol. 1, pp. 3–22). Greenwich, Connecticut: JAI Press.

Shadish, W. R. (1989). Science evaluation: A glossary of possible contents. *Social Epistemology, 3,* 189–204.

Shadish, W. R. & Neimeyer, R. A. (1989). Contributions of psychology to an integrative science studies: The shape of things to come. In Fuller, S., De Mey, M., Shinn, T., & Woolgar, S. (Eds.). (1989). *The cognitive turn: Sociological and psychological perspectives on science* (pp. 13–38). Dordrecht, The Netherlands: Kluwer.

Contents

PART III
DISCUSSION

PART IV
FOR FURTHER READING

Part I

INTRODUCTION, BACKGROUND, AND OVERVIEW

CHAPTER 1

Social Psychology of Science: A Conceptual and Empirical Research Program

WILLIAM R. SHADISH
STEVE FULLER
MICHAEL E. GORMAN

WITH
TERESA M. AMABILE
ARIE W. KRUGLANSKI
ROBERT ROSENTHAL
ROBERT E. ROSENWEIN

The psychology of science has finally arrived. Kuhn (1970) first broadly legitimized psychological studies of science. In particular, Kuhn (1970) noted the questions to which paradigm crises lead "demand the competence of the psychologist even more than the historian" (p. 86). Of course, psychological studies of science long preceded Kuhn (Campbell, 1989; Fisch, 1977). Still, prior to his work, such analyses were most often ignored in science studies. After Kuhn, even the least psychologically inclined scholars noted the contributions of psychological factors to science. Sometimes their aim was only to deny the relevance of those contributions to their own projects, but the debate itself helped legitimize the topic. Some scholars, however, valued the exploration of such factors in science:

> Perhaps the most important of Kuhn's insights was his emphasis on the socially embodied and socially sustained character of the traditions within

> which virtually all scientific work is and has been carried out. It is the activities of *persons* [italics in original], not disembodied ideas, concepts, theories, or "research programs," that constitute research traditions. (Rudwick, 1985, p. 445)

During the last 50 years, many topics have been studied under the rubric of psychology of science (Shadish & Neimeyer, 1989). Hence, the psychology of science is not one thing but a mixture of past and emerging studies. Early psychology of science focused mostly on studying the personal characteristics of individual scientists. Eiduson and Beckman's (1973) edited volume is an example, reporting the results of various personality, demographic, and biographical studies of scientists compared to other professions. Often, such studies were combined with an interest in scientific creativity (e.g., Chambers, 1964), with the latter viewed sometimes as a personality trait that scientists possessed to a greater or lesser extent and sometimes as a result facilitated by other personality traits. The underlying notion seemed to be that scientists might be qualitatively different from other people in such characteristics as intelligence, persistence, or sociability, and that more creative scientists might also be different from their less creative counterparts. If such tendencies could be identified, of course, this might illuminate how science works, and improve science by devising means to select personnel who might benefit most from scientific training. Problematically, clear trends were difficult to identify, and few psychologists seemed interested enough in the project to devote much time to it. Moreover, many philosophers of science ignored such work as part of the logic of discovery and so out of their province. This form of the psychology of science has received little extended attention in science studies, except occasionally as a foil against which others could present their own ideas. Nonetheless, such work continues (e.g., Diesing, 1991; Hart, 1982; Helmreich, Spence, & Pred, 1988; Johnson, Germer, Efran, & Overton, 1988; Simonton, 1989; Wilkes, Chapter 10, this volume). Much of it can be found in the ongoing "Psychology of the Scientist" series in the journal *Psychological Reports* (e.g., Frank & Rickard, 1988). Moreover, the topic of personality is in some respects closely related to the topic of this book, the social psychology of science (SPS), inasmuch as two of the main social psychology journals—*Journal of Personality and Social Psychology* and *Personality and Social Psychology Bulletin*—have long covered both topics. Space constraints preclude us from devoting much attention to the relationship of personality to social psychology, but a full treatment is needed.

A second, more prominent line of work has ensued from the mutual interests of cognitive psychologists and naturalistic epistemologists (Heyes, 1989). Part of the impetus for this work comes from epistemologists who are interested in such cognitive matters as the representation of knowledge in the mind (e.g., Fodor, 1981). They often see cognitive science as necessary for extending their work. Part of the impetus comes from philosophers of science. Houts and Haddock (1992) suggest that this is because philosophers see cognitive psychology as reasonably compatible with their own views, while providing philosophers with an empirical alternative to the onslaughts of the "Strong Programme" sociologists who often have "well-entrenched antipathies, in particular towards the notion of cognition" (Woolgar, 1989, p. 201) that is prominent in both psychology and philosophy. Part of the impetus comes from psychologists who have vigorously pioneered in the study of scientific cognition (Tweney, Doherty, & Mynatt, 1981). Consequently, the cognitive psychology of science has received much attention in science studies recently (Fuller, DeMey, Shinn, & Woolgar, 1989), so much so that cognitive psychology is not distinguished from psychology more generally in some authors' writings (Nickles, 1989; Woolgar, 1989). Perhaps the main finding in all this work is the limited reasoning power of human beings in general, presumably extending to scientists to some degree at least. Recent summaries of this area can be found in Faust (1984) and Fuller et al. (1989).

The cognitive psychology of science has largely remained distinct from the study of personal characteristics of scientists, but at least two areas of overlap exist. One is the extension of earlier interests in motivation (Maini & Nordbeck, 1973) to cognition through the rubric of "hot" or motivated cognition (e.g., Kruglanski, Chapter 5, this volume; Thagard, 1989). The other is studies of cognition and creativity, as with Kulkarni and Simon's (1988) computer simulation of the cognitive processes involved in discovering the urea cycle. In addition, both of these areas—cognition and personal characteristics—focus heavily on the scientist as an individual, paying less attention to social interactions, roles, institutions, and cultural factors that may influence science. This common focus is, of course, viewed as objectionable in some parts of science studies, especially in those parts of the sociology of science that have worked for years to eliminate "cognitive" explanations of science (Woolgar, 1989)—although we will demonstrate shortly that what sociologists mean by cognitive is not always the same as what psychologists mean by the term.

A third line of work consists of efforts to integrate past research in the psychology of science by summarizing its accomplishments and speculating on its themes. Early examples are by Maslow (1966) and Singer (1971), but neither seemed to spark sustained interest in the topic, nor did Grover's (1981) overview receive the attention it deserved. Mahoney's (1976) work, on the other hand, is one of the classics of the psychology of science, perhaps in part because it focused so much on psychological research to show the shortcomings of the scientific process, thereby challenging more rationalistic accounts of science (Ziman, 1984). More recently, Gholson, Shadish, Neimeyer, and Houts (1989) edited an overview of the psychology of science that aimed to review its history, explore its warrant, suggest its content, and set an agenda for theory and research. As a result of these integrative efforts, the psychology of science is emerging as a specialty in science studies (Gorman, 1992a). Its status is far from secure either epistemically or sociologically, however. In the latter case, the trappings of specialty status, such as specialty journals or chapters in annual reviews, have yet to appear—perhaps for the best, at least to the extent that further fragmentation of science studies is undesirable. In the former case, psychologists have yet to forward a statement of their agenda that is either widely accepted within the psychological community or of wide interest in science studies generally (Turner, Chapter 13, this volume). To judge from other specialties, however, it may take some time for such a statement to emerge. When writing about the empirical relativist program in the sociology of scientific knowledge (SSK), for example, Collins (1983) noted that "programmes, it appears, are best generated out of practice and example, and best proclaimed and systematized with at least some degree of hindsight" (p. 86). Generating programs in the psychology of science may take some time given how new its identity seems to be. We do not yet have much hindsight.

The present volume on SPS follows in all these traditions of applying psychological theories and methods to the conceptual and empirical study of science. More specifically, this book aims to anthologize various contributions to SPS, to examine some of the underpinnings of SPS, to explore the sorts of concepts and studies that SPS might suggest, and to criticize its potential and shortcomings. Further, while recognizing the existence of social psychological studies in the sociological tradition (Westrum's newsletter on SPS is a prominent example; see also Ziman, 1984, chap. 15; Westrum, Chapter 12, this volume), this book draws almost exclusively from psychology to discover what the unique contributions of psychological social psychologists to science studies might be. SPS differs from most previous psychology of science in incorporating an explicitly social component into psychological analysis. We explicitly endorse the assertions of macro- and microsociologists of science that social factors and social interactions are crucial components in science. However, to the extent that such sociologists wish to completely exclude cognitive explanations of science in favor of wholly social explanations (Woolgar, 1989), psychologists generally disagree (Houts & Haddock, 1992, is a partial exception, but for very different reasons than sociologists). Rather, this book explores the possibility that both social and cognitive factors can profitably be retained in science studies; in fact, SPS is dedicated to that project (cf. Fuller, 1992).

Efforts at integrating the sociological and psychological research traditions in social psychology have been few and far between (a recent and important exception is Doise, 1987, who writes in the Belgian tradition of Henri Tajfel). For an initial sense of the incommensurability between the two perspectives, consider how the phenomenon of "ideology" is defined. On the one hand, sociologists tend to treat ideology as linguistic behavior, or "legitimatory resource," used to justify particular actions within a larger normative framework, regardless of whether the ideology actually explains the individual's behavior, or, sometimes, even whether the individual believes that the ideology explains his/her behavior. On the other hand, psychologists interpret ideology as part of an individual's motivational structure, specifically a false or distorted belief that causally contributes to the individual's acting the way he/she does. As we descend from the macro- to the microlevel of inquiry, a useful way to see the obstacles to integrating sociological and psychological social psychology is to examine the treatment of "norm" in the two disciplinary traditions.

At the level of small-group interaction, sociological social psychology has tended toward constructivist and behaviorist interpretations of norms. In other words, a norm does not exist prior to what the members of the group do; indeed, the norm is none other than the degree of regularity and coordination that the members of the group can elicit from each other. Why is there one norm rather than another in a given situation? Typically, the sociologist wants to answer this question by examining the consequences that the norm-governed behavior has on the character of the group and its members. If the group survives and its members flourish (to their own satisfaction), the norm is presumed to be rational, or "functional," to use the sociologist's term of art. Individuals acquire a personal sense of the norm in a process that is variously called socialization, internalization, or acculturation. In each case, the individual learns the norm as a way of talking about behavior in which he/she has already been engaging to some degree. This approach unites schools as otherwise different as symbolic interactionism, ethnomethodology, labeling theory, and even certain forms of phenomenology.

By contrast, psychological social psychology has tended in a more cognitivist, even prescriptivist, direction, such that a norm may exist even when many or most people fail to abide by it. Consider two general psychological findings: the need for reducing "cognitive dissonance" and the inadequacy of beliefs and desires to explain and predict action. Both presuppose the existence of certain norms that people are said to violate routinely, respectively, the norm that one should be cognitively consistent and the norm that one's actions should be explicable in terms of one's beliefs and desires. Despite the violations, psychologists are comfortable calling these things norms because people have become concerned when they are informed that they have deviated from them. But where do these norms come from? Some psychologists, clearly influenced by the linguist Noam Chomsky (e.g. 1975), believe that the norms innately reside in the individual's subconscious, whose depths may someday be fully plumbed by neuroscience. The subconscious character of these norms is supposed to account for the intuitive sense that people have about the existence of such norms and yet their failure to have the norms adequately inform their behavior. Some psychologists have suggested that people can be trained to act more in accordance with the norm—or, as a philosopher might put it, to become "more rational." Other psychologists, such as those in the Tversky–Kahneman–Nisbett–Ross experimental tradition, make less robust ontological commitments. They are less interested in physically locating the norms in people's heads than in trying to account for how people can function so well in spite of violating the norms. This has led to the postulation of various "heuristics," rules of thumb that clearly do not produce optimal behavior in all cases but that nevertheless enable people to function effectively.

The tension between the sociological and psychological traditions should now be coming into focus. To put the contrast in boldest relief, in a way that will illuminate the existing tensions between contemporary microsociology and psychology of science: *For every situation in which a psychologist is prone to see an existing norm being violated, a microsociologist is likely to see a new norm being constructed.* What the psychologist sees as inconsistency and discrepancy, the sociologist sees as flexibility and innovation. Conversely, where the psychologist sees consistency and predictability, the sociologist sees mechanical routine. Although there is more overlap between the two disciplinary perspectives than these stark contrasts suggest (e.g., Aaron Cicourel, 1973 who espouses a cognitivist version of ethnomethodology), nevertheless the inherent tension between sociological and psychological social psychology becomes especially pronounced when the people being studied are scientists. For scientists are supposed to be *both* consistent and flexible, people adept at knowing when to abide by a certain norm of reasoning and when to abide by some other norm, people who know when to debate possible solutions to a standing problem and when to get down to the work of pursuing one such solution.

Kuhn (1977) originally cast this tension represented by the two disciplinary perspectives—that optimal mix of tradition and innovation—as *the essential tension* that distinguishes the scientific enterprise from other social practices. Nickles (1980) and Laudan (1984) have since pursued the philosophical side of this theme. From a social psychological standpoint, this tension can be effectively displayed by simulating the history of a tradition in the laboratory, specifically by studying successive "generations" of subjects who have been asked to transmit a norm. To put the matter simply, a norm that is intersubjectively constructed by an original group of subjects can acquire the status of an objective standard when it is used to evaluate the behavior of newcomers, especially if the newcomers are unable to renegotiate the terms of the norm. This experimental tradition, originating with Asch's (1951) studies of conformity and continuing in Jacobs and Campbell's (1961) work, is discussed more by Fuller (Chapter 3, this volume) and Campbell (Chapter 2, this volume).

For these sorts of reasons, then, we do not expect this volume to resolve the differences between the sociological and psychological traditions in social psychology. Our goal is more modest, to lay some groundwork on which debates about those differences can proceed in a more informed way. Our chapter begins this work with an extended exploration of conceptual, substantive, and methodological issues in SPS. The first part of the chapter briefly discusses nine conceptual features of SPS that help to clarify its nature and suggest how its emphases may differ from those of other science studies disciplines. The second part of the chapter reviews a wide array of social psychological theories and findings, suggests some substantive hypotheses about science that might be implied, and notes illustrations of those findings in science. The third part of the chapter describes some methodological issues in SPS. Some of those issues are metamethodological, having to do with the warrant for SPS's endeavors—one such issue concerns identification of exactly what it is that the social science of SPS is investigating, and the second issue concerns the reflexivity problems involved when science is used to study science. Other issues concern more specific methods that SPS is likely to use—including experimentation, quasi-experimentation, and meta-analysis—and problems raised by these methods, in particular the problem of causal generalization. The chapter then concludes with a brief overview of the rest of the book.

Conceptual Issues in Social Psychology of Science

The general character of SPS reflects certain prototypical features, partly derived from the parent discipline of psychology and partly emergent in ways we cannot fully know yet. These characteristics become apparent throughout the course of this book, but we anticipate them briefly here.

The first characteristic of SPS is that the "individual scientist in social context" is usually the basic unit of analysis. Tweney (see Chapter

14, this volume) discusses this in more detail. He notes that the engine that drives science is not just the individual or the social. It is both interacting together, the social psychological. Of course, some part of science occurs inside the skin of the scientist, and the skin makes for as clear a separation between inside and outside as one could wish for. However, what goes on outside—the social—is not just one other external cause such as, say, gravity. In particular, the social context is capable of being modified by the psychological, and, in the cases of most interest to social psychology, this modification is a continuous interaction, an ongoing inseparable process in which the proper unit is "person in context." Hence, we outline the rudimentary beginnings of a theory that moves from individual cognition to social cognition through other topics to end up at intra- and intergroup relations.

This explicit recognition of the importance of the social aspects of science should ease at least some concerns about the psychology of science. Nickles (1989), for example, rightly criticized some psychology of science as appearing to "wrench 'phenomena' out of the rich social contexts which largely define them, strip them of everything necessary to regiment them into uniform, homogeneous domain of natural objects, and then proceed to seek general causal laws" (p. 239). Where Nickles errs is in assuming that all psychology of science must hew this line as much as he perceives the *cognitive* psychology of science doing so. SPS recognizes the social embeddedness (indeed, the partial social constitution) of all the behaviors, thoughts, and emotions of the individual; and it tries in its own way to reflect this in its theories and methods. In its particular theory and method choices, SPS will inevitably incur some problems. Nickles (1989), for instance, claims that the key sense in which science is social is its "conception of human beings as members of moral and cultural communities rather than as mere natural objects or machines" (pp. 242–243). Psychology as a discipline, and therefore SPS, has often done poorly in recognizing this sense of the social, particularly in recent years (Cook, Chapter 15, this volume). But this simply means that we have much to learn from each other. SPS is not inherently inconsistent with this conception of the scientist as member of a moral and cultural community (Sieber, Chapter 9, this volume).

However, social psychologists do not wish to emphasize context to the point of losing the individual scientist's role. So, while this emphasis on the individual is cousin to a focus on the individual in the microsociology of science as embodied in, say, *Laboratory Life* (Latour & Woolgar, 1979), SPS will emphasize the individual more than sociologists would do. Giere (1989) criticizes science studies that move too far from the scientist as the causal locus of scientific activity, but he goes so far as to say that the social is no more than the sum of the individuals. Social psychologists agree with the first part of Giere's criticism but hold judgment in abeyance on the second part. In fact, social psychologists of science consider the question, for example, whether groups are qualitatively different from the sum of individuals as an empirical matter, not one to be decided beforehand (Gorman, Chapter 4, this volume).

Still, what all these cases (SPS, microsociology, and Giere) share is the centrality of the scientist to the enterprise we call science. To elaborate, Fuller (1993) points out that "the *locus* of causal powers (i.e., individual organisms) must be distinguished from the causal powers themselves (i.e., genes)" (p. 95). This statement has two implications for SPS. One is that the focus on the individual scientist is at least partly a matter of the "situatedness" of SPS in a parent discipline with particular strengths and interests in studying individuals. That is what psychologists do well. The fact that the individual scientist also may happen to be the causal locus of scientific activity is a happy coincidence that affords SPS the chance to contribute in some interesting and unique ways, but it does not make it the only or the best way to view science. Too narrow a focus on the individual, in fact, causes exactly the problems that socially oriented critics cite—but this is, of course, exactly why a *social* psychology of science is needed. The other implication is that SPS needs some method of moving from the scientist as causal locus to an understanding of the causal powers themselves. This is, of course, exactly where SPS's emphasis on experimentation comes in handy (Berkowitz & Donnerstein, 1982; Greenwood, 1989); we return to this several times later in this chapter.

An important corollary question is raised by the notion of "individual scientist in context": Does the individual in question have to be a human being, or can, say, a "computer in context" be substituted? Shrager and Langley (1990), point out that computational simulations are stripped of the social context in which science is embedded. It was this very lack of embeddedness that Slezak (1989) cited as a strength rather than a weakness. In brief, Slezak (1989) argues that: the fact that computer programs such as BACON can discover scientific laws independent of any kind of social context refutes the new SSK and, by implication, undermines the rationale for a social psychology of science. If machines can discover, why do we need to study all of this social context?

The answer is that the machine discovers because we recognize its act as a discovery—a phenomenon that could be studied by attribution theorists. Perhaps an analogy to the controversy over whether chimpanzees can learn language will help. Critics claimed that what the chimps learned was how to imitate their trainers; it was the trainers who decided that these imitative gestures constituted language. In the same way, it is the programmers who decide that when BACON stops, it has discovered

Kepler's laws. The program knows nothing about orbits, or planets; it merely discovers an arbitrary relationship between columns of numbers, similar to the sorts of relationships subjects try to find on numerical problems like Wason's (1960) 2–4–6 task. This sort of "data-driven discovery" does not emulate Kepler's path. Kepler had to squeeze the data out of the astronomical theory designed by his boss, Tycho Brahe, and come up with a new mental model of the solar system. Both students and program were given the sort of data that emerge clearly only after the controversy has been settled.

Shrager and Langley (1990) make the larger point that "the social organization of science in the laboratory and in broader contexts has a major influence on the nature of science, and future modeling efforts should move toward incorporating aspects of this structure" (p. 18). Apparently, these computational cognitive psychologists recognize the need for SPS. Gorman (Chapter 4, this volume) presents ideas about how to conduct experimental simulations that will address questions about embeddedness. In view of all this, we cannot preclude the possibility that some interest in computer models of science might emerge in SPS. On the other hand, this has not been an issue with which most social psychologists have been concerned. In that sense, computer modeling of the scientist has shown little sign of being prominent in SPS. Further, the outline later in this chapter of the substantive domains covered by SPS suggests that the problems with which such a model would have to cope to be comprehensive would be so daunting as to be unlikely to succeed. Nonetheless, psychologists are, on the whole, quite empirically inclined, and so the most commonly endorsed SPS answer to this question of the role of computers would probably be that the answer is best decided empirically rather than by fiat in advance.

A second characteristic of SPS is that it acknowledges the likelihood that both rational and irrational factors may play a role in the production, adoption, and change of scientific beliefs. Understood psychologically, rationality tends to imply reasoning in ways that are logical and consistent. Irrationality implies errors in such reasoning, and also factors that increase the likelihood of errors or that override rational factors, such as motivation, emotion, conformity, or wrong applications of heuristics. But these and similar terms—in particular, cognitive versus noncognitive—have been used in more diverse ways in science studies. In the philosophy of science, *cognitive* was originally a broad-gauged term to cover virtually anything that was not irrational or emotive. Thus, for the logical positivists, the following terms often appeared as synonyms for *cognitive*: *rational, meaningful, empirical, testable,* all generally being taken as descriptors of particular products of science. In each case, physics was presented as the paradigm case. Indeed, in early formulations of positivism, such as Ayer's (1936), it seemed that one needed to emulate physics in order to utter something with "cognitive significance." Ironically, then, Rorty (1979) is quite right in thinking that *cognitive* has often had little substantive meaning in its own right, as it just seemed to refer to whatever the positivists approved of. However, Fuller (1993) has noted that, in recent years, philosophers have increasingly used cognitive to refer to the process, as opposed to the products, of knowing. Moreover, unlike terms like epistemic, philosophers tend to use cognitive to describe actual mental processes, not simply processes that conform to some criteria of rationality. This shift in usage is more congenial to psychologists, and has enabled philosophers such as Dudley Shapere and Nancy Nersessian to recast in psychological garb much of the internal history of science that has traditionally interested philosophers. However, these self-styled "cognitive historians" (Nersessian, 1987) have been criticized by cognitive psychologists (e.g., DeMey, 1992) who study the history of science for themselves and believe that the cognitive historians are too beholden to philosophical prejudices to utilize the full range of findings from social and developmental research on cognition (cf. Fuller, 1991).

As a result, the science studies literature now contains much ambiguity in the usage of these terms. In some circles, rationality refers to whatever mechanism drives the success of science in generating knowledge of the world, whether or not that mechanism refers to logical variables. If we were to find that the generation of blind variations is a crucial mechanism, as suggested in some evolutionary epistemology, then that generation comes to be referred to as part of the rationality of science even though there may be no traditional logic to the process. In this sense, rationality refers more to rationale than to rational. Similarly, some in science studies use cognitive in psychologically recognizable ways, as when Woolgar (1989) uses cognitive to refer the mental operations of scientists. Others use cognitive in less psychologically familiar ways, usually to mean anything associated with how we construct true knowledge. Collins (1983) refers to "a cognitive goal—discovering the structure of a new compound, research the nature of a food processing operation" (p. 86). Rudwick (1985) refers to socially defined focal problems as a cognitive category (p. 427), and to the cognitive topography of a group of scientists as the way in which judgments of the validity of knowledge claims depend on the perceived competence of those making the claims (p. 425). Woolgar (1989) extends his description of what is cognitive to "the idea that mental or other inner processes enable the rightful perception of an already pre-existent world" (p. 206). In all three of these cases, the meaning of cognitive has subtly shifted from mental processes to mental processes that give correct answers—clearly more than psychologists typically mean by the term.

Third, SPS explanations can help clarify both beliefs we currently regard as true and those we currently regard as false; however, the same psychological explanation need not apply to true and false beliefs. This characteristic of SPS is partially sympathetic to Bloor's (1976) symmetry principle: that the same type of causes would explain both true and false beliefs of scientists. The latter may indeed prove to be the case, but SPS does not begin with the assumption that symmetry is necessarily true. Instead, SPS adopts what might be called an agnostic principle: We cannot know all the causes of true and false beliefs prior to studying the matter empirically. Simple, uniform symmetry may not emerge as the conclusion. Instead, different causes may emerge for true and false beliefs, both less rational than some philosophical depictions; beliefs may have multiple causes, both rational and irrational; or the answer may depend on the kind of belief. The possibilities are endless and should not be mandated in advance.

Fourth, to judge from what we know about the social psychology of humans generally, the behavior of scientists will never be simple or easy to predict. Rather, their behavior will reflect complex interactions among and exceptions to whatever general tendencies are present. In this sense, SPS is frustrating to those who want to reduce science to some relatively simple rules that would apply uniformly to explain or to improve science—for example, that science can be explained by its rationality, or that science is the product of social interactions. One aim of this chapter is to disabuse the reader of this notion. SPS does not provide simple "laws" that govern science. To the contrary, science is the product of very complex intra- and interpersonal variables that may not behave in a uniform way over time even in the same individual, that interact with other personal and situational variables to produce different results in different situations, that always seem to have exceptions, and that are highly contextual. What produces good science under some conditions may be irrelevant under others. Simple pronouncements are intuitively appealing, of course, but they are most often wrong if taken as literal descriptors of local instances of science. This does not mean that it is wrong to look for simplicity, or to reduce complex matters to simple statements on occasion. To judge from what we know of social policy generally, such oversimplification is sometimes necessary because policymakers almost never have the latitude to use their resources to dictate highly complex contingencies where some scientists are told to do something in some situations but not in others, and other scientists are given the opposite instructions in different situations (Cook & Shadish, 1986; Shadish, Cook, & Leviton, 1991). Scientists traditionally have too much discretion in the conduct of their work, and such policies would ultimately be so complex as to be largely unimplementable and unmanageable by policymakers (Shadish, 1989a, 1989b). In this sense, the main causal effects suggested by experiments are of more policy use than complex interactions between those causes and other variables even though those main effects oversimplify what we might know about the practice of science.

Fifth, SPS assumes that scientists first were and still are just people, subject to the same intra- and interpersonal processes as nonscientists. This assumption is partly dependent on the observation that all scientists first were not scientists and so presumably carry over some precareer psychological characteristics into their scientific career. It is also partly a matter of convenience in that this assumption allows SPS to take advantage of its basic research data base on ordinary people in generating hypotheses about science. It is also partly a matter of evidence, in the very limited sense that psychological studies of scientists tend to mimic findings from ordinary people in some essential ways (Griggs & Ransdell, 1986; Suls & Fletcher, 1983). But ultimately, SPS will treat as an empirical question whether results from nonscientists do generalize, whether scientists may eventually prove to have or be subject to unique characteristics as a result of their career training or situation (Ziman, 1984), or whether ordinary people behave in extraordinary ways when put into situations deemed scientific. These issues of generalization are crucial to the SPS enterprise, so we discuss them more extensively toward the end of this chapter.

Sixth, SPS rejects strong epistemological relativism as a necessary assumption. Just as occurred when discussing rationality or cognition, however, psychologists need to be aware of how terms like *realism* and *relativism* are used in science studies. Those who describe themselves as relativists are not usually denying the existence of a "real" world (but see Woolgar, 1983, 1989). Rather, strong epistemological relativists usually question whether the real world exerts any influence on knowledge, and in particular on theory choice in science. For example, Ashmore (1989) says that "the principled nonexistence of correct, best, or even better versions in any absolute sense—is, as you will realise, to embrace a form of *relativism*" (p. 1), and Laudan (1990) (himself not a relativist) describes strong relativism as "the thesis that the natural world and such evidence as we have about that world do little or nothing to constrain our beliefs" (p. viii). Strong relativists recognize that some constraints on scientific beliefs do seem to exist, but they describe those constraints only as social. Woolgar and Ashmore (1988) say that "scientific and technical knowledge is not the rational/logical extrapolation from existing knowledge, but the contingent product of various social, cultural and historical processes" (p. 1). SPS tends to reject strong epistemological relativism—while not rejecting weak relativism, the thesis that "there are some circumstances in which the available evidence fails to warrant a choice between certain rival perspectives" (Laudan, 1990, p. 56). In this weak sense, we are all relativists, wanting to retain a role for both social and empirical influences on scientific beliefs.

Pushing the relativist argument even further, Woolgar and others resolutely believe that cognition and the external world are not factors that can be brought in to explain science but rather are themselves phenomena to be explained in terms of social construction. Thus, they pursue questions such as: How do scientists (philosophers, psychologists) come to believe that there are distinctly "cognitive" processes? How do they come to believe that a world "external" to their local practices has made a difference to an experimental outcome? How do they come to believe that science is a "successful" enterprise. Jeff Coulter (1979) is the social psychologist (influenced by the Oxford philosopher Gilbert Ryle) who has influenced Woolgar here.

Adjudicating the differences among all these positions is notoriously difficult (Laudan, 1990). But despite all the attention garnered by strong relativist positions in recent years, weak relativism seems to be making somewhat of a comeback. In the history of science, for example, Rudwick's (1985) analysis of the relative roles of social and evidentiary matters in the resolution of the Devonian controversy in geology in the early 1800s suggested the following:

> It is possible to see the cumulative empirical evidence in the Devonian debate, *neither* as having determined the result of the research in any unambiguous way, as naive realists might claim, *nor* as having been virtually irrelevant to the result of the social contest on the agonistic field, as constructivists might maintain. It can be seen instead as having had a *differentiating* effect on the course and outcome of the debate, *constraining* the social construction into being a limited, but reliably and indefinitely improvable, representation of a natural reality. (p. 456; italics in original)

Similarly in the sociology of science, Latour (1988) says:

> Let us go back to the world, still unknown and despised. If you sneer at this claim and say "this is going back to realism," yes it is. A little relativism takes one away from realism; a lot brings one back. (p. 173)

(Earlier, Latour [1987] had made similar interpretations.) Weak relativism may be the dominant position in philosophy of science as well (Laudan, 1990), and it sometimes is seen under the guise of cognate terms such as *local realism* (Nersessian, 1987). SPS joins this chorus. Campbell (Chapter 2, this volume) presents SPS's case for weak relativism best, embracing much of what we have learned from SSK and the strong program within that tradition, but calling for what Chubin and Restivo (1983) refer to as a weak program in which we explore the procedures that might work to improve the validity of scientific knowledge if the world could indeed affect our beliefs. Campbell refers to his approach as neither realist nor relativist, but as competence enhancing—that scientists work to enhance the competence with which theories refer to their presumed ontological referents. Campbell (Chapter 2, this volume) also describes Jacobs and Campbell's (1961) autokinetic experiment as an early, fallible, but still paradigmatic exemplar for SPS's position that strong epistemological relativism is not necessarily warranted by experimental results.

Seventh, SPS helps clarify the micromediational processes of science, filling an important gap in current science studies. Some of the most important recent works in sciences studies can be interpreted as moving ever closer (substantively) to requiring a social psychology of science such as the one outlined in this chapter. Latour (1987), for example, pays close attention to transactions among scientists (and their "black boxes"), but often ends by saying *that* they occur, not *how* they occur. For instance, he says that successful scientists are those who win trials of attributions, but he says little about the sorts of variables that mediate the win. Similarly, Rudwick's (1985) analysis of the Devonian controversy can be read as a treatise about how the interplay of evidence, individual scientists, and intra- and intergroup processes led to changes in scientific beliefs, leading up to a social psychological theory of scientific belief change without explicating the processes that might underlie such a theory. What is offered in this chapter is not intended to be a complete theory of science, of course; it does not tap such important features as the political economy of science, for example. Rather, the claim is simply that this chapter helps fill a gap between more purely cognitive and social conceptions of science, by suggesting some micromediational interfaces between the two sides.

Eighth, SPS is empirical in the traditional social scientific sense, closest in substantive focus to the microsociological tradition of examining social interactions by observation (Knorr-Cetina & Mulkay, 1983), but closer methodologically to the quantitative approaches of macrosociology. More than these other areas of science studies, however, SPS uses experimentation, including analogue experiments in the laboratory and field and quasi-experimental field studies, as a legitimate means of exploring questions about science. The last third of this chapter discusses experimentation more fully, including the warrant for generalization from such work, but we should note a few points here. SPS's emphasis on experimentation is not exclusive. Nearly all of the chapters by social psychologists in this book are based in part on basic laboratory experiments, but the mix of methods is still broad. A close inspection of the methods reported in the chapters in this volume reveals that the experiments are embedded in multimethod programs of research, which is how it should be. Still, SPS's use of the experiment reflects the importance that social psychologists place on rigor of causal inference. While practitioners of SPS admit to shortcomings of the experiment, they also note that the experiment improves the warrant for causal-sounding assertions in science studies in some ways that other methods do not. Here is also where the observation comes home to roost about the difference between the scientist as causal locus versus the causal powers that may or may not flow partially through the scientist. The experiment elucidates the latter, using the former as the vehicle (Berkowitz & Donnerstein, 1982; Greenwood, 1989; see also the section Causal Generalization, later in this chapter).

Ninth, by providing information about what affects the conduct and outcome of science, the accumulated knowledge (always fallible and defeasible) from such work gives a locally based, quasi-normative function to SPS in the following form: "If you wish to have the following outcome, your chances will be increased if you do the following things." It is locally based because the inferences are limited to the historical situatedness of the actual experiment, yielding locally and historically valid, probabilistic conclusions rather than universal, deterministic laws, the generality of which must be explored carefully, tediously, and extensively through multiple methodologies that compensate for the weaknesses of the experiment in yielding generalizable knowledge. It is quasi-normative because the premise ("If you wish . . . ") is not itself justified by psychological research. By virtue of these causal focuses, however, SPS will have policy interests and implications. Some of that policy relevance will be at the governing level, to help state and federal science policymakers understand what works and does not work in broad strokes. But that relevance will also be for the local scientist who recognizes him/herself in some social psychological description and tries to take advantage of newly perceived strengths or to remedy potential weaknesses.

The above nine points have just begun to touch on the conceptual underpinnings of SPS, developing none of them in depth and omitting discussion of other important issues. In the next section, these points are substantively applied as we begin to develop hypotheses that social psychologists might want to test in science studies.

Substantive Issues in SPS

Here we review a wide array of theories and empirical findings in social psychology, organized into sections on social cognition, attribution theory, attitudes and attitude change, social motivation, social conformity and social influence, and intergroup relations. Even so, each area is discussed only briefly, and we omit many relevant areas studied by social psychologists, such as altruism, language use, leadership, power, mass communication, and deviance. Nonetheless, the hope is that we will begin to define the broad scope of what SPS might look like when fully developed. Although here we use examples from science as much as possible, the focus is primarily on presenting the theories so others can use them to generate studies of science. In places, therefore, the discussion is abstract; but subsequent chapters in this book provide more detailed examples of how some of the theoretical points might be used in the study of science.

At the same time, however, social psychologists of science must remember that the science studies community considers certain substantive issues as primary. Among these are the following:

1. Where do scientific ideas come from?
2. What factors play a role in the formation, revision, and persistence of scientific beliefs?
3. How are ideas diffused throughout the scientific community?
4. Do we need to provide separate accounts for beliefs that are currently regarded as true versus those currently regarded as false?

Fortunately, at least some of these problems have been historically important in social psychology, albeit with respect to humans generally rather than scientists in particular.

Nonetheless, to the extent that philosophers and sociologists are currently disputing these matters, or related ones such as the rational or social nature of scientific knowledge, psychologists find themselves unable to avoid these debates, a part of ideological and disciplinary disputes that are not of their making. The result is that at the same time that psychologists attempt to define their contributions to science studies, they are asked to take sides in existing disputes with which they are often only partly familiar; they find themselves befriended or attacked by sociologists or philosophers who, for one reason or another, may sense either congenial views or a threat to their interests; and they find their work characterized in terms they might not use, with which they might disagree, or which they simply do not understand. In these interdisciplinary debates, psychologists have a lot of catching up to do. The psychology of science stretches our limited knowledge of post-Kuhnian developments in science studies. On the other hand, scholars in science studies need to know more about recent developments in psychology—especially social psychology.

Each of the following sections is largely a summary of recent reviews by social psychologists, especially reviews from two of the classic sources in the field—the *Handbook of Social Psychology*, and the *Annual Review of Psychology*. While we impose our own organization and insert commentary on how these topics might relate to science studies, these sections are not original contributions to the social psychological literature. In places, in fact, the summaries simply paraphrase the reviews of these social psychologists. Rather than cite previous review authors constantly, we have cited the main sources at the start of each section; the interested reader is strongly encouraged to refer to these original sources for many more details.

Some Historical Background

[Excellent reviews of this topic are Allport (1985) and Hilgard (1987).] The origins of social psychology are diverse, found in a host of different intellectual traditions in France, Germany, England, Russia, and the United States, and in concerns such as hedonism, power, sympathy, imitation, suggestion, crowd behavior, and the notion of a group mind. The emergence of social psychology as a specialty is often marked by the appearance of two texts in 1908, one by the sociologist E. A. Ross and the other by the psychologist William McDougall. Psychologists have generally taken the definition of social psychology from Allport's (1985) statement that social psychology is "an attempt to understand and explain how the thought, feeling, and behavior of individuals are influenced by the actual, imagined, or implied presence of others" (p. 3)—one could paraphrase this statement for a concise definition of SPS, substituting "scientists" for "individuals." Psychologists also view social psychology as a consistent part of psychology generally, with the special implication that the center of study is always the person.

Social psychology is social in the sense that it recognizes the social influences on individuals that come from such sources as situational cues or the expectations of others. An example is the famous "autokinetic effect" experiments of Sherif (e.g., 1936; Campbell, Chapter 2, this volume). In that research, subjects exposed to a stationary light in a dark room see the light as moving, but their judgments of how much the light moves are influenced by the judgments they hear from other people. Social psychology is also psychological in its recognition of the diversity of human psychologies that seem to exist within similar social situations—that humans define social situations in different ways. In contrast, sociological social psychology tends to view its subject matter more from the perspective of the group. This is illustrated in the definition given by Ellwood (1925): "Social psychology is the study of social interaction. It is based upon the psychology of group life" (p. 16).

Cognitive Psychology and Social Psychology

[Much of the material that follows in this section is taken from Markus and Zajonc (1985).] In the last decade, the connection between psychology and science studies has perhaps been the strongest at the juncture of cognitive psychology and epistemology, especially naturalized epistemology. Hence, cognitive psychology is in many respects the most familiar psychological topic to those outside psychology. Cognition is currently so dominant a topic in social psychology that two leading researchers recently said: "Social psychology and cognitive social psychology are today nearly synonymous. The cognitive approach is now clearly the dominant approach among social psychologists, having virtually no competitors" (Markus & Zajonc, 1985, p. 137). Some of this work is not distinctly social, but rather closely resembles cognitive research done by other experimental psychologists. Because the cognitive psychology of science is not our main interest in this book, we comment on it briefly here and then quickly turn to more distinctly social psychological topics.

It may be difficult to distinguish cognitive research by social psychologists from mainline cognitive psychology when social psychologists study how information is represented, processed, stored, and retrieved for use—those events that putatively take place "inside" the individual's skin. Much of this attention is on the study of "schemas," those knowledge structures presumed to be "conceptually related representations of objects, situations, events, and sequences of events and actions" (Markus & Zajonc, 1985, p. 143). Individuals construct their own schemas through experience and social learning. That is, "events are not passively registered on the perceptual apparatus. They are organized in categories shaped by past experiences, and they take on their meaning as part of an active, constructive process in dealing with reality" (Jones, 1985, p. 83)—presumably in scientists, too. Rudwick (1985) says that during the Devonian controversy, "the relatively stable core of this body of practical knowledge was learned ostensively by newcomers to the science, while more seasoned practitioners were continually refining and modifying the application of even the most well-established categories" (p. 446). Because schemas mediate between "reality" and higher-order cognitive processes, and because we have no direct access to reality except through schemas, knowledge is always subject to whatever biases are reflected in the schemas we construct. Schemas influence initial coding and organization of stimuli, retention and retrieval of stimuli, subsequent judgments and inferences, and subsequent behavior. Schemas tend to be quite stable.

Much of this sort of cognitive psychology is familiar to the reader. For example, the inferential, judgmental, and heuristic biases outlined by such authors as Tversky and Kahneman (1974) or Nisbett and Ross (1980) have received much attention in the science studies literature, no doubt because of their all-too-obvious relevance to the conduct of science:

> Weak points include willingness to rely on small, biased samples; overreliance on initial data or vivid instances; too much protection of a theory against disconfirming data; failure to recognize and question background assumptions; too much exploration near an initial disconfirmed hypothesis and not enough exploration of more distant hypothesis space (anchoring bias); too much reliance on an analogy and failure to recognize disanalogies. (Diesing, 1991, p. 253)

It is possible to train people to do better in all these instances, but only to a point. Remedying these errors is a slow and unsure process at best, and some errors such as the fundamental attribution error (see the section Attribution Theory) seem mostly immune to training.

However, we often forget that many of these are functional heuristics that only become biases on those relatively infrequent occasions when people misapply them. The availability heuristic, for instance, suggests that people tend to estimate how often things occur in the world according to how easily they come to mind. Often cognitive availability accurately reflects frequency in the world, but availability can lead to bias when, for example, the sheer repetition of a statement increases its perceived truth value. Diesing (1991) speculates that confirmation bias occurs in part because of availability bias—the scientist's own theory is the one most available to him/her to explain the results. Other research suggests that people are inclined to perceive regularity and structure, which is functional as a means of finding stable cause-and-effect relationships to help us navigate the world. We also perceive regularity even where there is none—vividly demonstrated in studies of how researchers readily construct interpretions of results from statistical analyses of random data (Armstrong & Soelberg, 1968; Shaycoft, 1970).

Schemas do not have unimpeded power to constrain the nature of information processing. Some stimuli, for example, are particularly salient and so receive proportionately larger amounts of our attention than might otherwise be the case. Further, while much has been made of confirmation biases in the cognitive literature, evidence suggests that even the strongest schemas cannot overcome directly inconsistent stimulus information (Locksley, Borgida, Brekke, & Hepburn, 1980). This allows for the likelihood that the "data" really can speak to scientists in some independent fashion, at least some of the time. Rudwick (1985) notes that during the Devonian controversy, "even over the contentious and crucial issue of the alleged unconformity beneath the Culm strata, De la Beche's improved field evidence did eventually convince Sedgwick and Murchison, however reluctantly, *against* [italics in original] their theoretical expectations" (p. 455). In general, then, the biasing influence of judgment heuristics can easily be exaggerated:

> The review of biases and heuristics shows that none of them necessarily reflect the irrationality of human cognition in general or social cognition in particular, although there are several examples of quite faulty human problem solving and inference. The review also shows that none of the biases or heuristics have been demonstrated to be free of extracognitive factors and that if the purposes, intentions, and particular circumstances of the cognizer are taken into account, the picture that emerges is not always that of a misguided creature but often of one who is willing to suffer a few misses and false alarms for the sake of overall greater cognitive efficiency and general adaptation to a capricious environment. As was noted with respect to schemas, a failure to consider the question of these biases and heuristics in the larger context—to ignore the goals of the perceiver and the nature of the problem and of the surrounding situation—can easily lead to unrealistic and unwarranted fears about the power of biases and heuristics to distort the inference process. (Markus & Zajonc, 1985, p. 196)

Since our focus is on the more distinctive contributions of social psychology, we will not elaborate on these strictly cognitive matters. However, their superficial relevance to science studies is clear. Specifically, any psychology of science that invokes an active form of cognition must also have at least some constructivist component that arises in acknowledging that scientists construct schemas through which they interact with the world they study. A scientist's schemas may be based on such sources as formal scientific theories, expectations formed in past research, or interactions with colleagues, as well as schemas drawn from everyday life about social, economic, and political matters. Perhaps most relevant to science, "experts tend to have well-developed . . . cognitive structures for thinking about the issues" (Tesser & Shaffer, 1990, p. 484), and well-developed structures can be particularly resistant to change. The most general implications of all this are probably that scientific theory and behavior tend toward stability, given the structural and operational stability of schemas, and that the cognitive processes of scientists inevitably result in some distortion of the stimuli being studied.

Another implication is that any cognitive biases and limitations of scientists are telling against cognitively naive accounts of both philosophers and sociologists. After listing a host of cognitive liabilities of scientists, for example, Fuller (1993) notes:

> A point that often goes unnoticed in the recital of this litany is that our cognitive fallibilities are sufficiently deep to cut equally against standard internalist and externalist models of rationality. In other words, not only are the "philosophical" accounts affected, but so too are the "sociological" accounts that portray the scientist as making key epistemic decisions on the basis of political or economic interest, for the experiments seem to show that human judgment is no better when applied to self-centered matters (e.g., weighting personal utilities) than when applied to more self-detached ones (e.g., weighting hypothesis probabilities). (p. 106)

Of course, this is not to imply that such cognitive biases preclude any constructive cognitive contribution to science—only that these cognitive features sometimes fail us, equally so for all our decisions.

Social Cognition

[This section relies heavily on Higgins and Bargh (1987), Markus and Zajonc (1985), Schneider (1991), and Sherman, Judd, and Park (1989).] Some social psychologists make little distinction between cognition and social cognition (Hastie & Carlston, 1980). Others see large differences, with the most common delimiter being that the latter is about cognitions pertaining to social stimuli, including people, social events, and the social situations in which people and events are embedded. Markus and Zajonc (1985) claim that the distinction is important because social factors introduce four features that are not necessarily present in cognition about nonsocial matters. First is the more intense and regular participation of "hot" factors such affect and motivation, factors that are exacerbated as the target person and the cognizer interact with each other in ways that nonsocial targets rarely do. This was clearly obvious in Rudwick's (1985) analysis of the Devonian controversy:

> Open conflict or antagonism may only be sporadic, but there is an underlying structural conflict or field of competitive argument throughout . . . . "Pray *keep your temper*," Sedgwick had to instruct Murchison at one point, knowing how easily his collaborator's hasty anger could lessen the force of their case. (p. 437)

Second, we tend to categorize social targets in functional terms, especially terms that have implications for ourselves—for example, whether the target is one of our ingroup or a member of an outgroup will influence our judgments of the target and our allocation of resources to the target (see the section Intergroup Relations). Third, our social perceptions have consequences for the target. When some consensus develops that a scientist has made a discovery, for example, great rewards can acrue to him/her. In addition, the target can be oneself. This special case of self-perception clearly distinguishes social from nonsocial cognition, since in no other case are subject and object the same. Finally, social cognition and social communication are extensively involved with each other, providing unique access to at least part of social cognition.

An especially "social" stimulus is another person. Characteristics such as sex, race, age, and physical appearance are particularly salient in influencing judgments. Even in describing themselves, people are likely to appeal to and develop characteristics that differentiate themselves from others in the particular social context (McGuire & McGuire, 1982). In science, this might be a mechanism for generating diversity in theory as scientists try to differentiate their own work from that of others. Mulkay (1988), for example, points out that exact replications rarely occur in science in part because scientists want some room to display their own originality. Rudwick (1985) speculates that motivations to individuate one's contributions also played a particularly important role in the Devonian controversy.

Festinger (1954) postulated that other people serve as a basis for comparison about whether our own thoughts, emotions, and behavior are normative; and he also noted that such comparisons ought to increase the more ambiguous the situation is. For instance, Kuhn (1970) observed that the social sciences were less developed than the physical sciences, which should lead to greater uncertainty among social scientists on matters of procedures, data interpretation, and research directions. Combining social comparison theory with Kuhn's observation, Suls and Fletcher (1983) proposed that such uncertainty should lead to social scientists seeking out others' opinions more often than physical scientists do. They found this to be the case when judged on the extent to which consultations with colleagues were acknowledged in footnotes of publications. Collins (1985) noted a similar social comparison process in his studies of replication. In the case of the TEA laser, criteria for knowing whether it had been replicated in another laboratory were clear—either the laser vaporized concrete or it did not. In the case of gravitational wave detectors, however, no such clear-cut criterion existed for knowing whether the detector had been replicated, so that scientists could only look to each other to know—a social comparison process.

Just as there are prototypical cognitive biases, certain social inference heuristics also bias our inferences about social phenomena. For example, the *false consensus* bias suggests that people tend to believe that others are like themselves in beliefs or preferences. A host of *self-serving biases* suggest that we associate ourselves (and our ingroup) with positive things more so than we associate with other people (or with an outgroup). For example, many studies suggest that people attribute more credit to themselves for a joint project, possibly because one has more memory of one's own contributions than of those of one's collaborators. Many coauthors on a collaborative paper think that they made the most important contributions, and so deserve first (or higher) authorship. An example is the debate between Rudolf Diesel and his colleague over who really deserved credit for inventing and developing the Diesel engine (Latour, 1987). Such tendencies include recalling one's involvement in incidents in a favorable light. During the Devonian controversy in the 1800s, one of the main protagonists, Roderick Murchison, recalled his own involvement in a certain discovery more favorably than retrospective evidence suggests he should have. Rudwick (1985) describes it as follows:

> In retrospect he constructed a romantic myth around what he claimed had been a "Eureka! moment" in 1831, when near Builth he had found the conformable junction between the Old Red Sandstone and the Transition strata. He forgot or concealed the fact that this conformable relation was already well known, that the river Wye near Builth was not even the first place he himself had seen the junction, and that at the time he had paid no special attention to it. He tried to eliminate Lewis's priority by claiming that, even before seeing the local geologist's home ground, he himself had not only found the junction but had realized its significance. His mythical moment of discovery became the retrospective starting point for a research project he had not in fact envisaged at the time. (pp. 76–77)

And after even more time had passed:

> Murchison recounted what was now the established myth of its genesis in such a way that it too supported his emphasis on "the language of nature." It was, he claimed, "merely by seeing the letters of the alphabet spread out before him in a cabinet" that Lonsdale had solved the Devonian problem. (p. 385)

One of the strengths of Rudwick's (1985) analysis of the Devonian controversy is the detail it provides about social context—the political, economic, societal, organizational, and interpersonal settings in which the controversy occurred. Unfortunately, one of the least studied areas in social cognition is the effects of such social contexts on information processing. In social cognition research as it is typically conducted, subjects rarely expect to have to confront the target of their judgments, nor need they be concerned with being accountable or creating a good impression, or with how the other person is reacting to them, or with the social, economic, or political consequences of their actions. But available evidence suggests such matters greatly influence information processing:

> The act of understanding is not an individual achievement; in large part it is a consequence of a social process. . . . One cannot, then, determine the nature or the function of a given cognitive structure until it has been established that its influence is invariant across stimuli, goals, and social context. On the basis of the research thus far, this type of invariance seems to be the exception rather than the rule. (Markus & Zajonc, 1985, p. 174)

That is, the influence of cognitive structure changes in many important ways over situation; SPS predicts this but has not yet sufficiently explored its implications in practice.

The dynamic, changing nature of social cognitions has, however, been the topic of much research. Those dynamics often involve extracognitive factors such as affect, motivation, and action, often leading to a characterization of this approach as "hot" cognition in contrast to the "cold" cognition theories more typical of information-processing approaches. This sort of work has a cyclical history in social psychology. The 1940s and 1950s saw the domination of hot theories like Heider's (1946) balance theory or Festinger's (1957) theory of cognitive dissonance. To oversimply, the basic notion behind the latter two theories was that people dislike contradictions among cognitions, affects, behavior, people, or any combination of these. When they experience an imbalance or dissonance between, say, two cognitions, or between their thinking and some stimulus, they are motivated to change something to relieve the contradiction.

Presumably such accounts point to a mechanism by which scientific theories are adopted and changed, leading scientists to try to resolve inconsistent theories or contradictions between data and hypotheses. But like any mechanism, things often go awry. In the classic laboratory example of dissonance theory, subjects who performed an extremely boring and tedious task for a long time and little reward later reported themselves as being rather interested in the task compared to subjects given a larger reward. The notion is that subjects simply changed their cognitions to suit behaviors they had already done. Analogies to this process occur often in scientific theory testing, as scientists experience informational inconsistency, disconfirmed expectations, and insufficient justification for past actions. In such cases, the pressure to change a cognition solely to reduce the dissonance can be high. The resolution of such situations sometimes goes just as awry, especially since individuals actively seek information that reduces this dissonance, and avoid information that increases it (see the example of the physics student, Holt, in the section Social Conformity and Social Influence). Festinger and Hutte (1954) found that subjects felt dissonance about relationships in a group when people they liked happened to dislike each other. Presumably, then, scientists are similarly motivated. For example, it should be dissonance producing for scientists to have serious theoretical disagreements with colleagues whom they like. Dissonance reduction probably sometimes involves changing one's theoretical position in deference to the friendship, or eventually risk losing the friendship—examining which occurs more often, and under what circumstances, would be an interesting empirical project for SPS.

Some debate has occurred about what cognitive dissonance or inconsistency really means. The initial hypothesis is that it was arousing in such a way as to push the person toward reducing dissonance. More recent research suggests that a key variable is whether or not the dissonance-producing act implicates some aspect of one's own self-concept. If not, people seem to find it relatively easy simply to forget about the dissonance rather than, say, change their cognitions. Similarly, it seems to matter whether any of the possible consequences of the situation involve aversive outcomes; if not, motivation to reduce dissonance seems lower.

Dissonance is probably ubiquitous in scientific life, in no small part because all decisions and choices involve forsaking alternatives that usually have at least some attractive features and choosing an alternative that usually has at least one negative feature. It is difficult to believe that such choices are always resolved "rationally." More likely, scientists give little conscious thought to the matter, so that at least some of the resolutions go awry. One might hope that scientists would salvage the situation by judicious selection of which imbalances are most important in the context of their work as a whole, ignoring the other ones for practical rather than epistemic reasons. Without data, however, we do not know.

Since the end of the 1960s, social cognition has been more cold than hot. Yet the results of recent research suggest that one cannot account for all the variance without appealing to the hot factors. In that sense, the swing from cognition to motivated social cognition in social psychology has been gaining increasing momentum (Kruglanski, Chapter 5, this volume). Many of the sections that follow deal with such matters in more detail.

Attribution Theory

[This section benefited greatly from a review by Ross and Fletcher (1985).] To help make the world predictable, humans tend to make attributions about the causes of events. Collins (1985), in fact, suggests that part of the training of new student–scientists is to teach them to make the attribution that the world is orderly and lawful, and to teach them that any failures to detect such order are their own fault: "It is necessary for us to blame ourselves if we are to learn appropriate skills and learn to 'see' the world like scientists. An impression of chaos is the alternative; science cannot be learned that way" (p. 161). Collins rightly notes that it is often difficult and frustrating to train the student to think this way, since the world is often more chaotic and recalcitrant than simple pictures of science suggest.

Attribution theory is about the cognitive processes that lead to such attributions (it is not about the validity of the attributions). Kelley (1967) said that people look at three things in making attributions: (1) *consensus*, or whether others see the stimulus similarly; (2) *distinctiveness*, or whether other people respond distinctively to the stimulus or respond the same way to other stimuli; and (3) *consistency*, or whether others respond to the stimulus the same way over time. Each of these three kinds of information can be dichotomized as high (H) or low (L). People apply these criteria to judge such things as whether or not a causal attribution is valid, whether the cause of an event is internal (dispositional) or external (situational) to the actor, and whether the cause is stable or unstable.

Certain combinations of information typically encourage certain kinds of attributions. One is a combination of high consensus, high distinctiveness, and high consistency (HHH), which tends to lead people to infer that someone's judgment about a stimulus is based on characteristics of that stimulus. Suppose the stimulus is a scientific theory, and the judgment concerns the validity of the theory. The scientist is encouraged to judge the theory to be valid if (1) other scientists agree it is valid, (2) other scientists rarely agree on the validity of other theories in this domain, and (3) those validity judgments remain stable over time. Some process such as this probably also underlies the attribution that a scientist has made a discovery. Rudwick (1985) notes during the Devonian controversy that "the status of 'discovery' was not intrinsic to Murchison's insight, but was attributed to it, first by him alone but ultimately by competent consensual agreement" (p. 438) after Murchison persisently advocated his theory and its competitors slowly dropped away. By contrast, consider the combination of low consensus, low distinctiveness, and high consistency regarding this theory (LLH). If Scientist X judges a theory to be valid but (1) few other scientists judge the theory to be valid, (2) Scientist X tends to accept the validity of several approaches, and (3) this situation stays stable over time, other scientists are likely to attribute Scientist X's judgment not to the validity of the theory but to some personal characteristic. This latter situation might seem unlikely, but consider Campbell's (1979/1988b; Chapter 2, this volume) description of the Tolman–Spence debate in learning theory. Over the span of more than a decade, Tolman's cognitively oriented theory was not widely accepted across the entire community of learning theorists, and Tolman admitted that alternative learning theories such as that of Spence's had some positive attributes—an LLH situation (see Campbell, Chapter 2,this volume). Attribution theory predicts that, compared to the HHH situation, the LLH situation will encourage the community of scientists not to endorse the validity of Tolman's theory. Indeed, that is exactly what happened, at least over a period of a decade or two. Finally, an LLL attribution analysis helps explain why scientists tend to doubt some kinds of reports even from the most eminent of scientists. For example, Friedrich Kekulé's claim to have discovered the chemical structure of benzene in a dream was doubted by other scientists (Roberts, 1989)—no wonder, because Kekulé's dream could not be observed by others, because dreams tend to be classified by other scientists as part of an ephemeral class of entities that are not generally thought to produce scientific discoveries, and because dreams are rarely reported by others, or by Kekulé, as the source of their discoveries.

Some controversy surrounds the role of consensus in attributions. Consensus by itself cannot always overcome other influences. For example, scientists are less likely to be influenced by consensus about a stimuli with which they have extensive experience. But when they have little experience or are uncertain, information about others' opinions can have more influence. Further, people tend to assume that others think, feel, and act like themselves, fostering a false sense of consensus, but counterbalancing this is a tendency to perceive oneself as having some unique characteristics, too.

Kelley also postulated the discounting principle, which suggests that people will attribute less influence to a cause if other plausible causes are also present. Suppose one is doing a creative task in one of two conditions, either knowing that one will receive an extrinsic reward or knowing that one will not. In the former condition, the reward competes with intrinsic interest as the cause of one's performance. Indeed, research has found that intrinsic interest in fact decreases in the former condition compared to the latter. Amabile (Chapter 11, this volume) explores how this applies to scientists, finding that their motivation and their levels of creativity may decrease if they see themselves as working for extrinsic rewards rather than for the sake of their own interests in the task.

A clearly related extension concerns attributions for a person's achievements. Latour (1987), for example, says that winning trials of attributions for such achievements is a key part of scientific success, but his theory does not outline the mechanisms by which such attributions are made. Attribution theory helps fill this gap. In general, for example, a scientist's success can be attributed to either dispositional or situational causes. The former includes a person's ability and effort, and the latter includes opportunity and task difficulty. Notice that ability, effort, opportunity, and task difficulty vary on several dimensions—stability, internal versus external locus, and controllability. In general, people prefer to make attributions to stable causes of other people's behavior (ability and task difficulty) but will attribute to unstable causes (opportunity, effort) when current performance is discrepant from past performance. They are also more likely to expect continuing success when they make stable attributions, but their affective response stems more from the locus of attribution, feeling better about success and worse about failure under internal locus attributions.

Suppose, for example, a scientist consistently tends to receive negative reactions to his/her work, such as rejections from journals and failure to be included among elite scientists in a field. Over the long run, the tendency is for the scientist to attribute failure to his/her own ability and the difficulty of the scientific task, to expect continuing failure, and to lose motivation—the classic case of the burnt-out faculty member, for example. Conversely, this conceptualization also helps us understand why scientists are so often reluctant to admit to the role of serendipity in their work, for the externality of the attribution implies less control over success, and the instability of the attribution implies a decreased likelihood of future success. Finally, this conceptualization also applies to attributions of credit for scientific accomplishments. Roberts (1989), for example, notes the following:

> The discovery of oxygen is attributed to both Joseph Priestley, an Englishman, and Carl Wilhelm Scheele, a Swede. Scheele discovered oxygen more than a year earlier than Priestley, but did not publish his results until after Priestley, in 1774, announced his experiment and described the unusual properties of the new "air," as he called it. So Priestley received more credit for the discovery. (p. 25)

Attribution theory suggests that Scheele's discovery was probably discounted somewhat relative to Priestley's as being dispositionally unstable—if Scheele had really believed in his discovery, he would have published it, so to speak. Similarly, although Dr. Crawford Long apparently discovered and used a general medical anesthetic more than 2 years before Horace Wells did, the latter received credit because Long, "being a modest man, did not publicize his discovery" (Roberts, 1989, p. 40). More recently, we see attributions changing about who really discovered the AIDS virus as information becomes available to suggest that the discovery was situational in the U.S. laboratory (rather than dispositional in the U.S. scientists, and hence not due to them personally) since the material had been provided by the French laboratory.

The preceding example illustrates that attributions can affect motivation, but the opposite is also true. An extensive set of findings exists about self-serving biases as motivations for attributions. Most pointedly, for a variety of reasons presumably having to do with self-esteem and self-presentation, people tend to attribute to themselves more responsibility for their successes than for their failures. Failure may then require them to shift attributions in order to maintain self-esteem. So the burnt-out faculty members we previously described may not denigrate their own ability or effort. Rather, they may shift the attribution for failure to the task ("Psychology is too complex a field to study scientifically") or to opportunity ("Since I have been at State Tech all my career, I've not had the chances that my colleagues in the Ivy League have had"). They may then undertake different tasks in which they anticipate more success. Not only is self-esteem served, but the actor may increase the likelihood of being positively evaluated by others.

Attribution theory also postulates a *fundamental attribution error,* which suggests that when judging the causes of events external to themselves, people have a bias toward overestimating causes internal to the actor (e.g., cognitions, personality, and emotions) and to underestimate situational causes. Latour (1987), for example, discusses our tendency to blame drunk drivers for the 40,000 people who die in traffic accidents in the United States each year, rather than blaming, say, the cars or the alcohol manufacturers (p. 204)—but he does not explain why this attribution is made. From an attribution theory point of view, it is perfectly understandable as a case of the fundamental attribution error (several other attribution theory explanations converge on the same attribution to the drunk driver; it is an HHH situation, for example). An even better illustration of the fundamental attribution error is in science studies concerning attributions about whether the causes of science lie in the scientist or in the scientist's environment and social interactions. Those who study science are presumably prone to the fundamental attribution error—overestimating the contributions of scientists and underestimating social and environmental causes of success in science (another self-serving bias, by the way, is that the individual scientist is biased toward attributing his/her own failings to situational factors). In this sense sociologists are right to encourage examination of the external causes of science, and their constant cajoling to do so is a necessary antidote to biased inclinations to the contrary. Hence, SPS recognizes and even explains the correctness of the sociologists' external focus without throwing out altogether the notion of the scientist as cause. After all, the point of the fundamental attribution error is not that there *are* no internal causes, only that people tend to overestimate them. SPS keeps cognitions and emotions while seeking knowledge about the interactive role among these things and the social and physical environment.

Three questions thus need to be distinguished here: (1) Are there internal cognitive processes or mental states? (2) Do subjects have privileged access to them? (3) Do these processes or states significantly contribute to an explanation of their behavior? To answer yes to (1) is to leave open the answer to (2) and (3). For example, one could still answer no to (2) and yes to (3). Or, one could answer yes to (1) and (2), but no to (3). The answer to whether people have privileged access to all these cognitive processes has been hotly debated by social psychologists. The consensus among social psychologists seems to be: not necessarily. Subjects do not seem to be able to access all these processes on demand. For example, when subject behavior is experimentally manipulated, subjects do not necessarily report the manipulation as a cause of their behavior. On the other hand, people do have some privileged access to their own mental events, at least compared to observers, so that they have more information about the causes of their own behavior. Still, both observers and actors may be wrong despite all this information, since ultimately both are making inferences about causes. All this implies ambivalent support for the point of view that scientists do not necessarily have privileged access to the processes that generate science. The fact that there may be social or cognitive determinants of science does not imply that scientists are consciously aware of those determinants. Yet only scientists can provide access to certain mental events that may be involved in the production of science. This issue is discussed by Ericsson and Simon (1984) in their book on protocol analysis. Simply put, people can report their ongoing cognitive processes but not the causes of these processes—and retrospective accounts like Murchison's reconstructions in the Devonian controversy (Rudwick, 1985) are *not* reliable.

Attitudes and Attitude Change

[This section is based largely on recent reviews by McGuire (1985), Chaiken and Stangor (1987), and Tesser and Shaffer (1990).] The study of attitudes has been such a long-standing part of social psychology that some early theorists defined social psychology as the study of attitudes (Thomas & Znaniecki, 1918; Watson, 1925). McGuire (1985) claims that this research has passed through three phases, the first concerning attitude measurement in the 1920s and 1930s, the second on attitude change in the 1950s and 1960s, and the third emerging in the 1980s and 1990s concerning attitude systems. Through all three phases, a key question concerned the nature and structure of attitudes. For years, a dominant position was that attitudes had three components: cognition, affect, and behavior, all aimed at some object of perception or thought. However, although social psychologists have continued to study the role of each of these components, the increasing tendency is to view attitudes as evaluative dispositions that are expressed through cognitions, affect, and behavior toward objects. Of the three components, the inclusion of behavior has been somewhat controversial because many attitude researchers are interested in the effects of attitude as an independent variable on behavior as a dependent variable (we will return to this topic shortly). Hence much research on the structure of attitudes focuses on the cognitive and affective organization and expression of attitudinal dispositions to respond favorably or unfavorably to objects, and on the relationships among attitudes.

If attitudes are orienting evaluations toward objects that influence how we think, feel, and behave toward them, they may be key mediating variables in scientific interactions. We can expect such evaluations to be one of the first points of orientation between a scientist and such key matters as scientific theories, empirical studies, or participants in controversies. Blank (1988), for example, used the attitude literature to analyze reactions to Gergen's (1973) critique of social psychology; Blank claimed that social psychologists with positive attitudes toward Gergen's message were the ones who assimilated it, while those with negative attitudes rejected it. Such attitudes influence subsequent length of interaction, type of interaction, depth of processing, and social judgments and decision making—in Latour's (1987) terms, the decision about whether or not to open the "black box." Persuasion to change attitudes is therefore of interest as a precursor to scientific belief change: "If agreement in science is founded upon processes of persuasion and dissuasion . . . both the attainment and the dissolution of technical consensus in science can be taken to be social accomplishments" (Knorr-Cetina & Mulkay, 1983)—to which we would add, social psychological accomplishments, but not just social psychological accomplishments. Hence, a key topic in SPS is how scientific attitudes are formed and changed.

Many influences contribute to the initial formation of attitudes, but especially direct experience with the object and social institutions. Sometimes, a single favorable or unfavorable exposure to an object can be critical. A single crucial experimental demonstration of some phenomenon can sometimes positively or negatively dispose scientists toward a new theory or phenomenon. An example concerns the discovery of the general anesthetic, nitrous oxide (Roberts, 1989). Although one of the first discoverers was a dentist, Horace Wells, his initial demonstration in front of peers in Massachusetts General Hospital in 1844 accidentally went badly because Wells ordered the patient's tooth extracted before the anesthetic had taken effect. The patient screamed in pain, Wells left in disgrace, and the audience adopted such a negative attitude toward the incident that 2 years passed before they seriously attended to the discovery again. Collins (1985) suggests that a similar (but less clear-cut) demonstration took place in resolving the gravitation wave detector controversy when a single experiment by a scientist Collins refers to as Q swayed most attitudes toward the phenomenon.

More often, liking tends to increase with frequency of exposure (except for liking of outgroups when contact is forced and involves others who are disliked; see section Intergroup Relations). So, we should expect scientists to become increasingly favorable toward those theories (and theorists) with which they have had the most contact. However, to the extent that scientific controversies often result in opposing camps that rarely interact and take on personal animosities, increased contact between them may not always help the resolution. For example, at the turn of the century an extended and acrimonious debate took place between Karl Pearson and Almroth Wright about the efficacy of a typhoid vaccine. Roughly 10 articles back and forth between these two authors in less than 1 year did little to resolve either the animosity or the debate (Susser, 1977).

Social psychologists have studied the effectiveness of various factors in persuading others to change their attitude. Three often-studied factors are source credibility, attractiveness, and power, high levels of which increase the likelihood that attitude change will or will not occur. When the source is highly credible by virtue of expertise or trustworthiness, for example, and the subject's involvement in the topic is relatively low, the subject is especially likely to accept the source's conclusion without studying arguments in detail. This undoubtedly occurs often in science, especially since scientists cannot subject every aspect of their work, every assumption, every dependency on the work of others, to intense scrutiny—they are reluctant to open Latour's (1987) black boxes. So, for example, Peters and Ceci (1982) found that journal referees and editors may judge a manuscript more favorably when it is authored by someone who is famous or at a prestigious institution. Conversely, acceptance of Edward Jenner's discovery of a smallpox vaccine was delayed by opposition from a distinguished surgeon (Roberts, 1989). Latour (1987) postulates that such argument from authority is often used by scientists to silence dissenters. During the Devonian controversy, Rudwick (1985) describes how the central geologists in the debate accepted without much question the opinions of experts in related fields like botany when speaking on matters within their expertise:

> Just as no geologist, even in the elite, would have presumed to doubt the great Herschel on matters of astronomy, so likewise they accepted these

specialists' identifications of fossils almost without argument.... De la Beche tried to minimize the implications of the Devonshire plants.... Such comments were merely marginal reservations about conclusions that in the main were accepted on all sides as authoritative. (p. 422)

Such effects of persuasive, credible sources can be "used up," however. If one pairs a source repeatedly with undesirable messages, that source will gradually lose persuasive effectiveness as the receiver of the information learns to avoid the source of consistently unpleasant information. For example, a prestigious scientist who repeatedly proposes that serious scientific attention be given to parapsychological phenomena might begin to lose credibility in the scientific community and so be less influential in changing attitudes. Similarly, during the Devonian controversy, the amateur geologist David Williams made a number of cogent points but was ignored in part because of his continued assertion of a single incredible argument:

Among his many objections to the Devonian interpretation were some that his opponents would have found awkwardly cogent, if they had bothered to read the article.... But Williams forfeited any credibility he might have gained from such arguments, by his continued insistence that all the Greywacke of south Devon and Cornwall was younger than the Culm. (Rudwick, 1985, p. 388)

Certain characteristics of the message also affect persuasive impact. Persuasive impact is better facilitated, for example, by concrete case histories than by abstract statistical information, which may help explain why the liberal use of case studies in the history and philosophy of science has such intuitive appeal. Impact is also enhanced by presenting many arguments rather than just a few; by sustained use of analogies, similes, and metaphors; by acknowledging and refuting opposing arguments early; and by drawing conclusions explicitly rather than leaving them implicit. Rudwick (1985) describes such processes in his analysis of publications stemming from the Devonian controversy:

Formal publication in science is a ritualized means of persuading others of the *plausibility* of the author's conclusions, in light of his or her *credibility* as a competent practitioner.... Any scientific paper... presents the favorable evidence with maximum rhetorical effect; weak links in the argument are glossed over or concealed; any antagonists' cases is [sic] attacked, and the force of their evidence subtly undermined or openly dismissed or even ridiculed. (p. 434)

And also: "What belonged equally to the courtroom, although the legal imagery was not extended to it, was the use of analogues in the Devonian controversy" (p. 436).

On the other hand, resistance to attitude persuasion can be induced by several means. Training people in critical thinking seems to confer some resistance, as does modeling resistant behavior, increasing a person's public commitment to a position prior to persuasion attempts, anchoring the position to other important cognitions, and exposing the person to weakened arguments in favor of the attacking position prior to exposure to the full set of arguments (inoculation). One sees all these mechanisms occurring in scientific training within disciplines or specialties, probably making new scientists somewhat more loyal to the theories and methods of their discipline and more resistant to those of others from the outset of their career. In fact, Messeri (1988) showed that older geologists were quicker to adopt plate tectonics than were their younger colleagues.

However, all these factors seem to exert far less of an influence to the extent that people engage in detailed (central, systematic) information processing of persuasive messages, rather than heuristic or peripheral processing—the extent to which they open Latour's (1987) black boxes. Clearly we would have difficulty processing all messages in a detailed manner, carefully considering their arguments and thinking about their relationship to our own positions. Such central processing is encouraged when we are both motivated and able to do so. When we are not motivated or able, we tend to adopt attitudes that are not very sensitive to the issues presented in the message, or to the quality of the arguments in the message. At intermediate levels of motivation or ability, we scrutinize the message carefully only if superficial cues such as source credibility make it likely to be profitable to do so.

Many factors facilitate our motivation and ability to process information centrally. For example, central processing is encouraged when the message is of more personal relevance. On this basis, we might imagine a continuum of more to less personally relevant topics that range from detailed discussions of one's own recent work (e.g., a review of one's new book), one's favorite theory, one's specialty area, and one's discipline to far removed disciplines at the other extreme. Yet here again, complexity reigns. Research shows that we process information more carefully when we are highly involved in an issue but have not yet taken a strong position on the issue. Once strong positions have been taken, we become as concerned with salvaging our position as with seeking the truth. The picture we have drawn here is thus a complex one. Methods for persuading people are counterbalanced by methods for inducing resistance. Both undoubtedly occur frequently in science, but their mix and relative effectiveness in the context of science have not been studied.

A central question in science studies concerns changes in scientific beliefs. But the question of attitude–behavior relations suggests that formulation is too simplistic; rather, we could be interested in cognitions about a theory, attitude toward it, and behavior related to it such as writing about it. We learn different things by studying scientists' attitudes toward, say, a scientific theory versus observing their behavior toward it. These behaviors might include what the scientists write about it, their recommendations in peer reviews of articles submitted for journal publication in support of versus opposed to the theory, or whether they chose to pursue grant funding by submitting a proposal depending on the theory. In all these cases, the scientist's evaluative disposition toward the theory may differ somewhat from what one might infer from the behaviors. Many scientists have learned, for example, that it is risky to behave publicly in a way that implies even the most tentative support for parapsychological phenomena, no matter what one's attitude toward parapsychology might be (see Rosenthal, Chapter 6, this volume). Yet in many respects, behaviors are the key material features of science. In this respect, attitudes are important for the orienting function they serve in directing scientists' behaviors toward some theories and away from others. Scientists who are negatively disposed toward a certain theory are unlikely to engage in behaviors that foster that theory. If this negativism occurs on a wide enough scale, the theory itself may suffer decrements in the extent to which it receives a fair test.

The study of attitude–behavior relations has been controversial, mainly because of the apparently low empirical relationship between the two. However, more recent formulations suggest that behavior can be predicted reasonably well by assessing both attitudes toward a behavior and perceived social norms, both of which in turn affect behavioral intentions, which itself is then strongly related to behavior (Fishbein & Azjen, 1975). The only other variable that strongly predicts future behavior is past behavior (which predicts future behavior quite strongly in a host of domains—so much so that one author even suggests that the best predictor of future scientific discoveries is having made such discoveries in the past) (Barton, 1989)—although both perceived control over the behavior and expectations of success have also been implicated.

Conversely, social psychologists have also explored how behaviors affect attitudes. One germane set of findings regards the behavior of writing. Some research suggests that a subject's written description of a target influences that subject's attitudes toward the target. In a typical experiment, subjects assigned randomly to write a positive (versus negative) evaluation of a target later expressed more positive (versus negative) attitudes toward the subject. Interestingly, this effect held even when the subject knew of the manipulation. Especially since scientists engage in a considerable amount of writing in their work, it seems plausible to think that scientific writing then shapes scientific attitudes. At its simplest, this might result in reinforcement of existing attitudes, especially to the extent that scientists only write about what they already know. But this latter condition does not always hold, given the likelihood that scientific writing is partly a creative act rather than just a descriptive act (Bem, 1987), a process of creating a coherent narrative in which to fit the research one actually did. It seems likely that the scientist partly creates his/her own attitudes toward the topic through the process of writing about it.

Social Motivation

[This section summarizes work by Geen (1991), Pittman and Heller (1987), and Showers and Cantor (1985).] The earlier discussion of social cognition referred to the mutual interplays between cognition and motivation in social situations. Some social psychologists take this mutual interplay as their starting point in discussing social motivation. Showers and Cantor (1985), for example, note that a host of motivational elements can cause people to invoke more flexible cognitive strategies than do the stable, entrenched, conservative strategies that are usually the case. These elements include goals, mood, and expertise. When such elements are engaged, people seem more able to be responsive to the unique features of the situation, to generate multiple interpretations, to take more active control, and to change existing knowledge structures. The key seems to be that the perceiver, in one way or another, defines the situation as personally involving. This may occur naturally, by active search for opportunities in existing situations, or by taking control of situations to turn them to one's own advantage. Goals, for instance, are sometimes embodied in situations, for example, professional recognition goals are embedded in an annual convention but not in a family reunion; other times, perceivers bring their own goals to a situation. Induction of positive or negative mood states also make situations more involving. Positive mood, for example, seems to encourage more creative solutions to problems; and anxiety encourages people to increase their efforts to control the situation and their contribution to it. Finally, evidence suggests that expertise generally increases involvement in a domain, but not always; experts are more involved and use active control strategies in situations they see as important but use more passive processing strategies otherwise.

One of the oldest issues in this area concerns the distinction between intrinsic and extrinsic motivation. As suggested in the previous discussion of attribution theory, many studies show that when subjects work for extrinsic rewards or under extrinsic constraints such as deadlines or surveillance, their intrinsic motivation and creativity are decreased (see also Amabile, Chapter 11, this volume). However, it matters whether the extrinsic variable serves informational functions about one's competence or control functions about how one is expected to behave (Pittman & Heller, 1987). In the latter case, trying to control others' behaviors with rewards seems to decrease their intrinsic motivation. In the former case, information about quality of performance seems to enhance subsequent performance and intrinsic motivation. These effects seem to persist even after the rewards are removed. Even controlling variables can carry some competence information, however; to the extent that they do, as when reward is contingent upon performance quality rather than just activity level (e.g., upon publishing in the best journals rather than upon the quantity of works one has published), intrinsic motivation may increase. In addition, some interpersonal relationships seem characterized by mutual intrinsic motives (communal relationships), while others are based primarily on extrinsic motivations (exchange relationships). In the former case, partners are less likely to keep track of mutual contributions and may eschew norms such as immediate repayment of favors. We might speculate that some scientific collaborations are communal and others exchange based, which we should be able to distinguish by tracking such matters as exchange norms and contributions; the prediction is that communal collaborations would be more creative, and at least some tentative evidence in favor of the prediction exists (Kraut, Galegher, & Egido, 1987–1988).

Social motivation and attribution theory meet in asking why people make causal attributions. The answer seems to be that they do so to make their social world more predictable. Depriving people of control leads them to be more curious, to engage in attributional analyses that attend more closely to observations and data, and to increase the care with which they process data. Control motivation is also increased by unexpected information (e.g., anomalies in the data), unusually negative events (e.g., an unusually poor review of a submitted journal article), and expected future interaction with a person (e.g., interactions with a newly hired colleague). In science, of course, all this applies simultaneously at two levels—the scientist's desire to have a theory that predicts data generated by the phenomenon of interest and to control his/her own career track. Better science should be produced to the extent we can increase such motivations. One way of doing so is to increase the likelihood of encountering anomalies and new ideas by, for example, having more than one major line of scientific research in one's career (Gruber, 1989; Shadish, 1989a), conducting nonprogrammatic research, working with colleagues not directly in one's specialty (Watson, 1968), and conducting experiments in the informal sense of trying things to see what works (Roberts, 1989,), or by institutionalizing norms that favor debate and alternative opinion (Rudwick, 1985).

Simon (1976) distinguished between "satisficing" versus "optimizing" solutions, where in the former case only enough information is gathered to meet minimal situational demands initiated by control motivation. One way to understand better or worse science is to predict circumstances under which one of these two solutions will be sought. For instance, coupling motivation to a cognitive task increases depth of processing. Such coupling presumably occurs naturally in science; for example, social scientists often have more intrinsic motivation associated with learning about the substantive phenomenon they are studying and less with methodological or statistical matters. If so, one might predict that attention to substantive theory development is likely to be more rigorous and careful than is attention to methodological and statistical details of how that hypothesis is actually tested. In psychology, one set of authors presents evidence in favor of this prediction (Aiken, West, Sechrest, & Reno, 1990).

Geen (1991) approaches social motivation more in terms of task performance than cognition, and he is interested in the effects of the mere presence of others on motivational states. He focuses on three effects in the social psychological literature. The first is social facilitation, that the mere presence of others can improve performance. Apparently the presence of others causes an aversive arousal state that seems to be associated with evaluation apprehension. This arousal facilitates performance, but not uniformly, as when facilitation occurs more for simple than for complex tasks, and more for well-learned than for novel tasks. Geen also discusses social loafing, the observation that individual performance decreases as the number of group members increases in a group that is working on a task. The basic phenomenon has been demonstrated to decrease quality of written materials, creativity, and cognitive judgment. However, social loafing also seems to occur only if the task is not involving, or evaluation apprehension is low. A good example is a multiply authored scientific article; if some of the authors are less interested or see themselves as having less of their reputation at stake, their performance will likely be lower than the performance of those who are more invested or more concerned about how they will be evaluated. Finally, Geen (1991) discusses social anxiety, which is "created when a person is movitated to make a certain impression on any audience, either real or imagined, but doubts that this impression can be made" (p. 390). The failure to make the impression is hypothesized to decrease positive self-concept, an aversive outcome. Social anxiety tends to inhibit behavior, bring about disengagement from the situation, replace meaningful discussion with innocuous sociability, and cause the person to use strategies for avoiding blame or attributing failure to situational causes. Geen concludes by pointing out that social facilitation, social loafing, and social anxiety all share motives linked to evaluation apprehension and fear of disapproval—motives that have to be high in the extremely competitive environment of science. Rudwick (1985) says of the participants' motivations during the Devonian controversy: "The narrative of the controversy has shown . . . how dominant and obtrusive the issues of credit and priority were in the work of some of the participants, and evidently . . . in their emotional lives too" (p. 440). Rudwick speculates that those motivations may stem from both psychological needs and sociological sources such as the need to establish a viable career in an environment in which negative peer evaluations could lead to loss of livelihood.

Finally, we turn to a topic that falls on the border of social motivation and our next topic, social influence—expectancies and interpersonal processes (Miller & Turnbull, 1986). When we interact with others, we do so with some expectancies about their behavior that, in turn, affect their behavior. The most widely studied effect of this type concerns self-fulfilling prophecies, studied, for example by Rosenthal (1966; Chapter 6, this volume). The basic finding is that one's expectancies of other entities often cause the others to behave in ways that fulfill the expectation. In the classroom, for example, Rosenthal showed that teacher expectancies about students result in changes in student performance, even when those expectancies are randomly assigned on the basis of false information—Rosenthal dubbed this the Pygmalion effect.

The Pygmalion effect applies to science in several ways. One is that our expectations about other scientists might influence their subsequent behavior. Rudwick (1985) speculates that this occurred during the Devonian controversy with respect to attributions of scientific competence:

> Those to whom others attributed only a limited competence were generally content to accept that judgment and to work within their prescribed limits. Conversely, those who were looked up to as examples of outstanding competence were not unwilling to accept that attribution and to act accordingly. (p. 420)

Further, Rosenthal has demonstrated that experimenter expectancies can influence the results of experiments in such diverse areas as "animal learning, human learning, psychophysical judgment, reaction time, ink-blot tests, structured laboratory interview, and person perception" (Aronson, Ellsworth, Carlsmith, & Gonzales, 1990, p. 305). Aronson et al. (1990) note that "the effects seem strongest in some of the 'harder' areas and more variable in such areas as person perception, which might intuitively seem more susceptible to interpersonal influence" (p. 305). This effect should be of great interest in science studies on its own merits, but few seem aware of it. Ashmore (1989) cites Rosenthal's research as one of the few examples of reflexive practice in psychology. Similarly, Collins (1985) cites Rosenthal's work as a reason why replication might count for more if the positive finding is produced by someone *not* expecting to find it. Rosenthal and Rubin (1978a) further explore the limits of such an effect. As Rosenthal (Chapter 6, this volume) points out, his work is limited to experimenter interactions with humans and animals, not with inanimate objects; hence, it only clearly relates to the social rather than the natural sciences. Aronson et al. (1990) explain why the effect may not be as pervasive or troublesome as we fear. Still, it is a clear-cut example of how scientists can play a key role in creating their own data. Wilson (1952) provides examples of experimenter expectancy effects in the physical sciences as well. Much methodological work in the social sciences has been devoted to reducing such effects (Aronson et al., 1990; Rosenthal & Rosnow, 1991).

Social Conformity and Social Influence

[This section is largely drawn from reviews by Chaiken and Stangor (1987) and Moscovici (1985).] Perhaps the most dramatic studies in the history of social psychology are those of Asch (1951) and Milgram (1963). In Asch's study, subjects asked to judge the length of obviously different lines often conformed to the incorrect judgment of a majority. In Milgram's study, nearly all subjects instructed (but not coerced) to administer painful electric shocks to subjects continued to do so even after it became apparent that they might be causing the victim considerable physical harm. In both cases, subjects experienced considerable psychological distress at their own actions, but they still conformed, highlighting the dramatic extent to which humans seem willing to conform to the wishes of others. Throughout subsequent years, social psychologists elaborated many of the conditions under which such phenomena occur, and the reasons why conformity seemed such a dominant factor. For example, social psychologists found that conformity is reduced if the subject has a fellow dissenter, and obedience is reduced if the person giving the instructions appears to be of extremely low status or credibility. Similarly, theorists focused on such things as the need to reduce the dissonance caused by nonconformity as a motivation to conform.

If we extrapolate such findings to science studies, they clearly imply that the pressures to conform to existing norms and dogmas must be strong in science. After all, scientists often work in small theory groups, in disciplinary departments, or in research and development laboratories that involve considerable face-to-face contact with a small number of other scientists. Unpleasant disagreements among these scientists arise often. Authority structures exist in which one scientist is often subordinate to another, with the other often calling for obedience more or less explicitly. In all these cases, it must often occur that one scientist or a minority of scientists will disagree with a superior or a majority. In these cases, the pressure to conform is high, even when the intellectual merits of the dissenter's view are substantial. Diesing (1991) provides one example of a student named Holt whose results contradicted the laws of quantum mechanics. Instead of declaring the laws disproved, the student:

> Took the safer tactic of declaring publicly that his experimental result was erroneous. He then tried for two years to find the error, hopefully following up all the critical objections his teachers could invent. He failed; the result was consistently replicated. However, his efforts earned him the forgiveness of the community leaders. Holt had had back luck, they concluded; his experimental results were consistently wrong, but it wasn't his fault. He was a good experimenter. (p. 196)

At least some of the time, then, we can expect the dissenter to conform, and at least some of those times, this may be to the detriment of scientific knowledge.

In the 1950s and 1960s, research tended to emphasize conformity as the dependent variable and factors that increased or decreased conformity as the independent variables. This conceptualization began to change in the late 1960s, especially due to the work of Moscovici (Moscovici, Lage, & Naffrechoux, 1969). Moscovici (1985) attributed the change in conceptualization to scholars who noted that it could not be very functional in the long run for subjects to reach wrong answers for the sake of conformity, for in that case the ability of the group to adapt to its environment would be seriously jeapordized. He then claimed that the primary function of social influence was not to establish conformity but to foster innovation as a mechanism of change. In this change process, innovations virtually always begin as the achievement of one group member, or at most a small minority of a group, and it is the conflict between minority and majority that determines the outcome. Rudwick (1985) claims that this is the fundamental dynamic behind the successful resolution of the Devonian controversy:

> It is clear that the finally successful outcome of the Devonian controversy was not simply the achievement of one brilliant and innovative *individual*, as the older heroic tradition in the history of science might have portrayed it. On the other hand, to characterize it instead as the achievement of a *group* of scientists would be equally unrealistic, unless it is made clear that that

> group was a collection of highly unequal individuals in far from harmonious interaction. As the detailed narrative of the controversy has shown, the myth of the egalitarian research collective is in this instance as inappropriate as the myth of the lonely genius. (p. 411)

An enormous amount of work has subsequently explored and extended Moscovici's formulation, trying to integrate it with past theory and research. In a nutshell, the resulting formulation looks something like this. In all groups, disagreements and new ideas arise that cause conflict. The extent of the conflict seems generally proportional to the extent of the discrepancy among positions. With no discrepancy, the new idea may catch on quickly, as when Niels Bohr quickly disseminated the nuclear fission interpretation of experiments described by Otto Hahn, Lise Meitner, and Otto Frisch (Roberts, 1989). Conversely, when the discrepancy is too extreme, it is difficult to fit it into one's world view, in which case the discrepancy tends to be dismissed as irrelevant. Parapsychology probably fits this category. For cases in between, members begin to lose confidence in their position, and to think of new arguments they can use to reestablish consensus. We see this happen in science, for example, when authors whose work is criticized think up new arguments to defend themselves against the onslaughts of the critics. Presumably this dynamic also operated in the gravitational wave detector controversy (Collins, 1985), when Joseph Weber claimed to have built a device that successfully detected such waves. The possibility of such waves did not challenge the scientists' world views. Rather, because such waves were thought to be extremely weak, skeptics were not very confident that Weber had actually accomplished what he claimed. The subsequent debate introduced new ideas and new experiments in an effort to reestablish consensus, until gradually the community came to agree that Weber had not succeeded:

> By 1975, the vast majority of these candidate explanations for the discrepancies between one experiment and another had disappeared from the world of scientific discourse. The last four seemed quite bizarre and the range of discussion was restricted . . . to questions of statistical error and the like. This is exactly the sort of change we would expect to take place as the field reached consensus. As the disturbance brought about in the scientific community by the initial claims was smoothed away, there was no further need to dig deep into the background of "cherished beliefs" to try to bring a new order to physical reality. Weber was simply wrong. (Collins, 1985, p. 99)

Conflict is exacerbated when the discrepancy between positions is greater, when the disagreement is about categorizing (is it black or white) rather than about matters of degree (how gray is it), when commitment to the position is so strong that one does not fear rejection, and when excluding the deviant from the group is not possible. All these exacerbations occurred in the Devonian controversy, for example, where disagreements concerned opinions of the age of certain geological structures on the order of several million years, where they concerned to which geological category the structures belonged, and where the close-knit and open nature of the scholarly group made excluding a participant difficult. Consequently, the Devonian controversy was particularly conflictual (Rudwick, 1985).

Moscovici hypothesizes that strong conflict causes strong pressure on a majority to change, but weak conflict exerts pressure on the minority to conform. In both cases, conflict initiates a process of social negotiation in which the participants' behavioral styles play key roles in three ways. First, the more autonomous and independent is the source of the discrepancy is, the more influence the source will exert. The impression of autonomy can be conveyed in many ways, for example, when the source appears *not* to be trying to convince us, or when a low-status person advocates a position against his/her own interest. Second, and in some respects most important, consistency of position exerts tremendous influence, whether it is for or against group norms. So Rudwick (1985) concludes that his analysis of the Devonian controversy "suggests one important conclusion that is surely valid far beyond this particular case study. The final successful interpretation . . . was undoubtedly powered by Murchison's dogged conviction" (p. 411). This is not just a matter of credibility, for the effect even seems to be true in laboratory studies where the source was instructed to respond in an asocial, deviant, and incompetent way. In attribution theory terms, consistency seems to lead observers to make dispositional attributions that the deviant person must be certain and confident in his/her position. Third, taking extreme positions seems to increase influence; a most effective strategy is to take a consistently extreme position but make concessions at a critical moment. Yet even the nonflexible person has an influence, often indirect, on matters related to the topic of discussion, if not on the topic itself. Collins (1985) notes that the plausibility of an argument is increased just by one person speaking on its behalf: "Harvey (1981) saw similar things happening in experiments to do with quantum theory; a scientist's mere willingness to test a bizarre hypothesis rendered it far more plausible" (p. 150). Interestingly, Collins (1985) goes on to speculate that a majority wishing to silence the opinion should simply ignore it, for "even to criticize an idea in a devastating way is to start to bring about its institutionalization" (p. 150).

One might use this analysis to speculate about the pervasive impact that SSK has had in the last decades on a science studies literature in which sociologists of scientific knowledge were a distinct minority. Diesing (1991), for example, describes them as using minority influence tactics extensively:

> A subversion strategy, Kuhn's "revolution," is more promising if there is a small group willing to pursue it together. They they can announce a new paradigm, cite each other's writings, set up conferences published as edited volumes, and (they hope) make a big enough splash to attract disciples. If they succeed, they immediately become the community elite, accumulating symbolic capital rapidly in a stream of publications and citations and appointments and research grants. The microsociologists are an example of a group pursuing a successful subversion strategy. Their early articles are full of bold pronouncements that something new is happening, something important is being studied for the first time, a start is being made, science is now advancing. (p. 196)

We might add some other observations about their use of the three minority influence tactics mentioned previously. For example, at least some authors of treatises on reflexivity go out of their way to demonstrate that they are not trying to convince the reader of their position (Woolgar, 1988; Ashmore, 1989). Similarly, sociologists of scientific knowledge are consistent; they always follow the same game plan of deconstructing our common ways of seeing science. Finally, their positions on science studies tend to be extreme in many respects relative to the positions of other sociologists and philosophers of science. All this would lead us to expect SSK to have a particularly large impact on majority opinion in science studies—to get the attention of the majority, at a minimum (e.g., Laudan, 1990), and probably to change how the majority thinks about science at least in small ways.

These three influence tactics are also reminiscent of Mitroff and Fitzgerald's (1977; Mitroff, 1974) findings about the characteristics of Apollo moon scientists. In their research, they found three "types" of scientists. Type I scientists were rated by other scientists as bold, intuitive, speculative theorizers, and were most likely to be nominated by peers as the outstanding scientists of the Apollo moon program, but they were also rated as the most biased, controversial, aggressive, and committed to their hypotheses. In contrast, Type III scientsts were rated as obsessed with data gathering and disdaining of theory, and were nominated as the average scientists by peers. Type II scientists fell in between. Interestingly, most scientists rated themselves as closest to the Type III or Type II scientists, and they saw themselves as average or just better than average. It may be that Type I scientists are, in fact, exactly the influential minority Moscovici speaks of. They pose innovative ideas in an aggressive, committed, consistent, and even biased way, but have an influence and earn the respect of their peers for doing so.

The size of the minority and majority makes a difference in all this. If a group has a total of three persons, one in the minority and two in the majority, the minority member often will not yield. However, once the majority reaches three, minorities yield to pressure much more often, although increasing the majority beyond three seems not to matter much. The single minority group member then can be expected to search for at least one ally. This finding becomes complicated if one replaces individuals with groups. For example, two groups of three persons have more influence than one group of six persons in influencing a minority—especially to the extent that the groups are perceived as autonomous. In science studies, we speculate that this finding is relevant to relationships among theory groups (Mullins, 1973), so that as multiple theory groups rally around a position, pressure for remaining theory groups to conform to the new position increases. This dynamic is also apparent in Rudwick's (1985) analysis of the Devonian controversy—in fact, the physical arrangement of the meeting room for the Geological Society in which the controversy was often debated consisted of two sets of chairs facing each other, with conflicting groups encouraged to sit on opposite sides.

Moscovici claims that when conflict arises, groups resolve it in three ways. One is normalization by reaching compromises in order to avoid conflict. Early in the Devonian controversy, for example, both of the main protagonists in the debate made some efforts to avoid further overt conflict by going "through the motions of reconciliation" (Rudwick, 1985, p. 112). However, such efforts may be particularly successful when the object of disagreement is ambiguous, and when subjects are not too involved in the issue. Neither of the latter two conditions held in the Devonian controversy, so the reconciliation quickly broke down. A second resolution is to have the minority conform to the norms of the majority. For example, tasks that are especially difficult or complex encourage conformity. Especially since the tasks that scientists do are often complex, we might speculate that other scientists are often inclined just to go along rather than to question deeply—in Latour's (1987) terms, they are somewhat reluctant to open the many black boxes claimed by the original author.

The problem is that conformity, as we noted earlier, can eventually make a group dysfunctional in its environment if the group is wrong. At this point, the third style of coping with deviance becomes salient, as some members conform but others resist. To resist, however, means being willing to give up the benefits of group membership. In matters of scientific theory choice, for example, not all scientists are willing to do this, for it means giving up being part of a presumably successful theory group or research program in favor of a new program that may come up with nothing. Latour's (1987) story of Joao Dellacruz's eventual total isolation from the world of computer chip design is a particularly poignant example. Sometimes resisters have lower affiliation needs with the group and are willing to continue resisting (or to leave if necessary). Interestingly, some research suggests that more innovative scientists do have lower affiliation needs (Albaum, 1976; Buel, 1965; Fox, 1983). Similarly, Blank (1988) suggests that Gergen's (1973) deviant expression about the state of social psychology did indeed result in isolation and compartmentalization of Gergen and those who supported him. The latter people then formed their own coalitions both within and outside social psychology.

Moscovici postulates that minority influence occurs through different processes than majority influence. Majority influence operates by causing the minority member to focus on the characteristics of the majority—on whether the deviant still wants to be a part of the group, for example. Minority influence, on the other hand, causes the majority to focus on the characteristics of the stimulus being debated—to consider the issues. The latter process leads to deeper cognitive processing, and so results in more thorough and lasting changes. Majority influence, by contrast, results in compliance with the majority position, but relatively little cognitive processing of the issues involved. We return to this distinction between private internalization versus public compliance shortly.

When a single person starts off being the only deviant, his/her resistance to conformity is increased if a single other person supports him/her. Interestingly, it seems to make little difference whether the supporter is right or wrong, or agrees or disagrees with the specific point the deviant holds, as long as the supporter also disagrees with the majority. This is a search for allies—any allies. For example, Rudwick's (1985) description of the Devonian controversy clearly shows how often each main protagonist searched for someone to support his position—"being assured of Lyell's continuing support, Murchison now felt he could answer Buckland more defiantly" (p. 118). It did not matter that Murchison and Lyell did not agree on some important points themselves. In return, however, the deviant seems to feel increased pressure to conform to the later positions of the supporter—you support me and I'll support you, so to speak. Again, Rudwick (1985) describes this dynamic process of building and maintaining the alliance between Murchison and Lyell. Conversely, some research suggests that people conform to others to induce the others to conform with them at a later time.

A particularly influential deviant is one who has been accepted as a prominent member of the group for a long time, but who breaks away on some issue. During the Devonian controversy, for instance, "Sedgwick's change of sides began a more general swing from GRE to COA precisely because as an elite geologist his opinion carried such weight" (Rudwick,

1985, p. 421). Similarly, Latour seems to have broken away somewhat from the radical relativist position of some microsociology of science, as illustrated earlier in this chapter in Latour's discussion of a little relativism taking one away from realism but a lot bringing one back. His dissent should be particularly influential because he is both prominent and one of the flock. Conversely, the person who tries to remain neutral on an issue, trying to please all parties by remaining moderate or indecisive, often ends up rejected in deference to more extreme positions. One recent chapter (Shadish & Neimeyer, 1989), for example, spoke in a conciliatory manner about how the psychology of science might contribute to *integrative* science studies. The call was quickly rejected by a sociologist (Woolgar, 1989) and a philosopher (Nickles, 1989) in favor of more extreme positions, with the latter countering the olive branch with the extreme view that "the peace offering is not only presumptive but, in the end possibly fatal—not only to sociology but (hysterically stated) to the human race as we know it" (p. 244). Finally, of particular relevance to science, people with particular competence on the issue being debated seem more resistant to conformity. This process is apparent to anyone who has seen, say, a symposium at a convention in which a lone dissenter continues to object on the grounds of special expertise.

Some research has also related minority influence to the quality of resulting knowledge. In one study, when groups were given the task of generating original responses, the number of deviant responses increased and was more tolerated. Similarly, groups with a deviant minority member were more likely to find correct answers when the problem was a difficult one. Scientists recognize the need for deviant minorities, of course, as a source of new ideas and criticism of old ideas. During the Devonian controversy, for example, after nearly all participants agreed on the age of the geographical strata involved in one debate, one person still held out: "Only Williams declared he was unconvinced, and Buckland said he was glad there was still one opponnent to 'the theory,' implying that a little opposition was not unhealthly for its future development" (Rudwick, 1985, p. 319). Similarly, Kasperson (1978) found that more creative scientists are more likely to use other people as sources of information and to take information from a wider array of disciplinary sources. Pelz and Andrews (1976) reported that technical scientists were more effective "when their associates held divergent viewpoints" (p. xv).

A key issue in all this research is the extent to which any changes that occur are fictitious or genuine—typically referred to as compliance versus internalization. Clearly the minority sometimes say they agree with the majority, later to indicate they did so only under duress and had not really changed their minds. During the Devonian controversy, for example, a meeting took place at Bristol in which Henry De la Beche's position was called into question in a paper by Roderick Murchison and Adam Sedgwick. Taken by surprise, De la Beche made significant concessions to the argument. Later, however, he partially recanted those concessions. One of De la Beche's contemporaries, George Greenough, acknowledged De la Beche's claim in subsequent correspondence that he had merely complied rather than internalize the new arguments at the Bristol meeting, saying, "Your retreat under the circumstances was perhaps wise, but I did not at the time, nor do I now consider the question settled" (quoted in Rudwick, 1985, p. 181). Pointing poignantly to the pressure on a minority, Rudwick (1985) goes on to note: "More than ever before, De la Beche felt acutely the intellectual isolation that his employed status now imposed" (p. 181).

One problem, of course, is that we usually rely on subject self-report about attitude or belief change, and subjects can lie. To counter this problem, one of the most intriguing demonstrations that the changes induced by deviant minorities are often internalized was conducted by Moscovici and Personnaz (1980). The task was color perception, with subjects being presented with blue slides that a minority confederate consistently described as green. Instead of relying solely on majority members' reports of the color of the slide, the authors relied on their reports of the color of the afterimage—when we see a color, and then look at a white screen, the afterimage is the complementary color. Most subjects are very unlikely to know which color is the correct afterimage for either blue (yellow–orange) or green (red–purple), so they would have great difficulty falsifying the report since the assessments of afterimage were done in private. Indeed, the authors found that majority members did see the complementary color of green after a blue slide when exposed to a consistent minority member's reporting green. This finding did not occur, however, when the influence source was the majority reporting a green response to a blue slide; the minority being influenced reported green, but saw the afterimage of blue. The changes induced by minorities, then, seem to be fundamental, not just superficial, and suggest that the deviant minority foster internalization while the deviant majority foster only compliance. This example also lends force to claims that psychological and social processes can actually affect the very perception of reality.

In a sense, Latour's (1987) theory of science as the Leviathan could be viewed as a conformity theory of scientific change in the tradition of Milgram or Asch. Latour hypothesizes that scientists accumulate black boxes that become leverage for power over other scientists, and that changes occur as we engage in conflicts over these black boxes. One implication is that the sheer number of black boxes one has accumulated determines the outcome because it improves the ability of the Leviathan to impose its view. But work in Moscovici's tradition suggests otherwise. If power via black boxes were all that counted, there would be no negotiation; the powerful would simply impose their view. Sometimes this happens. But clearly not always, because negotiation occurs. The minority can always refuse to conform. Sometimes power has the opposite effect and can be seen as unfair pressure or abuse of privilege. Further, group change can also occur by winning the confidence of an individual or group and then exploiting that confidence so as to change existing relations in one's favor. Sometimes scientific change need not be a conflict but ingratiation.

We have spent much time reviewing social influence, mostly from the perspective of Moscovici's approach as an integrating theory. Yet his approach has not been without criticism. Chaiken and Stangor (1987), for example, suggest that Moscovici's approach does not sufficiently take into account social motivations such as the desire to get social approval, the desire to identify with those we like by genuinely adopting their views, and the desire to hold correct opinions. They also note that some individuals have more need for cognitive processing than do others—presumably scientists might rank higher on this tendency than do others—and that they might process majority opinions more than Moscovici suggests. Similarly, other theories of minority and majority influence exist, most notably Latane's theory of social impact (Latane & Nida, 1980). In that theory, the impact of a source on a target individual is a function of the strength (e.g., expertise and credibility), immediacy (e.g., proximity), and number of sources. Latour (1987), for example, argues that scholarly citations in an article serve this sort of function, increasing the number and strength of things that a critic has to attack to discredit an article. Unlike Moscovici's presumption that two different processes are at work in minority versus majority influence, Latane assumes one process for both. Finally, Chaiken and Stangor (1987) also point out that most social influence research has used "minimal" groups with no real history or likelihood of future interaction, so that generalization is still open to question—although Nemeth and Staw (1989) summarize research in real-life organizational settings that seems supportive of at least some degree of generalization of Moscovici's work (see also the discussion of causal generalization later in this chapter).

Intergroup Relations

[Much of this section is drawn from Brewer and Kramer (1985), Messick and Mackie (1989), and Stephan (1985).] Social psychologists have approached the study of intergroup relations from the perspective of the individual, so that group-to-group relations, for example, have received less attention than how the individual perceives, processes, and acts on intra- and intergroup relations. Like most of social psychology, a cognitive perspective has been heavily dominant in recent social psychological study of this topic. Such studies are particularly concerned with group stereotypes, and with how those stereotypes affect attention, coding, and memory of data about groups and their members. Stereotypes can be thought of as schemas about groups and their members, and have the same general effects in processing information about groups that schemas have on information processing regarding any other stimulus. For example, especially in the presence of ambiguous data, the perceiver tends to make judgments about the stimulus that are heavily affected by the stereotypes. One group of scientists, for example, may perceive another as generally less competent in the area, and so may tend to dismiss the data they present, especially if the data are unclear in implications or difficult to understand. More to the point, scientists may attack the credibility and competence of colleagues who advocate opposing theories, as when Rudwick (1985) notes the following about the Devonian controversy:

> The rhetorical devices that pervaded both formal and informal exchanges throughout the Devonian controversy were merely the surface expression of the agonistic field, on which the relatively *plausibility* of alternative explanations, and the relative *credibility* of their proponents, were contested ceaselessly. (p. 437)

Such stereotypes profoundly affect how we process and act upon the information in social interactions. Brewer and Kramer (1985) suggest the following:

> The knowledge structures . . . that individuals bring to an intergroup contact situation may be as important as the structure of the situation itself in shaping the behavioral and attitudinal outcomes. Information gained from direct experience with outgroup members may be rejected or assimilated to prior expectations unless alternative knowledge structures or schemata have been provided in advance of the contact experience. (p. 237)

Further, "This bias toward collecting expectancy-confirming evidence is followed by a bias toward selectively attending to expectancy-confirming information during social interactions" (Stephan, 1985, p. 604). People often expect and perceive behavior associated with a stereotype even if the behavior did not occur. These expectations can influence the perceiver's behavior in such a way as to elicit the expected behaviors from the person with whom one is interacting—Rosenthal's interpersonal expectancy effect is an example. That person, in turn, comes to believe that he/she possesses the expected trait, and acts in ways consistent with that belief in many later interactions.

Initially, social psychologists focused mostly on the negative connotations of such stereotypes, but more recent work attends to the heuristic function of stereotypes for reducing cognitive overload, just as all schemas do. In this light, group stereotyping ought to increase under conditions that increase overload. Kruglanski and Freund (1983), for example, found that stereotyping increases as time pressure increases on individuals to make judgments or predictions. In science, given that they do not have the time to read in all related fields, scientists therefore inevitably have to rely on some degree of stereotyping of the contributions of those fields.

When we encounter a person who might be a member of another group, we compare that person to a prototype for the group. Features such as age, gender, skin color, attractiveness, or obvious handicaps are particularly salient and so widely used in social categorization. Of course, the defining features of scientific groups are less salient and often have to be solicited during interactions—theoretical commitments, applied versus basic interests, or discipline, for example. Once stereotypes are established, however, they are difficult to modify in part because a good deal of behavior that may be inconsistent with the prototype can be tolerated; low frequencies of such behavior are simply discounted.

A persistent finding in this literature is the "illusory correlation"—the perception of a strong relationship between group membership and some other stereotypical feature when the relationship either does not exist or is fairly small at best. A good example is the illusory correlation once asserted by Shadish, Houts, Gholson, and Neimeyer (1989) between being a sociologist of science and emphasizing "institutional, societal, and cultural factors that impinge on science" (p. 2) when such factors are not the primary concern of some important microsociological branches of that field (Ashmore, 1989; Diesing, 1991). Often, when perceiving a group to which one does not belong (an outgroup), the correlation involves a variable that is negatively evaluated, whereas illusory correlations with ingroup categories are often positive. An example is the illusory correlation between being a psychologist and holding a psychoanalytical (e.g., Freudian) point of view, as perceived by nonpsychologists who view psychoanalytical theories negatively (e.g., Popper, 1970). In fact, relatively few psychologists would describe themselves as psychoanalytically oriented today (or even in 1970), and the proportion of psychologists who do so might be no greater than, say, the proportion of psychiatrists, biographical historians, or some groups of literary scholars.

A related finding combines the fundamental attribution error identified by attribution theorists with these findings on group stereotyping to produce what Pettigrew (1979) called the "ultimate attribution error": that positive attributes of ingroup members tend to be attributed to internal, dispositional causes, but similar attributes of outgroup members are given transitory or situational attributions, and negative attributes are more likely to be given dispositional attributions in outgroup members than in ingroup members. For example, Nickles (1989), a philosopher, in criticizing the notion of integrative science studies proposed by Shadish and Neimeyer (1989), both psychologists, speculated about the possibility that their notion "is just a mask for more standard imperialistic ambitions of their own" (Nickles, 1989, p. 226), but he makes no such dispositional attributions about his own field's conquistadorial history in science studies.

People also tend to draw finer distinctions among members of groups to which they belong (ingroups) than they do among members of groups to which they do not belong (outgroups). So, for example, SSK insiders can more easily draw distinctions within SSK than can non-insiders, as Ashmore (1989) does when he says, "I am not and I have never been a 'Bath relativist' . . . . However, nonmembers of SSK would be much more likely to see SSK membership itself as the relevant category" (p. 136). Important causes of this tendency to make distinctions seem to be that we have more contact with ingroup than with outgroup members, anticipate more future contact with them, and are more dependent on them. In fact, evidence suggests that ingroup norms may discourage both prosocial and antisocial acts against outgroup members in many situations, so the net result is to discourage the overall amount of contact with outgroup members.

A well-established finding is that members allocate more resources to their ingroup than to an outgroup. Collins (1985), for example, describes how physics laboratories that had knowledge of how to construct TEA lasers often withhold part of that knowledge from members of other laboratories trying to construct one. Ingroup members also give negatively biased evaluations to the outgroup. Interestingly, this finding persists not only in natural social groups such as political parties but also in so-called minimal groups—groups formed for a short period of time solely for laboratory research and on the basis of minimal discriminators such as the preference for paintings by two different artists. Moreover, this bias is stronger when the ingroup is a minority and the outgroup is a majority.

Social psychologists have speculated about the motivations underlying such biases. Social identity theory, for example, posits that our membership in groups or categories provides information about our *distinctiveness* from others, and that we wish to be perceived as distinct in positive ways. Perhaps the most amusing example of this is from Woolgar and Ashmore (1988), who first comment that "the reflexive project in the social study of science, to which the papers in this volume are an essential contribution, is the inevitable next step in—indeed the culmination of—some of the most exciting intellectual work currently being undertaken anywhere" (p. 9), and then reflexively criticize this statement as including "several passages which attempt to propagandize the reflexive project as the most important intellectual development since SSK! Obviously, such hyperbole has little place in an enterprise which justifies itself in terms of the abandonment of realism" (p. 10). But in both passages, the reflexive project is portrayed as distinct in a positive way—in the former case directly stated, and in the latter by exemplar while showing that reflexivity is now what "we" do better than "you." In fairness, however, we admire much of the reflexive efforts, for they might well serve to help reduce ingroup and outgroup biases in science studies.

Similarly, we often adopt interpersonal strategies that might enhance our own group. This tendency is increased in competitive situations, or when status differences between groups are threatened. For example, some authors have suggested that competition exists between psychologists of science and sociologists of scientific knowledge concerning whose explanations of science should be accepted (Nickles, 1989; Woolgar, 1989). Our prediction is that both groups are motivated to perceive their own groups in ways that are positive and the outgroups in ways that are negative. An additional prediction of social identity theory is that acts of discrimination against outgroups increase the actor's self-esteem; and experimental data seem to support the prediction. So, for example, the prediction would then be that if one takes a "Strong Programme" stance against outsiders, one gains esteem within one's own group.

Category structures are associated with affect, both positive and negative. When a stereotype is associated with negative affect, the perceiver tends to make increased use of heuristics and biases. The most prominent result is intergroup prejudice and hostility; social psychologists have long studied methods for reducing such problems. By far the most frequently studied method is the hypothesis that contact among group members reduces these problems. Results have been somewhat positive. Successful cooperation among members of different groups on a joint project is especially helpful in reducing conflict. In science, for example, conflict should be reduced when investigators from different theory groups or disciplines collaborate with each other on a project—although only when the project has a successful outcome, which may not bode well for collaborations between social psychologists from psychology and sociology given the failure of the so-called golden age of interdisciplinary social psychology in the 20 years following World War II (Sewell, 1989).

But there are important caveats and exceptions to the success of the intergroup contact hypothesis. A key caveat is that unequal status between groups makes positive intergroup relations more difficult to achieve. One might speculate, for example, that psychologists of science rarely go to philosophy meetings and vice versa, partly because when they go they are usually in the minority with low status, and the contacts they make do not necessarily result in intergroup acceptance or harmonious agreement. Similarly, if the environment is designed to promote intergroup cooperation, members can attribute cooperation to the environment rather than to the disposition of the outgroup members, and thus group relations are not as much helped. For example, in some public forums it is simply in bad taste to attack one's scientific opponents publicly, so an air of cordiality prevails that does not convince many participants that relations are really smooth.

Another variable that helps reduce intergroup conflict is the presence of cross-cutting category distinctions. For example, in Rudwick's (1985) analysis of debates between geologists in England during the 1800s, he noted that participants could be categorized simultaneously on four variables: (1) involvement in the controversies (major, middling, or marginal involvement); (2) field of specialty (geology, botany, biology); (3) competence in their field (elite, accomplished, amateur); and (4) national origin (English vs. Continental). Elite geologists were prone to disparage the contributions of, say, amateur geologists, but they were less prone to do so for, say, an elite botanist even though the botanist might have low status in terms of general involvement in the controversy. Subcategorization also helps reduce the prejudicial effects of category-based processing; it should help to reduce stereotyping of psychology of science in general if we can portray the subcategories that are beginning to exist in the field among psychologists interested in cognition, personality, social psychology, creativity, and behavioral approaches.

Stephan (1985) listed the following variables that produce favorable reactions to intergroup contact:

1. Cooperation within groups should be maximized and competition between groups should be minimized.
2. Members of the ingroup and the outgroup should be of equal status both within and outside the contact situation.
3. Similarity of group members on nonstatus dimensions (beliefs, values, etc.) appears to be desirable.
4. Differences in competence should be avoided.

5. The outcomes should be positive.
6. Strong normative and institutional support for the contact should be provided.
7. The intergroup contact should have the potential to extend beyond the immediate situation.
8. Individuation of group members should be promoted.
9. Nonsuperficial contact (e.g., mutual disclosure of information) should be encouraged.
10. The contact should be voluntary.
11. Positive effects are likely to correlate with the duration of the contact.
12. The contact should occur in a variety of contexts with a variety of ingroup and outgroup members.
13. Equal numbers of ingroup and outgroup members should be used.

One can easily think of situations in science that have at least some of these qualities—small interdisciplinary conferences focused on a particular topic of mutual interest, for example. Sedelow (1976) speculated that computer networking of scientists would also be a means to reduce intergroup conflict. Gorman (Chapter 4, this volume) reports simulations of these sorts of characteristics in his studies of interacting groups.

We close this section by noting that one of Moscovici's observations about minority group influence should apply to intergroup relations as well. We have, in some respects, been assuming that reducing such conflict is good, but one might argue that science benefits from conflict through the generation of new ideas. Just as groups need deviants to generate new, potentially more adaptable ideas, so, too, intergroup conflict can be a source of productive problem solving. Katz and Tushman (1979), for example, found that when several separate research and development groups in the same organization were working on the same difficult task, the more successful groups were those that communicated effectively with other groups (see also Katz, 1982). In general, the likelihood of successful communication patterns between groups seems lower than within groups given that negative stereotypes and outgroup biases seem less of an obstacle in the latter case. But this likelihood may be changing over time in science as a whole. Brannigan and Wanner (1983a, 1983b), for example, suggest that multiple discoveries of innovations in science have been decreasing over time in part because communication among scientists as a whole has been vastly improved over time, thus scientists learn about each other's work more quickly and are less likely to duplicate it.

Summary

To summarize, this review suggests a host of questions that are both interesting and practical to pursue in SPS, including:

1. Investigations of the effect of specific social factors (e.g., the size of majority opinion, expectation of evaluation, time pressure, weight of previous consensus, reward structures, attitude, and stereotypes) on *each* aspect of scientific inquiry (e.g., question formation, data gathering, interpretation, and communication).
2. Comparison of group versus individual performance on scientific tasks, under a variety of conditions.
3. The study of group processes, especially minority influences, as they influence science.
4. Closer examination of the role of social influences and social motivation as mediators of scientific thought.

In the next section, we consider methodological issues that will arise as social psychologists examine such questions.

Methodological Issues in SPS

In general, this section covers two kinds of methodological issues. One is highly practical, concerning the methods that social psychologists of science are likely to use in the actual conduct of research. Those methods will be heavily influenced by the methodological conventions that are dominant in social psychology generally. Hence, this section aims to introduce the reader to those conventions (especially about experimentation, quasi-experimentation, and meta-analysis); to explore those conventions in more detail, and to discuss various conceptual advantages and disadvantages that are likely to ensue from following any such conventions. The discussion is often elementary in places, aiming to establish a common vocabulary accessible to nonpsychologists in science studies; but some of the discussion may be of interest even to those social psychologists who are quite familiar with methodological matters, especially the parts describing generalizations from experiments.

Prior to discussing these methods, however, we discuss certain metamethodological issues that address the very legitimacy of the SPS endeavor. Such issues have a long history in science studies. For example, anyone who asserts an interest in using psychology to study science quickly encounters questions about the warrant for such study. The kinds of questions seem to vary depending on who is doing the asking. Some philosophers may criticize the work as psychologism (Popper, 1970), and both philosophers and sociologists may ask about reflexivity, although they may each mean something different by it, each with different evaluative connotations (Ashmore, 1989; Barker, 1989; Woolgar, 1988). Some of the questions are generated by psychologists themselves while exploring the basic tenets they wish to use in studying science. A number of authors have written about such matters (Barker, 1989; Campbell, 1989; Houts, 1989; Houts & Haddock, 1992; Neimeyer, Shadish, Freedman, Gholson, & Houts, 1989). In the next two subsections we continue this metamethodological effort on two topics: adequacy conditions for SPS as a social scientific inquiry and the reflexivity problem.

Adequacy Conditions for a Social Science

If SPS is a social science (i.e., a science that investigates "social" phenomena such as "society"), it must satisfy certain "adequacy conditions" for social sciences in general. These include (1) a definition of the "societies" under investigation, (2) criteria for identifying the parts of the society under investigation, and (3) methods for analyzing and integrating the parts of the society under investigation. Let us consider each, in turn.

Definition of "Societies" under Investigation

In the case of SPS, this definition need not involve specifying the "essential" features of a scientific community, although such a definition would be useful for simulating the social aspects of scientific inquiry in an experiment. However, for ethnographic or historical work, it is better that the definition specify empirical criteria for distinguishing one scientific community from another, and scientific communities from the society at large.

Much here turns on whether science is defined as an extension and refinement of our ordinary problem-solving ability or as an activity quite unlike anything people normally undertake. The former position, closely associated with the pragmatist tradition in philosophy, generally trusts human cognitive powers, and hence is comfortable with the idea that experiments on nonscientists working on, say, simple number series tasks can simulate the basic processes involved in scientific reasoning. By contrast, a positivistic philosophy stresses the uniqueness of science as an institution that overcomes the profound biases and limitations in human cognition. Consequently, such a philosophy tends to emphasize the organizational and technical features of scientific inquiry that can typically only be examined by studying actual scientists at work.

In the pragmatist view, all our encounters with the world are protoscientific, whereas, in the positivist view, science occurs only in special locations, such as laboratories, under special conditions, such as during an experiment. Clearly, these are polar views between which SPS needs to maneuver. However, there is an intermediate viewpoint that complicates the boundary between science and the rest of society. This is what sociologists call the "scientization," "modernization," or "rationalization" of society at large, that is, the remaking of society in the image of science by social engineering, bureaucratization, and, most generally, quantified accounting schemes. Critical sociologists of knowledge, following Weber, Mannheim, and the Frankfurt School, have spoken of this process in terms of nonscientists adopting a scientific attitude toward—or imposing a scientific method on—ordinary social interaction. The tendency in this literature has been to portray the process as mildly pathological, causing people to become alienated from the natural disposition of the practices that are scientized. A microinstance of this phenomenon occurs whenever people judge the quality of a good primarily on the basis of its price, size, or other quantitative measure.

In the social psychology literature, scientization is discussed in two separate contexts. One, associated with attribution theory, portrays ordinary individuals as "intuitive scientists," which is to say, as persistently trying, but failing, to adhere to scientific standards of reasoning, especially standards pertaining to causal and statistical inference. The other, associated with Moscovici's work on collective representations, focuses on the role of popularly reported science—ranging from psychoanalysis to chaos theory—as resources that people increasingly use to represent everyday features of their life. These two bodies of research sit uneasily with what public opinion surveys often reveal, namely, the suspicions that people officially have about the applicability of scientific theories and methods outside of scientific contexts.

Criteria for Identifying Parts of Society under Investigation

The task here is typically much trickier than it first seems, as it raises thorny issues of ontology. Initially, the task appears quite simple: Because a society consists of people, it automatically follows that individuals are the relevant units for a scientific analysis of society. In a certain sense, this is obviously true, but it still leaves room for ontological pitfalls. One pitfall that is especially relevant in this context is what the sociologist Dennis Wrong (1961) has called the "oversocialized" conception of human beings. People are made to look oversocialized if a given individual can be made to stand as an exemplar for an entire society. Thus, a social scientist has an oversocialized conception of physics if he/she thinks that the best way to study physicists is to compile the typical characteristics of individual physicists taken in isolation of one another. Historicist philosophers of science, such as Kuhn, who tend to identify a scientific community with the world view or mind-set that its members supposedly share, are susceptible to having oversocialized conceptions of scientists. While such conceptions seem prima facie friendly to social science, their net result is to discourage research into the role that social interaction and other collective processes play in the coproduction of knowledge and knowers. A bias to oversocialization also leads one to underestimate the presence and necessity of diverse perspectives, skills, and tasks in a community of scientists.

Having disposed of oversocialized individuals, there remain two fruitful general strategies for capturing individuals in one's ontology. These strategies, in turn, reflect general solutions to the philosophical problem of relating "universals" and "particulars." To put the problem in a form that has some purchase for SPS research, what do such general properties (or universals) as "being scientific" and "being rational" have to do with particular individuals? Here are two possible answers, framed in terms of the SPS research agenda.

The first answer (the A strategy) is that an individual may be envisaged as a unique point of convergence for several general properties. For example, an individual may be simultaneously a physicist, a Catholic, a member of the middle class, middle age, an American, and so forth. In ontological terms, the individual is seen as a particular *instantiation* of several universals. Individuals, in this view, are typically portrayed as experiencing what Robert Merton calls *sociological ambivalence* in the course of trying to balance the conflicting role expectations associated with literally being several things at once. Each general property, such as "being a physicist," implies (either logically or probabilistically) characteristics of its own. However, a particular physicist may possess some other properties that either inhibit or promote the appearance of the ones associated with being a physicist; hence, the ambivalence. Nevertheless, general properties can be articulated and studied in abstraction of particular individuals, which enables inquirers to avoid many of the effects brought on by ambivalence. In fact, this has been a popular strategy in the history of the social sciences, which will be explored next.

The second answer (the B strategy) is that an individual may be envisaged as an incomplete part of some greater whole, literally a member of a body politic or a social organism. In this view, to identify someone as a physicist is not to attribute to him/her certain cognitive and motivational dispositions that are common to physicists; rather, it is to say that he/she regularly participates with other such identified people in activities that together produce the corpus of physical knowledge. Thus, when people train to be physicists, they do not acquire a competence that is shared by other physicists; instead, they simply learn how to work with others who are involved in the production of physics. This orientation to the study of scientists has been championed by one of the few contemporary psychologists who is favorably cited in the science studies literature, Jean Lave (see, e.g., Law, 1986). Lave's popularity is due to her heavy reliance on the anthropological theories of Pierre Bourdieu and Jack Goody in interpreting the difference in skill required for, say, formal arithmetical calculations and determining whether one can afford an item in the supermarket. For Lave, the latter is not simply an "application" of the former, but rather a completely different cognitive process that is dictated by the nature of the situation involved (Lave & Wenger, 1991). Lave goes to great lengths to externalize and socially distribute the sorts of competences that psychologists would be inclined to see as properties of individual minds. For example, by characterizing the learning of a skill as "legitimate peripheral participation," Lave means to convey the idea that the novice is not trying to acquire a property that others already possess but rather that he/she is acting in a way that is strategically related to another form of action in which the novice ultimately wants to engage.

Now it turns out that the general properties identified in the first answer are the names of institutions in the second answer that periodically require an ensemble of individuals to act in an appropriately coordinated fashion in some location. There may be considerable variation as to what the individuals actually do on these occasions. This viewpoint has met strong resistance throughout the history of social science because it casts aspersions on the autonomy of the individual, that is, the idea that individuals have clear identities above and beyond the groups in which they participate. Nevertheless, it does address a social psychological phenomenon that increasingly appears in the multiply-authored-paper world of "big science," namely, "collective deindividuation," the tendency of individuals to do whatever it takes to help their partners get the job done, which involves identifying themselves more closely with the final product than with any fixed role they might have played in the process.

Let us now see how these two ontological strategies have played out in the history of social science, starting with the clearest case of the A strategy, namely, the attempt to articulate and study "economic man," *homo oeconomicus*, in abstraction of particular individuals. We will then be in a position to draw some telling comparisons with the attempts so far to study "scientific man," as it were, *homo scientificus*. This, in turn, will enable us to see where the B strategy may be of particular relevance to SPS.

Homo oeconomicus is the self-interested, utility-maximizing, labor-saving creature who is the unit of analysis in neoclassical economics. Economists typically study the properties governing economic man in an abstract environment of perfect competition, with relatively little attempt to complicate the analysis by introducing noneconomic character traits or market interference. While many mathematically sophisticated normative principles of rationality have been derived from economic research, their empirical status remains highly controversial. In an early attempt to make economic research more accountable to the emerging body of empirical sociological research, Max Weber portrayed *homo oeconomicus* as an "ideal type," that is, a theoretical construct whose virtue would ultimately lie in its empirical applicability: Under what conditions do people's behavior most closely approximate the neoclassical economic ideal? Under what conditions do they deviate from it? How is the variation to be explained? Weber himself used this strategy to study why the "capitalist spirit" arose in Protestant rather than in Catholic countries.

One problem that immediately arose with Weber's ideal type methodology was that it seemed to presuppose a "nominalistic" (or "instrumentalist") solution to the problem of universals in the social sciences. In other words, for Weber, economic man was *merely* a theoretical construct whose reality depended entirely on the extent to which the behavior of individual human beings could be explained by that construct. Such nominalism went against the grain of many social scientists, especially the early political economists such as Adam Smith, David Ricardo, and John Stuart Mill, who saw *homo oeconomicus* as a set of innate human dispositions that are displayed, to varying degrees, depending on the disposition of the environment. Thus, prior to Weber, it had been common for political economists to define the relative "progressiveness" of a society in terms of the degree to which its members displayed an economic disposition. The social conditions that Weber would identify as empirically falsifying *homo oeconomicus*, political economists would either simply call "backward" (especially if nothing can now be done to change the conditions) or suggest that something be done to transform those conditions so as to enable the backward peoples to act more like *homo oeconomicus*. The slide from abstract science to "white man's burden" is plain enough. Indeed, it is the theme that dominates most histories of Victorian social science.

The debate over the ontological status of ideal types became complicated in the course of this century, as it was generally realized that social scientists were not alone in their use of such types. The social phenomenologist Alfred Schutz (1962) elaborated this point in examining the role of "typifications," or what social psychologists nowadays call "stereotyping" in everyday life. Schutz argued that both Weber and his opponents failed to appreciate the role that, say, *homo oeconomicus* played in people's conceptions of one another during economic exchanges. After all, in modern society, the people one encounters in a transaction are likely to be strangers. Yet, people are able to structure their transactions by assuming that others will act much as *homo oeconomicus* would. Of course, successful encounters do not automatically follow from such assumptions, because other qualities of the person (or the situation) may dominate the purely economic ones. However, even under those circumstances, the atypical person would be described as "irrational" or the situation as somehow "unusual." Schutz's point has been central to such broadly "phenomenological" approaches to the social sciences as symbolic interactionism, constructivism, and ethnomethodology: to wit, that in the course of socialization, people acquire ready-made accounts of how others act in certain situations, accounts that serve both explanatory and normative functions. They are used to explain people's behavior when they conform to the typification, and used to evaluate people's behavior as excusable or not when they fail to conform. Because the same typification is made to serve both functions, there is no reason to think that they will be particularly good predictors of other people's behavior. Thus, we may typify each other as economically rational agents during an exchange, even if there is little evidence that we have either the ability or the inclination to behave as economically rational agents. It is here that social science plays an important role in testing the empirical applicability of such "folk" typifications.

Thus, with regard to the study of general properties, the social scientist needs to distinguish the following:

1. The properties that define an ideal type, and their consequences in certain ideal settings.
2. The empirical application of the ideal type to explain a range of human behavior.
3. The normative desirability of applying the ideal type to maintain or transform a range of human behavior.
4. The availability of the ideal type as a resource for human beings to account for each other's behavior.

Now let us move the discussion from *homo oeconomicus* to *homo scientificus*. One can easily imagine a similar debate arising over the attempt to restrict SPS to the study of the distinctly "scientific" properties that people, especially scientists, might possess. The economists' quest for *homo oeconomicus* is largely analogous to the project of positivist philosophy of science, with its sense of the history of science as having an "internal" trajectory largely autonomous from the rest of cultural history. The positivist ideal is *homo scientificus*. Its distinguishing features are a preference for an inductive approach to concept formation and a deductive approach to theory testing, a tendency to measure the worth of a concept by the cases it can empirically cover, and an ultimate interest in explaining the widest variety of phenomena by the fewest general principles. The history of logical positivism (including contemporary analytical philosophy of science) is very much like the history of neoclassical economics in that we also see here the development of more mathematically powerful ways of representing, say, inductive relations between cases and alternative conceptual formulations—indeed, often with the aid of computer simulations, again just like the economists.

Strictly speaking, *homo scientificus* has never met its Max Weber. In other words, no one has tested the applicability of the positivist ideal across a variety of cultural historical settings in which science has been done. Interestingly, what has happened is that some philosophers, influenced by Karl Popper, have tried to articulate *homo scientificus* in such a way that it becomes a special case of *homo oeconomicus*, which is, in turn, taken to be the base model for all forms of rationality. Radnitzky (1987) has been the strongest proponent of this viewpoint. It presumes that all rationality is "instrumental" (*zweckrational* in Weber's sense), which is to say, means–ends oriented. In that case, *homo scientificus* is engaged in the efficient search for truths, which are characterized as "ends" that must be presupposed by all other instrumentally rational pursuits. Even in a strictly economic case, in order to decide what he/she ought to buy, the consumer needs to know the truth about the quality of the goods available, as well as the truth about what he/she really wants and needs. Influenced by Popper's falsificationist criteria of theory choice, Radnitzky believes that scientific rationality is an especially efficient form of instrumental reasoning because it enables the scientist to take calculated risks on supporting novel positions that end up paying off in the long term. However, Radnitzky does not go beyond a philosophical explication of this suggestive idea. It has yet to be subject to empirical test.

Sometimes it is thought that Kuhn empirically refuted the positivist ideal. However, a more apt description of his achievement was to propose an alternative ideal by extending the Piaget–Bruner model of individual cognitive development (popular in Harvard in the 1950s) to the social history of science, precedent for which could be found in the work of the influential French historicist philosopher Gaston Bachelard (1985), from which Louis Althusser and Michel Foucault developed their own views of "epistemic rupture" slightly after Kuhn did. The absence of a Weber-like figure willing to test the applicability of *homo scientificus* reflects something curious about the model's fate in the hands of social scientists. They have either presumed its realizability, if not its actuality, or they have claimed to have refuted the very possibility of *homo scientificus*. As will become clear, SPS is designed to steer a middle path of research between these two extremes.

On the one hand, both political theorists and sociologists of science—especially before the rise of SSK—have simply taken for granted that the positivist ideal was the way in which people reasoned when placed in a democratically organized community. Although their opinions varied as to how many people in today's societies actually conformed to this ideal, such liberal philosophers as John Stuart Mill, John Dewey, and Karl Popper agreed on *homo scientificus* as apt credentials for membership in the "open society." And although they are proposed as part of empirical sociology, Merton's four norms of science—disinterestedness, organized skepticism, communalism, and universalism—are little more than expressions of the sociological preconditions for the positivist ideal to thrive. Merton's presentation is striking in that it draws indiscriminately from accounts of the scientific enterprise made by philosophers, sociologists, and natural scientists, without checking whether they have accurately characterized actual settings in which science was done.

On the other hand, both ethnographers of "science in action" and experimental psychologists who study scientific reasoning have found little evidence for the presence of *homo scientificus*. Unfortunately, the tendency in these literatures has been toward offering "conceptual" reasons for concluding that further research on the matter would likely turn up more of the same. For example, experimental psychologists in the Tversky–Kahneman–Nisbett–Ross tradition operate with an image of human beings as possessing trenchant cognitive biases and limitations that make us ill-disposed to reason scientifically. The best one can hope to do under the circumstances is to anticipate and compensate for the errors in judgment that inevitably result. The ethnographers, for their part, shift the blame from individual incapacities to the very ideal of *homo scientificus*, as someone who reasons according to a set of explicit rules. Such an ideal fails to appreciate the contextually sensitive character of rule application, which requires that one attend to the nonexplicit "tacit dimension" of scientific practice (but see Wagner, 1987). Given the conceptual nature of their rejection of *homo scientificus*, neither the experimentalists nor the ethnographers have been especially moved to consider a range of conditions under which the ideal might be more or less approximated.

SPS aims to mitigate the excesses on all sides by conducting the type of comparative–historical, ethnographic, and experimental work that is needed to show exactly what sorts of environments enable *homo scientificus*, to come into his/her own. However, the extreme tendencies in the preceding lines of research do raise some interesting issues for the SPS research agenda, which will force us to incorporate more of the ontological B strategy. In particular, exactly what is the relation between the quality of scientists' reasoning patterns and the products of scientific research? The positivists conceived of *homo scientificus* in much the same spirit as the economists conceived of *homo oeconomicus*: Both were proposed as creatures equipped with the most efficient means to attain their ends. In the philosophy of science, this spirit is expressed in the largely taken for granted relationship between, on the one hand, rationality and methodology and, on the other, knowledge and truth. Yet, to accept even some of the experimental and ethnographic findings alluded to above requires that we revise our understanding of this relationship. But rather than denying the relationship altogether, a more fruitful empirical strategy would be to examine somewhat larger units of analysis than individual scientists. After all, as historians have begun to point out, when philosophers first proposed theories of rational methodology, they envisaged the scientific community as being on the order of the Athenian *polis* or a British society of gentlemen, that is, a small, relatively intimate group of peers. But what works for groups of that character may have little hold, not only on isolated individuals but also on collectivities with the dimensions of today's big science. Consider the case of Popper's method of "conjectures and refutations." Individuals may be ill-disposed toward refuting their own conjectures, but a supportive small-group environment may enable each member of the group to function as his/her own conjecturer and his/her neighbor's refuter, although both will probably be seriously constrained by the shared interests and biases that brought them together as a small group in the first place. However, once the group grows beyond a certain size, and members must communicate by more formalized means under intense competitive pressure, Popper's method may once again become inapplicable.

The point, then, is that philosophical theories of rationality may be best seen as abstract blueprints for organizing social interaction among inquirers. The exact scope of these theories—covering groups of what size, over how long a period, consisting of members with what range of skills that are integrated by what means—is to be determined empirically by SPS research. We see here a shift to the ontological B strategy, which implies that the interactive whole of the scientific enterprise is greater than the sum of scientists' particular endeavors. However, this shift may have moderate or radical consequences for the way in which individual scientists are portrayed:

1. *Moderate deindividuation.* Each individual has a set of competences and motivations that remain relatively constant across the different group activities to which he/she contributes. Nevertheless, the individual still needs to participate in those activities in order to bring any of his/her capacities to fruition.
2. *Radical deindividuation.* The competences and motivations that each individual has are manifested only during social interaction, and thus may change substantially as the nature of the interaction changes.

Roughly speaking, moderate deindividuation corresponds to the "cognitive sociology" advanced by the ethnomethodologists Aaron Cicourel and Karin Knorr-Cetina, while radical deindividuation corresponds to the ultra-contextualist social constructivism of Michael Mulkay and Steve Woolgar. According to the definition of moderate deindividuation, the physicist may display the traits associated with being a physicist, even when he/she is outside a strictly scientific context. But according to the definition of radical deindividuation, the physicist is only a physicist in virtue of certain standard situations in which he/she is called on to behave as a physicist. Whereas the first makes the idea of a physicist extending his/her knowledge into a new domain an intelligible course of action, the second denies the intelligibility of such a move, arguing instead that, in moving into the new arena, the physicist will need to negotiate the extent to which his/her actions can be attributed to his/her knowledge of physics.

Regardless of how radically one pursues the ontological B strategy, it nevertheless is likely to have salutary consequences for SPS research. One traditionally intractable area in which the B strategy might help is examining the relationship between scientific creativity and the impact of science on the rest of society. Loose talk about ideas being "applied" in technology or "embodied" in artifacts has tended to mystify the process of knowledge transmission. A certain popular way of telling the social history of science starts with a genius from whose head emerges a full-blown theory, which the rank-and-file scientists then try to comprehend for themselves in order to fulfill the promise of the theory in a variety of arenas. The presumption here—that the entire scientific community and relevant onlookers basically aim to copy and apply the ideas of its brightest members—is simply bad social psychology. Rather, as the ontological B strategy would suggest, a theory must appear to others in a scientific community as somehow incomplete or flawed—but, in any case, requiring the supplementation of the other members.

Methods for Analyzing and Integrating Parts of Society under Investigation

As highlighted in the discussion of Schutz's account of ideal types, social analysis and integration are common to both social scientists and the people they study. SPS needs to be sensitive to this duality, or what Anthony Giddens (1984) has called the *double hermeneutic* that defines the epistemology of the social sciences. There are two ways of thinking about this issue: (1) in terms of what the social psychologist of science needs to assume about the social world in which science is done; and (2) in terms of what the scientists, whom the social psychologist studies, need to assume about the social world in which they do science.

The first way of thinking (what the social psychologist of science needs to assume about the social world in which science is conducted) is perhaps best seen as a special case of the "micro–macro" dispute that currently dominates sociological theory. In that case, the question becomes one of identifying the combination of actions and interactions that result in the production of scientific knowledge. This involves integrating different levels of social psychological analysis, from small face-to-face group processes ("micro") to large, impersonal international networks ("macro"). Moreover, integration must be achieved not only at the level of theory but also at the level of method, since micro- and macrosocial realities have traditionally been studied by methods that do not naturally fit well with one another. Part of the problem here is that historians, philosophers, sociologists, and psychologists have each typically wanted to explain certain purported facts or features about science that some of the others have wanted to deny. In recent years, this problem—the site of most of the incommensurability that afflicts science studies—has been the need to identify and explain the sense in which modern natural science has been a distinctly "successful" cognitive enterprise: Do particular scientists, research teams, or disciplines succeed? Are specific ideas, techniques, or technologies especially successful? Failure to agree on the micro–macro dimensions of what needs to be explained has hampered interdisciplinary collaboration—any collaboration, for that matter.

The second way of thinking is in terms of what it is that the scientist, the subject of interest of the social psychologist, needs to assume about the world he/she studies in order to attain knowledge of it. Philosophers of science typically see this issue as having to do with the degree of "realism" that scientists need to invest in their theories: Can science be done without supposing that successful theories capture some real features of the world? It is in the context of addressing this question that issues pertaining to the "ideological" character of science arise. After all, like anyone else, scientists frequently make claims about their activities that turn out to be false or misleading. Thus, there are two empirically separable questions about realism that concern the social psychologist of science: To what extent do scientists claim to be realists? To what extent does realism explain what scientists do? In answer to the last question, psychologists may be in an especially good position to distinguish between the role of realism in explaining the sustaining *motivation* of scientists and its role in explaining the more lasting *products* of scientific endeavor.

As we have remarked throughout this chapter, differences in the methodological preferences of psychologists and sociologists have often blocked the kind of SPS needed to address the double hermeneutic involved in scientifically studying science. These differences often embody quite distinct assumptions about what the scientist is able to do. Consider the different assumptions embodied in the ethnographic method preferred by sociologists of science and the experimental method preferred by psychologists of science. The ethnographic method aspires to provide an account of naturally occurring, intersubjective processes. It thus tends to presume that all the real possibilities for action are eventually realized in normal practice. Ethnographers tend not to believe in what Max Weber called "suppressed possibilities" and Marx, before him, "structural contradictions" in history. Consequently, ethnographers are skeptical of the existence of competences that are not normally elicited in the everyday life of the individual. Called into question by this skepticism is any theory of *power*. In particular, the ethnographer has no way of studying how people may realize latent possibilities for transforming their current social situation. Instead, the existence of such possibilities can be demonstrated only once they have been actualized. In contrast, it is just this aspect of power that the experimental method directly addresses when the experimenter constructs situations, often quite unlike the normally occurring ones, in order to generate some hitherto hidden behavioral tendency. The more the experimental method is credited with eliciting such latencies, the more the experimentalist is compelled to posit capacities that enable people to act differently in different situations. Nevertheless, the experimental method is not sufficient for a complete theory of power because it typically leaves open the representativeness of a given experimental intervention. After all, it is one thing to say that people will change their behavior if their environment is changed in a certain way; it is quite another to show that an environmental change of that sort is feasible outside the laboratory. The prospect of new norms for governing social behavior is hollow if no mechanism is in place for distributing the relevant laboratory conditions to society at large.

Reflexivity

Although "reflexivity" is currently a popular topic in SSK, it has been integral to most dynamic accounts of the history of science. Put most simply, a system is reflexive if it applies something it has learned about its environment to its own internal workings. SSK differs from earlier accounts of science in terms of the general character of what it describes science as having learned. In many ways, these differences capture what is at stake in the *modernist–postmodernist dispute* that cuts across all the human sciences today. On the one hand, such 19th-century modernist theorists as Hegel and Comte took science to be discovering order in the world, which could then be applied to regulate the development of science itself. On the other hand, the postmodernist science practiced by SSK has discovered disorder in the world, *especially* in the world in which science is practiced. The reflexive histories related by Hegelians and positivists are of ever better methods that enable greater prediction and control of the environment. These are stories of increased *closure* and *inclusiveness*. In contrast, the reflexive histories related by sociologists of scientific knowledge are of ever greater discrepancies between universal principles and situated practices that are patched up in ever more opportunistic ways. These are stories of increased *openness* and *dispersion*. There are three dimensions along which the reflexive consequences of modernism and postmodernism may be compared, namely, in terms of the ways they *broaden, deepen,* and *limit* the scientific enterprise.

To broaden science reflexively is to apply to all of science what has been learned about one science. This is another way of stating the "unity of science" thesis associated with logical positivism. Crudely put, if a model or method enables order to be elicited in one domain of inquiry, it should be extended to all domains; hence, the ubiquity of mechanistic models and experimental methods, once they were shown to succeed in physics. In this way, the positivists and other modernists thought that laws governing each domain would be forthcoming. The postmodernists also have their version of the unity-of-science thesis, which may be called pan-constructivism. If the appropriateness of a given attribute to a given case must always be negotiated by social actors, then, in principle at least, there is nothing that prevents a nonhuman from being socially constructed as, say, a scientist or even a person. Conversely, it is just as much a political act to withhold rights and responsibilities from a computer as it would be from a human. The only difference, according to the pan-constructivist, is that whereas suppressed human voices often find someone to speak on their behalf, computers typically do not. However, a general strategy for granting nonhumans voices would be to treat the technological interfaces between ourselves and the nonhumans as media of communication instead of control. A cloud chamber would thus be a means for communicating with microphysical particles, and not simply for tracking their motions. One paradoxical consequence of this version of broad reflexivity is that postmodernists now face the prospect of investing everything with personhood—a veritable "sociology of things"—and, in the process, diminishing the value of being a person. Indeed, this consequence seems to be deliberate, insofar as postmodernists have often remarked on the discriminatory character that ontologically inflated criteria of personhood have had not only for nonhumans, but also for humans who failed to meet those criteria by not looking or acting in the right ways. In short, devaluing personhood might be one of the best things to happen to people.

To deepen science reflexively is to divide a domain into parts that are then analyzed by the same principles that had originally been used to study that domain. This process is familiar to modernists as the division of cognitive labor into special disciplines. This process typically involves adapting general principles, techniques, and instruments to ever more specific objects. For example, an experimental task that was originally used to test the problem-solving ability of humans in general can be refined in various ways to capture differences between, say, scientists and nonscientists, or men and women. The experimental method is not abandoned; rather, it is intensified. For their part, postmodernists become reflexively deep by intensifying the openness of their inquiry, typically by highlighting the discrepancies in perspective that are already latent in any situation that is defined by more than one person. This goes beyond, say, the ethnographer noting discrepancies in behavior. Indeed, it is here that the ethnographic method itself comes under attack for privileging the ethnographer's account of an episode at the expense of silencing the perspectives of those who participated in defining the episode. In that regard, the classical ethnographer is no less authoritarian than the experimentalist who dismisses his/her subjects' accounts of their behavior during an experiment. The reflexive remedy is to articulate the alternative perspectives without any attempt at resolving their differences. Ironically, this evenhandedness of representation often costs postmodernist research some credibility as contributing to a body of factual knowledge, as opposed to how some of the postmodernists themselves describe it, a "new literary form."

Finally, to limit science reflexively is to consider the practical and moral implications of the inquirer's knowledge claims as contributing to the validity of those claims. Certain modernist attempts to link social scientific knowledge with emancipatory politics, most notably Marxism, have been put under enormous strain by this reflexive requirement: What if our best science shows that our favorite ideologies are unfounded? Or, to put it more subtly, what if we learn that the scale of social transformation demanded of most schemes to liberate humanity could be realized only at a very great risk to our current quality of life? Do the value commitments of Marxism, say, override the scientific research that casts doubt on its validity? Whereas modernists have increasingly let the canons of scientific research constrain their value commitments, and have hence moved away from Marxist positions, postmodernists have stressed people's ability to live with the image of humanity that is projected by social science research. If, according to the dominant paradigm, the terms in which people make sense of their life turn out to be untenable, a new science is in order, not a new way of making sense. Given the constructivist premises of postmodernism, this piece of advice is meant to be taken quite literally. As a result of taking this advice, SSK practitioners tend to portray the discrepancies they observe in scientists' behavior as marks of cleverness (e.g., exploiting the openness of a situation) rather than of stupidity (e.g., failing to notice inconsistency). This tactic has the morally salutary consequence of the practitioners of SSK ascribing to the scientists only those qualities that they would not mind being ascribed to themselves.

Are there some concrete courses of action that the practitioner of SPS can take to reveal his/her sensitivity to the dimensions of reflexivity just outlined? The practitioner of SPS wants to remain committed to the modernist ideal of an empirically informed, theoretically progressive social science, but at the same time the possibility of integrated science studies depends on his/her addressing the postmodernist challenge posed by SSK's sense of reflexivity. Some brief suggestions cover each of the three senses of reflexivity.

Broadening Science Reflexively

Instead of conceptually prejudging the issue whether computers can be scientists, decide the issue empirically by placing, say, an expert system in a scientific setting and seeing how often, and under what circumstances, people who are recognized as scientists come to rely on the computer's judgment. As in the case of measuring the credibility of human scientists, the key indicator here is less a matter of whether the scientists consult the computer and more a matter of whether they actually follow its advice—especially in situations in which there is competing advice from a recognized colleague.

Deepening Science Reflexively

Multiple perspectives undermine a piece of research only if they are allowed to diverge indefinitely, rather than being forced to enter into dialogue with one another, specifically to encourage each perspective to articulate in its own terms the differences that it can detect in the others. The ultimate goal would be a more inclusive discourse that found a place for each distinctive position. Thus, the multiplicity of perspectives on a piece of research should be encouraged—but so too equally should many attempts at their integration.

Limiting Science Reflexively

If social scientific theories are to have moral import, they must empower the people who believe them. In the first instance, this means that the theories need to be comprehensible to their intended audiences. Failure to address this reflexive concern led the Frankfurt School to be cynical about Marxism's ultimate ability to liberate the masses. Realism has worked so effectively as a scientific ideology in large part because of its cognitive simplicity: It projects real-world objects from theoretical terms, and it promises that in the long term, all the theories will come together to explain everything by the fewest principles possible. If part of SPS's mission is to change scientific practice, it must seriously come to grips with limited reflexivity.

Experimentation and Causal Inference

Like all psychologists, social psychologists use many specific methodologies in their empirical research. These include experiments, surveys, case studies, meta-analyses, and structural equation modeling, to name just a few. Yet it is fair to say that experimentation is the ideologically dominant methodology in social psychology, and, as a corollary, to assert that this dominance reflects an implicit endorsement of the importance placed on valid causal inference about the effects of social psychological variables. In this section, we describe the approach to experimentation that dominates social psychology and address certain implications of this domination.

Randomized Experiments

To the general science studies community, the notion of experimentation connotes the deliberate introduction of some putative causal agent to see what results, at the same time controlling for other possible causes. This notion fits with a commonsense understanding of experimenting, as when we experiment with different spices in a recipe to see how they might improve taste, introducing them one at a time in different trials so as not to confound their effects. More to the point, the notion fits with our understanding of the meaning of simple experiments in the natural sciences, as when we deliberately introduce some chemical to see what it does, or deliberately vary temperature to explore how it affects the pressure of some gas in a closed container. In the latter cases, the scientist can usually ensure that the extraneous variables that might influence the results are held constant. In short, such procedures give us a plausible causal inference because the manipulated cause precedes and covaries with the effect, and the covariance cannot be explained as due to some variable other than the cause—a rough approximation to Mill's canons (Cook & Campbell, 1979). Clearly this is a great oversimplification of the nature of causal inference in science, but it sets the stage for exploring social psychology's use of experimentation.

The methodology of social psychology is in many respects faithful to this notion of experimentation. But the notion of experimentation described above only works well in limited circumstances, especially when the substances being introduced are relatively uniform, and when both putative causes and extraneous conditions (e.g., temperature) are easily controlled. Often that is not the case, certainly not in social psychology where the subjects are usually people who are anything but uniform and easily controlled, and where many important extraneous variables constantly covary with the causes of interest in a manner that is also not easily controlled. To deal with these difficulties, social psychologists mostly use the *randomized* experiment—a variant of experimentation that dates back to the work of statistician R.A. Fisher (1925, 1926, 1935) in agriculture (see Cowles, 1989, for historical background). In brief, subjects are assigned to treatment or control conditions using some random procedure so that at pretest, on average, those scheduled to receive treatment are similar (formally, on expectations) to those not scheduled to receive treatment within known probabilities. If differences emerge at posttest between treatment and control conditions, they are then attributed to the treatment as the probabilistic cause. Notice that the treatment itself need not be manipulated because subjects can be assigned to existing treatments that the experimenter does not control, such as to classrooms of different size. Further, no effort is made to hold subject characteristics constant; doing so would often be impossible, anyway. Rather, these things are randomly distributed over treatment-control conditions so that they do not introduce any systematic bias. In this sense, criticisms of experimentation in social sciences that are based on difficulties in holding such factors constant are misplaced. In fact, such criticisms are doubly misplaced because the process of randomization is sometimes necessary in physical sciences as well, as Wilson (1952) described in the case of testing the strength of explosives.

Several points are worth noting about social psychologists' reliance on the randomized experiment. First, of all the methods devised for facilitating causal inference in the social sciences, the randomized experiment is generally regarded as the most rigorously justified, providing the best available basis for saying that some putative cause made a difference to some outcome. In the jargon of the discipline, the randomized experiment yields causal statements with high *internal validity* (as contrasted with *external validity*, or the degree to which the causal inference generalizes) (Cook & Campbell, 1979). Social psychologists acknowledge some exceptions to this rule, of course (e.g., Campbell, 1975), and recognize the existence of debate about other methods that might serve the same purpose (Cook, 1991). These exceptions notwithstanding, however, the fact that social psychologists rely so heavily on the randomized experiment suggests that they place a premium on establishing the most rigorously defensible inferences about cause-and-effect relationships. Such a premium seems to reflect an implicit belief that the most important features of social psychological theory are the causal connections it posits—as opposed to, say, the descriptions it offers of phenomena. We might further speculate that social psychologists would tend to be very critical of the empirical warrant for *causal* statements offered by those in the science studies community who rely on other methods, most especially ethnomethodologies, but even the quantitative survey methods of some sociologists. For all their strengths, case-study methods are particularly prone to cognitive biases:

> Insofar as social scientists use induction—for instance, in the various case study methods—the experimenters have pointed out characteristic biases that need to be guarded against and counteracted. They include anchoring bias, availability bias, matching bias, representativeness heuristic, analogy heuristic, and the fundamental attribution error. (Diesing, 1991, p. 268)

Of course, this criticism is just as true in the present chapter to the extent that we have used case examples extensively to illustrate desired points.

Second, the randomized experiment tests a very particular kind of causation, what Campbell (1986) called local molar causation: (1) whether a molar treatment package that probably consists of a very large number of molecular elements (2) had an effect in a particular setting, at a particular time, with particular subjects, (3) on a molar dependent variable that itself probably consists of an amalgam of more specific operations and constructs. No pretense is made that the experiment yields universal causal laws. In fact, Campbell (1986) suggested relabeling internal validity as local molar causal validity to highlight this feature of the randomized experiment; the new label faired poorly, but the intellectual point was well taken. In any case, the randomized experiment provides no special evidence about another important kind of causation, causal explanation—about how the treatment produced the particular effect, about whether particular components of the treatment are more or less important in producing the effect, and about whether the effect itself can be broken down into component parts, some of which are more or less affected. Yet these latter processes are often of central interest in scientific theory. Hence, the randomized experiment requires certain adjuncts to explore such processes. We return to this point in the discussion of causal generalization later in this chapter.

Third, the report of an experiment is a statement about whether A (the treatment, whatever it might be) caused B (the outcome, whatever it might be) in the setting and time of the test. The design features that are essential to the randomized experiment—random assignment of units to alternative conditions with subsequent administration of a posttest—provide no special evidence about about what A or B should be called (what psychologists call construct validity), or about whether the relationship between A and B would generalize to other persons, settings, times, or variants of A or B (external validity). Again, certain features can be added to the randomized experiment to provide at least partial remedies to these lacunae, as when multitrait–multimethod measurement (Campbell & Fiske, 1959) provides more detailed information about how effects vary as a function of method and construct. But still, these features are mostly omitted, and are almost always regarded as adjuncts to the basic randomized design, implying that the questions about which they provide answers are themselves of adjunct interest to the basic causal question.

Fourth, social psychologists have long been aware that the randomized experiment is no panacea, even for local molar causal inference (Cook & Campbell, 1979; Rosenthal & Rosnow, 1969). Some problems are technical, as when treatment-correlated differential attrition of subjects from conditions vitiates initial group equivalence. Other problems concern biases to causal inference that random assignment may cause (Cook & Campbell, 1979), as when the assignment of subjects to a less desirable condition causes them to become demoralized and so to perform worse than they might have otherwise. Still other problems are practical, as in ensuring that random assignment is executed properly (Dennis, 1988). Of course, such obstacles to experimentation are often greatly overstated (Boruch, 1975). They are often more feasible than opponents acknowledge, and may even be ethically warranted when scarce resources make some unequal distribution of resources inevitable. Moreover, a common misconception is that *randomized* experiments must also be *laboratory* experiments. To the contrary, they can be and have often been implemented in field settings (Boruch, McSweeney, & Soderstrom, 1978). Particularly in matters of public policy, in fact, randomized field experiments have been used extensively this century to test potential policy interventions, especially during the last 25 years (Reicken et al., 1974). Although they can indeed encounter significant problems of execution and interpretation (Cronbach et al., 1980; Haveman, 1986), our understanding of how best to use them has increased greatly (Boruch, 1991; Rossi & Wright, 1984; Shadish et al., 1991).

Fifth, the endorsement of experimentation is not an implicit endorsement of positivist interpretations of experiments as revealing nature in some simple or direct way. Nickles (1989), for example, agrees that not all psychologists are "self-conscious positivists; but the latter do tend to accept their experimental results as being stated in nature's own language" (p. 241). But contrast this with the view of Donald Campbell, probably the foremost authority on experimentation in social psychology. Campbell (1988a) likens the experiment to tribal divination rituals in which:

> Caribou hunters roast a shoulder blade on the fire and use the cracks resulting to choose the direction the hunting party should take. The details of the ceremony contain many features designed to prevent human hunches from influencing the outcome, thus providing an uncontaminated channel through which the supernatural powers can speak if they will. Similarly, Norwegian fishermen once located new fishing sites by building a shoreline and island map of sand in a pan, filling it with water, and watching for the first place where bubbles rose. (pp. 501–502)

Human history, construction, and interpretation are ever present in these rituals, of course, but humans do not have much control over where the bone cracks appear or where the bubbles first rise (Fienberg, 1971, identifies the Caribou hunters as the Napaski Indians in Labrador). In this latter, very narrow sense, the experiment allows a role for nature to speak. To be sure, the very idea of allowing nature any role at all in experimental evidence is objectionable to some critics. Such differences are grist for the mill. But tribal divination rituals can hardly be fairly characterized as simple logical positivism. Diesing (1991) goes a step further, claiming that it would be difficult to conceptualize the actual experimental practices of psychologists as being logical empiricist at all, and "those psychologists and sociologists who describe their research in logical empiricist terms simply don't know what they are doing. Like Freud, they misunderstand themselves" (p. 328).

Campbell (1982) also describes experiments as arguments, not demonstrations of fact:

> A dialectical perspective does more justice to the history of experimental physics than does an image of the experiment as a window through which nature is seen directly. At each stage the "experimental variables" and the "outcome measures" are never "defined" for out-of-context or all-context meaningfulness. Instead, they are historically and dialectically indexical, acquiring their transient meaningfulness in the context of previous experiments and theories. In this important sense, *experiments are arguments* [italics in original] in a historical dialectic for the physical sciences and perhaps potentially for the applied social sciences. (pp. 120–121)

This conceptualization resembles Latour's (1987) concept of "trials of strength to evaluate the resistance of the ties that link the representatives to what they speak for" (p. 79). Such trials are always fallible and heavily socially influenced, but still influential in scientific argument.

In summary, then, social psychology brings a special tool to the study of science, one that places great stress on the importance of valid causal connections. The spirit in which social psychology brings the randomized experiment to science studies is captured by a quote from Brown (1986): "When social psychologists are so naive and arrogant as to suggest that our experiments 'generate' the answers to great questions, other disciplines necessarily bridle, but when we suggest that an experiment makes a distinctive contribution, that contribution is welcome" (p. xiv). We hope the randomized experiment does make such a distinctive contribution, but encourage debate about the matter to provide a still broader perspective on what experiments do and do not have to offer.

Quasi-Experimentation

It is not always possible to do a randomized experiment. When the social scientist is nonetheless interested in causal inference, two options are popular in the social sciences, although both options are generally acknowledged to be inferior to the randomized experiment. One is characterized chiefly by the use of statistical adjustments that aim to reduce biases and confounds in the data by properly modeling them or by "adjusting them away." Examples include path analysis, structural modeling, and econometric selection bias modeling (Bollen, 1989; Trochim, 1986). Social psychologists rarely use this statistical option, even though its joint use with quasi-experimentation could be a salutary development.

The other option was developed by Campbell (1957; Campbell & Stanley, 1963; Cook & Campbell, 1979) and is called quasi-experimentation. To oversimplify, quasi-experimentation refers to designs that are like experiments in all ways except that they lack random assignment of units to conditions. That is, some units experience the putative cause, others do not, and both groups are measured at posttest. But the process by which units are selected (or select themselves) to experience the cause is unknown (and so is called selection bias). The problem, of course, is that the putative cause is now confounded with the selection bias in unknown ways, and thus confident attribution of the effects to the cause are thwarted. Quasi-experiments attempt to remedy these problems by the use of design features that rule out or render implausible at least some alternative explanations for the cause. For example, when subjects are not assigned to conditions randomly, those who are to be exposed to the treatment may differ from those who are not to be exposed to it before the experiment ever started (at pretest); and if differences are observed after treatment (at posttest), we do not know if they were due to the treatment or merely reflect the existing differences at pretest. To help assess this possibility, a quasi-experiment might add a measure of the outcome variable taken at pretest and then examine whether groups differed at pretest in a way that mirrored the differences at posttest. To be sure, this does not rule out such pretest differences as alternative explanations of observed effects, but it at least allows partial assessment of the plausible likelihood of such differences, and of the direction of the biases they might introduce.

The use of pretests is just one of many quasi-experimental design features that function under this general logic of ruling out or assessing the plausibility of alternative explanations for the observed effects (Cook & Campbell, 1979). Unfortunately, the causal inferences yielded from quasi-experiments are generally weaker than those yielded by randomized experiments, so the former are encouraged in social psychology only when the latter seem impossible to use. Quasi-experiments might be of some use in SPS. Many of the questions of interest in science studies involve causal inference. Investigating them with randomized experiments would be ideal, but for practical reasons, many if not most practicing scientists would have little patience with being randomly assigned to some treatment that might affect their work. Instead, we could use other subjects who agree to participate. For example, Gorman (Chapter 4, this volume) used undergraduate college students as subjects in a series of randomized experiments. The obvious criticism of this practice has been made often (e.g., Brown, 1989)—that there is no obvious reason to believe that the results of experiments on undergraduates have any relationship to the results that would have been obtained had real scientists been the subjects. In the jargon of social psychology, this is a criticism of the external validity of the study.

We discuss some responses to this criticism in the section on causal generalization, but for present purposes the point is that quasi-experimentation may have a constructive role to play in this kind of situation. Many quasi-experimental design features (e.g., the use of nonequivalent control groups, cohort designs, predicted interactions, nonequivalent dependent variables) (Cook & Campbell, 1979) can be implemented without the subject's even knowing that they are being studied. Other features (e.g., staggering introduction of some change over time and setting) can sometimes be justified on practical or ethical grounds, as when limited resources make it impossible to implement the change everywhere at once. This makes quasi-experiments far more feasible than randomized experiments.

A good example is the quasi-experiment by Suls and Fletcher (1983) that was briefly discussed earlier. Recall that they started with Kuhn's (1970) observation that the social sciences were less developed than the physical sciences, which they posited should lead to greater uncertainty among social scientists on matters of procedures, data interpretation, and research directions. Based on Festinger's (1954) work, Suls and Fletcher (1983) proposed that such uncertainty should lead social scientists to seek out others' opinions, which they found to be the case when judged on the extent to which consultations with colleagues were acknowledged in footnotes. The putative cause is uncertainty due to being in the social rather than the physical sciences. Obviously it is not possible to randomly assign scientists to those two options; so Suls and Fletcher (1983) simply compared existing groups of social and natural scientists—a quasi-experiment using nonequivalent comparison groups. Especially since we cannot fully know all the reasons why people choose to become social and physical scientists, this design is replete with selection biases that make causal inferences problematic. Hence, Suls and Fletcher (1983) discussed alternative explanations, such as that physical scientists are more secretive and competitive than social scientists, or that social scientists are more gregarious. Sometimes evidence can be brought to bear on these alternatives; for example, Suls and Fletcher (1983) described an unpublished study by Hagstrom (1967) to the effect that there were no differences among five social and physical sciences disciplines in how afraid the scientists were to share their work—so the "secretive and competitive" hypothesis is then somewhat less plausible for now.

Of course, the causal inferences yielded by quasi-experimental methods are always suspect because it is rarely possible to rule out all possible alternative explanations, and because new evidence is constantly emerging about the plausibility of rival hypotheses. But as one part of an array of methods, each flawed in different ways but all yielding some unique information, quasi-experiments might well facilitate more generalizable causal inferences. Gorman, for example, might be able to extend his program of research by studying falsification processes quasi-experimentally in naturally occurring groups of scientists. If the key results seem to be the same in both randomized experiments with undergraduates and quasi-experiments with scientists, our confidence in the validity and generalizabilty of the causal inferences is incrementally increased. Similarly, by itself the Suls and Fletcher (1983) quasi-experiment is far from convincing, but when coupled with laboratory research showing very similar findings, their results gain more credibility and extend the warrant for generalizing results from the laboratory to the field (Cook, 1991).

Meta-Analysis

Meta-analysis is a set of quantitative techniques for cumulating and exploring the results of multiple primary studies. The term was coined by Glass (1976), an educational researcher, who also popularized the main techniques for conducting meta-analyses. Social psychologists helped develop this methodology, have used it often, and have made unique contributions to it (Cooper & Lemke, 1991; Rosenthal, 1984). Even though meta-analysis is not SPS per se in the sense of being about social psychological variables in science, we include a discussion of it for two reasons. First, the chapters by Rosenthal (Chapter 6, this volume) and by Miller and Pollock (Chapter 7, this volume) discuss meta-analysis extensively, so readers unfamiliar with the technique will need a general introduction to it. Second, and more to the point of this book, meta-analysis speaks to at least two key issues in science studies. One concerns replication, that is, the extent to which an experimental effect can be reproduced under various circumstances. The other concerns scientific evaluation, because meta-analysis can be used to quantify characteristics of studies that are thought to bear on study quality, and those characteristics can be related to study conduct or outcomes (Shadish, 1989b). Yet, except for cursory mention in a few sources such as Collins (1985), meta-analysis seems relatively unknown in science studies. This is especially unfortunate because meta-analysis, as much as any quantitative innovation in the social sciences during the last several decades, has caught on rapidly as the method of choice for doing certain tasks in literature reviews. Its impact on the literature review in the social sciences may be nearly as great—and as controversial—as the impact of the cognitive revolution on primary theory and research in psychology.

The function of meta-analysis is clearer if we think of science as having two complementary tasks. One is gathering primary data using such methods as the randomized experiment. The other is the integration and interpretation of primary data from multiple primary studies to summarize what is known about some topic. Until recently—at least in the social sciences—this function was usually served by the *narrative* literature review. The researcher would use whatever bibliographic sources were available to locate the studies done on a topic, gather and read the reports of those studies, and then write a narrative review that summarized and criticized the results of the studies, complete with commentary about implications for theory and future scientific work. Meta-analysis is not really distinct from these literature review procedures but rather complements them. Meta-analysis is a set of quantitative techniques that aid in doing parts of the review by quantifying certain information from the primary studies so it can be described and analyzed more efficiently. This is especially advantageous when the number of studies is large, for the reviewer then has difficulty remembering all the outcomes and variables without some numerical aid—just as in primary research.

Specifically, whenever the goal of the review includes summarizing the magnitude of some effect or relationship (called effect size), meta-analysis can be helpful. Some common measure of effect size is necessary because each primary study often reports results on several different outcome measures, and those outcome measures often differ from study to study as well. This diversity makes it difficult to compare outcomes over studies (notice also that it is difficult to interpret such studies as simple "replications"). Meta-analysis converts them into a common metric so that all outcomes are scaled identically. Doing so allows one to compare the magnitude of effect obtained on different measures both within and between studies, and to compute the average effect size over any subset of studies. Meta-analysts use a few standard statistics as the common outcome metric, such as the correlation coefficient, the standardized mean difference statistic, the rate difference, and the odds ratio (Shadish & Haddock, in press). In addition, meta-analysts usually quantify many other study characteristics such as whether the study appeared in the published or unpublished literature, the kind of treatment administered, what sorts of methodologies were used in conducting the study, and how the dependent variable was measured. The resulting data set allows the meta-analyst to conduct extensive analyses to examine not only the magnitude of the effect but also how effect size varies over other methodological and substantive variables.

Since meta-analysis was first introduced in 1976, its use has expanded geometrically, to the point where hundreds, perhaps thousands, of meta-analyses have now been done. The following example is particularly instructive, since it provides an introduction to one of the chapters in this book. Rosenthal (Chapter 6, this volume) discusses two topics. In the first half of his chapter he describes the origin and development of his interpersonal expectancy research (see also the section Social Motivation, earlier in this chapter). He then devotes the second half of his chapter to a discussion of how meta-analysis might be useful in coping with certain problems in social psychology. But he does not provide an example of meta-analysis, and neither do Miller and Pollock (Chapter 7, this volume). A relevant example is the meta-analytic work that Rosenthal and his colleagues have done on the interpersonal expectancy effect itself. Using meta-analytic techniques, Rosenthal and Rubin (1978a) showed that in over 345 studies "the overall probability that there is no such thing as interpersonal expectancy effects is near zero" (p. 377), and that "the average magnitude of the effect of interpersonal expectancy is likely to be of practical importance" (p. 377). They also showed that the effect was highly variable in magnitude across studies of reaction time, inkblot tests, animal learning, laboratory interviews, psychophysical judgments, learning and ability, person perception, and everyday situations, and also over dissertations versus publications. This article was followed by a large number of critical commentaries, to which Rosenthal and Rubin (1978b) then responded. The meta-analysis clearly did not solve all controversy; in fact, it created some. It did, however, help to sharpen the focus of the debate by providing a more concise and numerically precise source of information about how effect size varied in this area.

A second example is a set of several meta-analyses of psi ganzfeld research. In the latter field, subjects try to guess which answer has been transmitted to them extrasensorially by another person. Hyman (1985) located 42 studies and used meta-analytic techniques to summarize them. His analysis suggested that the actual rate of successful replication of the effect was only 30%. He then ran subsequent analyses to adjust studies for procedural flaws and found the effect diminished to virtually zero. Hence he concluded that available studies failed to support the existence of the effect. Honorton (1985) responded by using meta-analytic techniques to examine the same data set again, and he concluded that the effect is present and reliable. Hyman and Honorton (1986) then wrote a joint communiqué in which they discussed some of the possible sources of their differences, and the agreements they could reach on how future research ought to proceed to provide more useful information. Rosenthal (1986) then reanalyzed the data one more time and concluded that the effect was probably smaller than Honorton (1985) claimed but larger and more reliable than Hyman (1985) claimed. This example shows that different meta-analysts who review the same literature can reach somewhat different conclusions depending on exactly what assumptions they make and what analyses they use. Rosenthal (Chapter 6, this volume) discusses this example in the conclusion of his chapter.

One reviewer of a draft of this manuscript asked why, if meta-analysis was such an important contribution to science, the physical or biological sciences did not seem to have a need for it. The underlying implication is that meta-analysis is popular in the social sciences because it gives the illusion of making sense out of the inconsistent results that the social sciences seem to yield. But the question is based on a false premise. What may be the first use of these techniques (not under the name meta-analysis, of course) came in public health when the biometrician Karl Pearson (1904) examined whether a newly developed vaccine for typhoid was effective in preventing infection or death. Perhaps even more to the point, Hedges (1987) showed that the physical sciences, including physics, use cumulation techniques that are nearly identical to meta-analytic techniques. More strikingly, his examination of reviews in physical sciences suggested that "the results of physical experiments may not be strikingly more consistent than those of social or behavioral experiments. The data suggest that even the results of physical experiments may not be cumulative in the absolute sense by statistical criteria" (p. 443).

One can interpret Hedges's result in many ways, of course—as complimentary to the social sciences, as deleterious to the physical sciences, or as deconstructing the boundary between the social and physical sciences. But no matter how one interprets it, it is clear that meta-analysis has some potential for contributing to (and causing) debate about key issues in science studies like replication and cumulativeness. For example, much discussion of replication in the science studies literature reflects naive conceptions of how science is actually done. Mulkay (1988), for example, says:

> Replication in science is usually taken to contribute to the process whereby experimental claims are validated. Valid claims are supposed to be "reproducible" by other competent experimenters and it is often assumed that experimental observations come to be accepted only after they have been successfully replicated, that is, repeated exactly, by numerous independent observers .... But recent work by several sociologists has begun to show that the process is much more complex than may at first appear. (p. 81)

But to whom did the process appear simple? Not to most practicing scientists. For one thing, the notion that multiple studies in an area are a sign of attempts to replicate is wrong. In many areas there is no "first experiment" that everyone is trying to replicate. Psychotherapy research, for example, reflects the work of hundreds of scientists who may have common interests, but who probably rarely see themselves as trying to replicate any particular work. These psychotherapy studies are not replications; they are more or less close to some prototype of what psychotherapy studies might look like, but that prototype is always subject to change and may vary for each scientist. Even when replication is in some sense the goal, in the social sciences such exact replications are logically impossible if for no other reason than the fact that time has passed so that even if one could use the same subjects again, they have changed. More important, exact replications are rarely even the goal of social scientists, who instead almost always intentionally vary numerous parts of a past experiment to suit their own needs and interests. One must even doubt the extent to which exactly replicated results are expected in physics, given Hedges's (1987) work. Mulkay (1988) himself quotes biochemists as saying "But if you look, no two people have done exactly the same experiment" (p. 83). We might glean from all this that most practicing scientists probably do not have some naive expectation of exact replication in mind when they refer to multiple studies in an area.

Meta-analysis leads us to expect to find a distribution of effect sizes routinely in any given literature. The task is less to decide what is to count as a replication than it is to explore the many features of the study and the phenomena that give rise to variation in study outcome—including the possibility that the effect itself is driven by forces that are at least partially random (Hedges, 1983). Compare this to Collins's (1985) seven-level sorting scheme for deciding if something is a replication of some results "r":

> *Level one*: Eliminate all activities not to do with the subject of "r."
> *Level two*: Eliminate all activities that are not scientific.
> *Level three*: Eliminate all activities where the identity of the experimenter is inappropriate.
> *Level four*: Eliminate all activities that are not experiments.
> *Level five*: Eliminate all activities that are not competent copies of the original.
> *Level six*: Divide the remainder into those which are positive and those which are negative.
> *Level seven*: Decide whether "r" has been replicated. (p. 39)

Collins's goal is dichotomous, to determine what counts as a replication—it either is or it is not. Approached meta-analytically, however, these levels need not reflect sorting tasks with a dichotomous end point, but rather sources of hypotheses for empirical study that assume a continuum of degree of replication—for example, do variations in defining the subject of "r" lead to variations in outcome, and if so, how and with what implications for theory? From this perspective, Collins's critique of replication in science simply points to hypotheses to be studied empirically. Rosenthal (1990) elaborates this view of the success of replication as properly being a continuous rather than a dichotomous dimension. Moreover, it is continuous not only in study outcome, but in all study characteristics such as conception, execution, writeup, and entry into the literature. We discuss this again later in this chapter (see the sections Causal Generalization through Generalization and Robustness).

Meta-analysis also offers a slightly different perspective on a matter raised by Latour (1987). He notes that one way to render a scientific claim safer from attack is to embed it in a scientific paper that depends on a large number of citations as underlying foundations. In the case of research on a growth hormone, for example, Latour (1987) notes:

> The control of growth hormone by the hypothalamus could be disputed, it has been, it will be disputed; but to do so the dissenter will be faced not with one claim in one paper, but with the same claims incorporated in hundreds of papers. It is not impossible in principle; it is just enormously difficult in practice. Each claim comes to the future author with its history, that is, with itself plus all the papers that did something with or to it. (p. 42)

There is a great deal of truth in this observation, but meta-analysis could change the understanding of such claims in two respects. One is that the meta-analyst marshals an army of other studies on the same claim that can be used to confront the army of supporting references and claims cited by the original author—"Yes, you said the claim holds, and you said others agreed, but the empirical evidence in forty other studies does not seem to support it." The other change is that the meta-analyst abstracts numbers from the article in which the claim was made, combines them with similarly abstracted numbers from like articles, and makes statements about the aggregated numbers rather than about any one author's claims. This process to some degree strips the supporting references and arguments from the claim and leaves it to stand in a new context. Latour is right to point out that such literature reviews often add more fuel to the fire—but that is the point, that they do so by stripping away some of the support for the original claim and putting the claim in a new context not always of the original author's choosing. Of course, the results of the meta-analysis itself could be subject to Latour's analysis, so that they are themselves now less subject to challenge because of the supportive context in which they have been put—although we will see shortly that, in fact, such meta-analyses have been challenged by other meta-analyses on the same topic that reach different conclusions.

Causal Generalization

Gorman and Carlson (1989) defended the use of experiments to study causal inferences in science studies. Gholson and Houts (1989) followed that with an empirical example—two studies using undergraduate students. In the first experiment, subjects were assigned randomly to work either in groups or alone on a cognitive problem-solving task, and they found that subjects working in groups solved the task more often. In the second study, changes in the cognitive structure of students over the course of a semester were mapped. Although Gholson and Houts (1989) explicitly labeled these as analogue studies, they also claimed that these studies had at least some implications for the conduct of science.

Brown (1989), a philosopher of science, argued that, to the contrary, the Gholson and Houts study "does not make any contribution to the study of science" (p. 129). Brown summarized his argument as follows: The studies in the Gholson and Houts "paper deal solely with subjects who are scientifically naive and who are . . . engaged in activites that are so different from those that characterize scientific research that these studies cannot throw any light on the 'relevant practices of working scientists'" (p. 129). In the jargon of social psychology, Brown argues that these experiments have low external validity (generalizability). We use this example as a springboard to consider some of the issues involved; the arguments here draw heavily from Cook (1990), to whom the reader is referred for a more detailed exposition than current space allows.

Trade-offs between Internal and External Validity

It has long been thought that there is a trade-off between internal and external validity in randomized experiments—that is, obtaining a valid causal inference in a particular place and time, with particular subjects, may come at the price of obtaining a generalizable causal inference to other places, times, and subjects. Brown essentially plays off this presumed trade-off. A classic example of the trade-off stems from the fact that those subjects who are willing to accept randomization to conditions are quite unlikely to be similar to those subjects to whom we would like to generalize. After all, randomization is an interference in one's routine, forcing one to risk being assigned to some condition that may be inconvenient at best and noxious at worst. Small wonder, then, that Gholson and Houts did not even try to assign real scientists randomly to conditions. More to the point (and even Brown misses this), even if Gholson and Houts had approached real scientists to gain their consent to participate in the experiment, those few who might consent would undoubtedly be unrepresentative of the population of scientists to whom we wish to generalize. They might be the least busy, or be at Gholson and Houts's institution, for instance. While such a sample is superficially more representative of scientists than it is of undergraduate students, a generalization problem would still remain to some nontrivial degree.

We can elaborate on this trade-off in at least three ways. First, the peculiarities concerning which subjects are willing to be randomized are not the only trade-off. Some trade-offs are logistical. Social experiments are often resource intensive, for example, so that paying for their implementation at one site often leaves little left for implementing the experiment elsewhere. Other trade-offs are conceptual. For instance, experiments require some manipulable cause, but the very act of manipulation can interfere in the natural ecology of the setting of interest, thus producing unrepresentative results. Second, the trade-off works both ways (Brown misses this point too); methodologies that facilitate external validity lack the key features of the randomized experiment that facilitate strong causal inference. Hence, the humanistic methods that Brown (1989) advocates are as unlikely to yield strong causal inferences as the experiment is to yield generalizations. This is not a matter of whether quantitative or qualitative methodologies are used, for a similar weakness in causal inference is true of quantitative sample surveys in the sociological tradition. Third, although the focus of the present discussion is on the generalizability of results from the randomized experiment, in principle the trade-off that Brown is claiming need not be tied to the experiment. After all, Brown's (1989) essential criticism of Gholson and Houts (1989) was that they failed "to make use of scientifically trained subjects" (Brown, 1989, p. 136). By implication, Brown would probably question the generalizability of any other methodology—the survey or a case study—that departed far from the use of scientists as subjects.

To the extent that this trade-off exists, it is not limited to an experimental psychology of science. Many sciences have both basic and applied research components. The former are usually conducted in laboratory settings under purified, controlled conditions; the latter are often conducted in field settings with more realistic or naturally occurring conditions. Generalizability is always an issue of debate in all such sciences. In this sense, then, laboratory-based experimental social psychology might be viewed as the basic laboratory component of science studies that are conceived of broadly to have both basic and applied components. Distinguishing between basic versus applied components does not lessen the trade-offs involved, but it does suggest that they may not be unique to SPS in science studies. If we reject these laboratory experiments in SPS, we must also reject them in other fields as well, and few observers are likely to do so.

Nonetheless, Brown has identified a problem that is not only important but probably widely viewed by the science studies community as a plausible threat to the contributions of an experimental psychology of science. Specifically, how do we generalize causal connections, especially those that arise in experiments, and even more especially those that arise in contrived laboratory situations with apparently unrepresentative subjects and tasks?

Generalization and Proximal Similarity

Brown implicitly uses a principle that Campbell (1986) calls proximal similarity to criticize the generalizability of the Gholson and Houts study. That is, when we have found an effect in an experiment, "we will apply it with most confidence where treatment, setting, population, desired outcome, and year are closest in some overall way to the original program treatment" (Campbell, 1986, pp. 75–76). According to this rule, Gholson and Houts's results are not very generalizable since the target is not proximally similar to what was studied—practicing scientists in real laboratories who are working on real tasks with real colleagues around the world today bear little apparent resemblance to undergraduates in laboratories working on contrived tasks in contrived groups in the Psychology Department at Memphis State University several years ago. Houts and Gholson (1989), in responding to Brown (1989), call this a criticism of the *ecological validity* of their work (Gorman, Chapter 4, this volume, uses the same term). We prefer proximal similarity to ecological validity to keep clear the distinction between mechanisms for enhancing validity versus types of validity to be enhanced. What has been called ecological validity might more accurately be thought of as a mechanism (proximal similarity) to enhance external validity.

Houts and Gholson (1989) then reject Brown's criticisms as missing the point, suggesting that Brown focuses only on the superficial features of their work (mundane realism) whereas their experiments really probed the true underlying processes at work in scientific-type thinking. This response, however, assumes that *either* proximal similarity or probing true underlying processes must be the sole arbiter of external validity, when in fact both (and much more) are relevant. Like it or not, proximal similarity is indeed one proper mechanism for enhancing generalizability of experiments, and Brown is perfectly correct in pointing to this limitation of the Gholson and Houts studies. Where Brown errs is in implying that proximal similarity is the only mechanism to facilitate generalization, and that studies that fail this test cannot have any external validity. This is overstated because other mechanisms to facilitate generalization do exist.

But before going on to these other mechanisms, let us examine proximal similarity a bit further. Following Cronbach (1982), experiments are composed of *units* being treated, such as undergraduates or scientists, the *treatment* conditions to which they are assigned, the *observations* that are made on those units on pretest and posttest measures, the *setting* in which all this takes place, and the relationships among these four components (from the first letter of each of these four words, Cronbach coined "utos" to indicate the study as done). Experiments enhance their external validity via proximal similarity to the extent that these components as operationalized resemble the targets of interest. Hence, it is too limited to criticize the Gholson and Houts studies solely on the basis that the units (students) were not very similar to the targets (scientists). To some degree those studies were proximally dissimilar to all their respective targets. Of course, this judgment would vary somewhat depending on which component was examined. The setting, a research laboratory in a research-oriented university psychology department, was probably the component that was least distant from a reasonable target because much science does take place in such settings—although the social roles involved were probably quite different. Hence, clearly this judgment is not all or none, as the component is similar in some ways (e.g., general physical location) but dissimilar in others (e.g., not providing extensive funded time to the subjects to do research), so the dimension on which similarity is judged will make a difference.

Equally clearly, the judgment of similarity depends on which target is chosen. Results are not generalizable either to everything or nothing, but more or less to different targets. Cook and Campbell (1979), for example, distinguish between generalizing *to* versus *across* targets. In the former case, one generalizes from the study as done (utos) to the study as intended (UTOS)—the kinds of units, treatments, observations, and settings to which the investigator designed the study to be relevant. In the latter case, one generalizes from the study as done across unintended targets (*UTOS) that later prove to be of interest even though were not originally the focus of study. Generalizing "across" is always more difficult than generalizing "to"; however, generalizing across is often the most frequent use of research results. Consider, for instance, the generalizations from Gholson and Houts's second study on how the cognitive structures of psychology students change over the course of a semester's study of cognitive psychology. Presumably generalizing from the undergraduates in this study to other psychology undergraduates is most proximally similar, across domains to psychology graduate students next most, but across domains even further to physics graduate students or across practicing biologists the least similar.

Generalization and Sampling

Proximal similarity requires that one make a difficult and fallible judgment about how similar the sample and the target might be. One can reduce the need for such judgments by using some form of sampling. Under the classical statistical model, generalization is facilitated by systematic sampling with known probabilities. Of the relevant methods, random sampling of a subset of units from a known population is the best known instance, and it allows one to set error limits on the likelihood that the obtained results generalize to the population. Many public opinion polls use variants of this technique. Similarly, in studying scientists' perception of citations, Shadish (1989a) randomly sampled one citation from each article that appeared in the 1985 volume of each of three journals in psychology; the warranted generalization in this study is from the sample findings to all citations that appeared in those articles.

In principle, then, one can easily generalize from studies to populations when an appropriate sampling of units, treatments, observations, or settings is used. But two problems thwart the use of this solution on a routine basis. One is the previously described trade-off between random sampling and random assignment. The Shadish (1989a) study was a survey, not a randomized experiment, and so did not encounter this difficulty. The first Gholson and Houts (1989) study did encounter it. To use sampling as a solution, they would need a population from which they sampled randomly, and those subjects would have to agree to be assigned randomly to conditions. But in a randomized experiment with real scientists and with treatment conditions that might involve distribution of resources or other elements that might affect science, randomly sampled scientists may not be willing to allow themselves to be assigned randomly to treatments. They may, for instance, prefer to fight to get the best available treatment rather than risk getting an inferior one. If they drop out, the random sampling scheme is vitiated. The second problem is that it is differentially feasible to randomly sample units, treatments, observations, and settings. Ideally, such sampling requires knowledge of the identity of each population member, which is sometimes possible for units and settings but rarely possible for observations, and probably never for treatments. Even when it is possible in principle, the practical costs of sampling units and settings are far greater than would be the case if the study could be done in one setting with convenient units. Thus, we must anticipate that statistically sophisticated methods of sampling will rarely prove to be a widely practical means of generalizing from studies.

Still, however, two options can be used to approximate the benefits of random sampling when generalization is a desirable goal but better sampling methods are not possible. One is that even when systematic sampling with known probabilities cannot be achieved, fallback methods may still warrant generalization. Two such fallbacks are deliberately heterogeneous sampling and sampling modal instances. In heterogeneous sampling, one samples units that are known to be different from each other on putatively important dimensions. For example, if the performance of laboratory scientists in the nation is at issue, random sampling may not be possible because we lack a list of all such scientists, or even of all such laboratories. But we do have reason to think that laboratories in the public sector differ in important ways from those in the private sector, and that university-affiliated laboratories differ from those not so affiliated. Designing one's sample to take cases that vary on these dimensions increases the warrant for subsequent generalization. In modal sampling, one samples cases that are thought to be similar to the most commonly occurring type of case in the population to which one wishes to generalize. For instance, if we know that private-sector research and development laboratories are the source of most laboratory science, sampling from such laboratories at least increases the likelihood that one's results will be of some relevance to the largest possible subset in the population. Of course, data are often not available to describe the characteristics of the modal instance in a population; under this condition, the researcher simply makes his/her best judgment about how the case that was studied might reflect different target cases—back to proximal similarity again.

A second option for approximating the benefits of better sampling procedures occurs in data analysis. First, the use of random-effects data analyses allows us to place some confidence bounds on findings under the assumption that they were sampled randomly from some population of like instances. In essence, those confidence intervals are widened considerably, reflecting the greater inherent variability likely to exist in the population than in the obtained sample. Second, one can divide the data from the available sample into subsets that vary from each other in ways that approximate population variability (on units, treatments, observations, or settings), and then examine whether results generalize over these subsets. For example, in their first study Gholson and Houts (1989) used undergraduate students to test their hypothesis about the effects of working in groups on solving the 2–4–6 problem. Undoubtedly some of those students were science majors while others were not, some were advanced undergraduates but not others, and within the sciences some were in the natural sciences and others in the social sciences. Examining study results by such subsets is another way to explore generalizability, although as progressively finer subsets are created the sample size in each subset decreases rapidly.

Ultimately, however, there is another practical limit on the extent to which sampling can be used to facilitate generalization from experiments. Although one often can sample multiple units and observations, it is costly to conduct the research in multiple settings, and difficult to implement multiple variations of the treatment. Doing all this simultaneously is usually beyond the scope of any experiment. Any given experiment will almost inevitably be restricted in the range of its sampling. Thus, we rarely expect a single experiment to have truly widespread generalization through sampling mechanisms.

Generalization and Robustness

Fortunately, we need not limit ourselves to generalizing from just one experiment. On any given topic, more than one experiment has often been done, each differing somewhat from the other in units, treatments, observations, and settings. If so, one route to causal generalization is to demonstrate that some effect has occurred robustly over different variations in each component. Consider, for example, the many laboratory experiments that psychologists have done over the years on cognitive strategies and biases. In general, they suggest that undergraduate students are not good at using disconfirmation strategies—although they can achieve some limited success if they are instructed to disconfirm (Gorman, Chapter 4, this volume). Over all these experiments, the effect seems to generalize in the sense that it has been demonstrated not just with the 2–4–6 task but with a host of other tasks that differ from each other on such dimensions as content, technology used, and duration. Similarly, this research has been conducted in a variety of universities in the United States, suggesting that it generalizes over this aspect of setting.

The goal in this search is twofold, to examine the variability of the effect and to examine factors that may moderate that variability. The former search is based on the premise that even under the best of circumstances (e.g., pure replication, perfect measurement, and no errors of implementation), the results of multiple studies on a topic will vary somewhat due to sampling error, that is, because each study is based on a sample of people who differ from each other and from the population in important ways both within and between studies. Examining the distribution of effects helps us know whether the effect is more or less variable and whether we can expect sometimes to find no effect or even effects in the opposite direction from the central tendency. Such sign reversals are more threatening to causal generalization than are simple fluctuations in the size of an effect that still remains in the predicted direction. However, we rarely work under the "best of circumstances"; studies are rarely intended to be strict replicates, for example, at least in the social sciences. Hence the search for moderators of effects is also crucial to test the limits of causal generalization when units, treatments, observations, or settings are known to be different.

Meta-analysis has a contribution to make here as a quantitative tool for exploring the robustness of some observation or some hypothesis. After a point, summarizing the mere existence of an effect is of limited interest, as many critics said of Rosenthal and Rubin's (1978a) summary of the effect across 345 studies (Rosenthal & Rubin, 1978b). Exploring the robustness of the effect over variations in units, treatments, observations, and settings is still worth doing, though, in the interests of causal generalization. For example, Rosenthal and Rubin (1978a) showed that the interpersonal expectancy effect seemed to vary considerably depending on which research area was studied, with relatively large effects being obtained for animal learning or psychophysical judgments but with smaller effects for human reaction time. They also showed that the effect was considerably smaller in dissertations than in publications. Such variations not only suggest the limits of generalization but also prompt us to develop hypotheses to explain their occurrence.

Collins (1985) makes brief mention of the Rosenthal and Rubin (1978a) meta-analysis in his book on replication in science, reaching skeptical conclusions about both. But it may be less useful to think of meta-analysis as a tool for confirming replication in some dichotomous fashion, as Collins does, than as a tool for exploring generalization through the mechanism of robustness assuming a distribution of study characteristics and outcomes. Consider the disconfirmation literature, for example; Gorman's (Chapter 4, this volume) research suggests that he can get disconfirmation strategies to work under tight laboratory constraints. Meta-analysis could suggest how disconfirmation strategies worked across a wider range of situations. Similarly, few if any social psychologists would view the 345 studies in the Rosenthal and Rubin meta-analysis as strict replications in which "the variation among experiments is entirely accounted for by random fluctuations in unknown background variables" (Collins, 1985, p. 42). Nor is this latter assumption necessary in meta-analysis, although a relevant statistical test is available (called the test of homogeneity of effect size or of model specification) (see Hedges & Olkin, 1985). Experience has shown, however, that this test is usually rejected, indicating (roughly) either that some nonrandom background variables are contributing to effect size variability or that the effect itself may result from random variance in the population (Shadish & Haddock, in press). Indeed, as Collins points out, Rosenthal and Rubin go on to explore such potential sources of variability as described in the previous paragraphs. This does not imply that the worth of meta-analysis turns "on the possibility of distinguishing between competent and incompetent experiments" (Collins, 1985, p. 42). Rather, all such sources of variation, including but not limited to what is counted as a competent experiment, are grist for the mill of exploration. In fact, different meta-analysts often reach somewhat different conclusions when synthesizing the same research literatures, sometimes tending more toward agreement as in the case of the effects of professional training on psychotherapy outcome (Berman & Norton, 1985; Durlak, 1979; Hattie, Sharpley, & Rogers, 1984), but sometimes tending more toward disagreement as in the case of the effects of school desegregation on children in the United States (Wachter & Straf, 1990) or of the psi ganzfeld research discussed previously.

Hence, one way to think of meta-analysis is in terms of categories of studies like some prototype rather than as replications of some "first experiment." When one considers various instances of members in a category, one tends to have in mind certain instances that best represent that category. The latter are prototypes. For example, apples are probably a more prototypical instance of fruit than are lemons or figs; carrots are a more prototypical vegetable than are pickles. Such categories mostly do not have clear boundaries, nor do their instances have uniform characteristics. Rather, they are bound together by "some set of attributes, none true of all instances, but each true of some, a collection of overlapping short-range similarities" (Brown, 1986, p. 472). Prototypical instances are not necessary better or preferable—just because a robin is a more prototypical bird than a penguin does not mean the latter is a worse bird. In this terminology, meta-analysis could be thought of as a form of prototypification and protodifferentiation—that is, exploring how studies do and do not resemble some prototype on dimensions that include but are not limited to study results. Note that the prototype study itself need not even exist; it may be a concept, only minimally explicit, that the scientist has in mind in deciding what to include in some category of studies; and the prototype can change over scientists and within scientists over time.

The major problem with the robustness approach to causal generalization is the presence of constant biases across multiple experiments so that the effect has not been shown to be robust in important ways. Some constant biases are known. With some exceptions (Kahneman, Slovic, & Tversky, 1982; Tversky & Kahneman, 1971), for example, research on the cognitive psychology of science seems mostly to have been conducted with undergraduate students, so we do not know much about how robust the effect is in other kinds of subjects. Similarly, observations of the effects of these studies have mostly been concerned with whether subjects reach the correct conclusion. But Gorman (Chapter 4, this volume) reports that the treatments in these studies seem to generate much discussion among subjects about such matters as better or worse ways of proceeding toward a solution. Few if any researchers have systematically examined such variations in outcome, so we have little idea about the robustness of these treatments across different dependent variables.

Even among the dimensions that have been varied somewhat—treatment and setting—constant biases still remain to be examined. Most of the settings, for example, have been universities.

More worrisome are the constant biases that are undetected, especially those features of the units, treatments, observations, or settings that are particularly different from the real world of science in crucial ways. Criticism such as that of Brown (1989), Nickles (1989), or Woolgar (1989) is invaluable in ferreting out such biases; in fact, it is not clear that better ferrets exist than one's sharpest critics. Indeed, some of Brown's criticisms of Gholson and Houts (1989) would undoubtedly characterize the bulk of this sort of laboratory research—for instance, the tasks in most of these experiments may not ask for much real conceptual innovation from the subjects relative to the task facing many scientists. But this, too, is a matter of debate, since Brown makes assumptions about what most scientists do with their time that may be quite unwarranted—they may spend much more of their time on mundane tasks than we know because we have not really looked at it closely.

Some of these constant biases, of course, arise from the nature of the randomized experiment, as when it seems unlikely we will be able to generate many randomized experiments on how scientists themselves use disconfirmation strategies. But again, programs of research that incorporate randomized experiments, quasi-experiments, and other methodologies can provide some information about robustness. We return to this matter shortly in the section on Programs of Research.

Generalization and Theoretical Explanation

Finally, causal generalization can also be facilitated by theoretical explication of the more and less essential elements and processes for an effect to be produced. This involves decomposing the molar cause-and-effect constructs into subsets and explicating the processes that may mediate causal relationships among all these subsets. Sometimes common causal processes underlie apparently different phenomena. Knowledge of those processes would facilitate our ability to reproduce the effect with other places, times, and people by focusing our efforts on transferring the essential components only. Cook and Campbell (1979) use the example of a light switch to illustrate the point. Knowledge that flicking the switch results in light is knowledge of manipulanda of the sort experiments give; knowledge of switch mechanisms, wiring and circuitry, and electricity and how they are combined to produce light is explanatory knowledge. The latter is more dependable than the former for facilitating causal generalization because it ensures that we can create the proper conditions for lighting in a wide array of circumstances where light switches currently do not exist. Knowledge of manipulanda is only useful for generalizing to places where light switches already exist.

This sort of rationale has long undergirded basic theory development in social psychology. For example, Jones (1985), in his review of major developments in social psychology over 50 years, said:

> Without some kind of bridging theory of interpersonal process, would-be experimenters were thwarted by what might be called the generalization question. How could an experimenter claim that his findings on thirty male college sophomores were in any important sense generalizable to a broader population of human beings or even to college males? Sociology essentially confronted the generalization question by abjuring the laboratory experiment and constructing or testing theories through survey methods in which sample representativeness was an important consideration. Psychologists, on the other hand, developed and refined experimental techniques that would test the plausibility of general process theories in restricted concrete contexts. In the late 1920s and 1930s this effort increased particularly in studies of animal behavior in which psychologists like Hull and Tolman attempted to theorize about general learning processes from data produced by rats in mazes. Thus there developed a broad context in U.S. psychology nurturing the importance of theory as a bridge between concrete experimental findings. As the experimental tradition developed in social psychology, researchers became more preoccupied with the conceptual generality of their findings than with the representativeness of their samples. Theories were useful to the extent that they predicted superficially different but conceptually similar relations in a variety of contexts. (p. 67)

A classic example in social psychology concerns the Kitty Genovese murder, where the victim was stabbed repeatedly for over half an hour while nearly 40 neighbors watched and did nothing to help. Latane and Darley (1970) hypothesized that a process underlying this apparent apathy was diffusion of responsibility: Everyone thought someone else would surely help. In a series of subsequent experiments, they demonstrated that this process was common to a host of other apparently diverse situations. Their general approach was to manipulate the number of bystanders as the independent variable; the results showed that when there are more people, responsibility is diffused and each person feels less obligation to act. This diffusion seems to occur in situations as diverse as a woman falling and spraining her ankle, smoke coming into a room from a fire, someone having an epileptic seizure, and a cash register being robbed (Brown, 1986).

Another example concerns Rosenthal's expectancy effects. Rosenthal (1973a, 1973b) proposed a four-factor theory of the mediation of teacher expectancy effects—at least some of these are likely to mediate interpersonal expectancy effects in general. The four factors are (1) cli-

igh expectations lead to a warmer socioemotional climate;
that high expectations leads to more differentiated feedback
cy of student responses; (3) input, that teachers will teach
more difficult, material to high-expectancy students; and (4) output, that teachers will give high-expectancy students more opportunity to respond. In 1985, Harris and Rosenthal (1985) located 135 studies that examined relationships between expectancies and these mediators, or between the mediators and outcome. All effect sizes were converted to the common metric of correlation coefficients and then aggregated over studies separately for each of the four factors. In general, the results suggested that all four factors helped mediate the effect. Armed with such knowledge, we are better able to know exactly how to produce (or minimize) the effect across diverse situations, thus fostering causal generalization.

This mechanism for facilitating the generalization of experimental results also has its problems, not the least of which is the previously mentioned fact that the essential components of the randomized experiment (random assignment of units to conditions with posttest) give us no particular information about the processes mediating the cause-and-effect relationship. Hence the addition of methods for generating and exploring causal mediation is required. These can be as qualitative, simple, and exploratory as Gorman's (Chapter 4, this volume) ad hoc reports of the comments of subjects about how they were solving problems, or they can be as quantitative, complex, and confirmatory as linear structural equation modeling to test a priori hypotheses about causal mediation. Social psychologists generally take two approaches to this problem, however. One is that certain features can be added to the randomized experiment to aid in causal explanation, such as the assessment of mediators that can be modeled in path analysis (e.g., Fiske, Kenny, & Taylor, 1982). The other is that certain aspects of causal explanation can be addressed in randomized experiments through careful control of the conditions to which subjects are assigned, or through the use of additional causal agents in factorial designs. To give just one oversimplified example, if one hypothesized that Milgram's obedience paradigm had its effects through the credibility of the experimenter, one could hold other conditions constant but vary credibility to see if the effect remains. The limitation of this latter strategy is that relatively few explanatory mechanisms can be investigated at one time using this method, so it leads to slow cumulation of knowledge. But Milgram's (1974) extended experimental research program on obedience illustrates the process involved.

In any case, this sort of causal explanation is what Houts and Gholson (1989) seem to have in mind when they responded to Brown (1989) that "concerns about ecological validity are misleading precisely because they focus attention on surface features rather than on underlying processes" (Houts & Gholson, 1989, p. 143). They point to Kulkarni and Simon's (1988) computer simulation of the discovery of the urea cycle, in which about half of the processes used in the simulation were independent of the domain being investigated. The hypothesis that these are common underlying processes across a range of scientific activities is, of course, fallible so that future research may change our opinion of them. Further, the claim to have found a generalized mediating process is an empirical one that cannot be established in a single study. However, identifying such processes would indeed improve our ability to generalize from the results of the Gholson and Houts experiments.

Conclusion about Generalization

Brown (1989) was correct to call attention to the serious issue of generalization from the Gholson and Houts (1989) studies to science in general. His criticisms of the proximal similarity of the Gholson and Houts studies to science may have referred only to the superficial aspects of their studies, but the criticisms were not themselves superficial on that account because proximal similarity is one legitimate mechanism for facilitating causal generalization. He was probably wrong, however, to claim so extremely that Gholson and Houts offered nothing to the study of science. As an exploration of possible underlying cognitive processes, and as one study of many that explore the robustness of such phenomena, the Gholson and Houts studies are a small contribution to the possibility of generalized knowledge about scientific cognition from a psychology of science.

On the other hand, this discussion should not be taken to imply that causal generalization from social psychological experiments is automatically ensured through the mechanisms we discussed. To the contrary, we have emphasized that the randomized experiment, by itself, rarely includes these mechanisms. Hence, active adaptation on the part of social psychologists will be required by (1) facilitating proximal similarity through studying real science when possible; (2) attending more carefully to nonredundant and representative sampling of units, treatments, observations, and settings in single studies and in programs of research; (3) seeking out multiple experiments on a topic to examine robustness of effects across diverse operationalizations, or pointing to obvious gaps in those operationalizations; and (4) including the necessary qualitative and quantitative means of explaining observed effects through their mediating processes. This is an ambitious agenda, and no single experiment is capable of doing it all well. Still, these are the steps that need to be taken.

Social Psychological Methods in Context

We have made a broad sweep of the methodological contributions of SPS. Social psychologists should and undoubtedly will use these methods in much of their work because these are methods that they do well. But such methods are not sufficient by themselves for science studies generally, or for a viable SPS by itself, or even in the research of a single practitioner of SPS. In this section we attempt to place such matters into a somewhat larger context of programs of research that build on multiple methods.

Using Case Studies

Social psychologists rarely use case-study methodologies, but such work is crucial to the other science studies disciplines. Hence, social psychologists need to be conversant with existing case studies to draw connections between the cases and their own work. At the end of Chapter 8, for example, Rosenwein (this volume) calls for the establishment of a well-worked data base of case studies and the formulation of criteria for those case studies to be developed that would be most useful for social psychological analyses. For example, the style of sociologists of science is to study in microscopic detail the routine practices of scientists. But Rudwick (1985) feels that these ethnographies are often static in character and do not give a sense of the way in which routine procedures are used in a temporal process to develop new scientific beliefs. He calls for studies that do the following:

1. Focus on a specific *problem* that brought together some group of individuals in an interacting network of exchange.
2. Attend to the role of all participants, however minor their contributions seem.
3. Follow the dynamics of interaction with equal attention to exchanges both public and private, formal and informal, and ritualized and spontaneous.
4. Explore the *meaning* of the activities for the participants themselves.

These case studies can be either historical or about current controversies; and they can describe a variety of levels of analysis from fine grained to macroscopic.

A wide array of such case studies are already available so that social psychologists will rarely need to create them anew. Rudwick's (1985) description of the Devonian controversy is one of the most salient recent ones, but one can hardly pick up any work in sociology, history, and philosophy of science without encountering them (e.g., Laudan, 1990). We also see the occasional development of clearinghouses for some cases that keep tabs on current work in the field, produce bibliographies, and stimulate research. For example, historians at Cornell are funded by the National Science Foundation to gather materials on the cold-fusion controversy. Social psychologists of science should know about this archive, know how to access it, and build it into their work. It will help, in turn, if case-study researchers include information about the kinds of concepts and theories that social psychologists study; thus, one rationale for this chapter is that it will help guide interested case-study researchers to include relevant information.

Experiments in Context

Experiments, by themselves, are limited statements about the effects of a molar cause on a molar effect in a particular historical context; and the context is often a university social psychology laboratory with undergraduate psychology students. The relevance of such experiments to science studies is not obvious, so it is no wonder that Brown (1989) had the reaction he did. Ideally, of course, social psychologists would modify their practices in the ways discussed previously to produce work that is more obviously generalizable to science studies. But one cannot be hopeful that this modification will be widespread because it is expensive and time-consuming, and because probably relatively few social psychologists have sufficient interest in science studies to make the relevant changes. Hence, much of social psychology will remain laboratory-based efforts conducted by social psychologists who have no apparent interest in science studies.

Yet even these very limited laboratory experiments have a role to play in science studies in two ways. One is that basic laboratory experiments provide a new vocabulary for analyzing the conduct and context of science. That vocabulary may point to new directions for such analyses to take. An example is the extent to which Moscovici's work on social influence processes seems to lend itself naturally to analysis of how minority ideas eventually come to be accepted by a majority of scientists. Turner (Chapter 13, this volume) notes this possibility, but warns social psychologists that convincing the science studies community to adopt any new vocabulary will be difficult at best. We agree. It will take careful analysis to show that any new vocabulary carries implications not already carried in existing vocabularies; and it will take convincing empirical demonstrations to show that any new vocabulary is a difference that makes a difference. But in this regard we appreciate the comments of

Serchuk (1989)—another respondent to and critic of the Gholson and Houts (1989) study—who shared Brown's (1989) uneasiness but who preferred "to use this uneasiness as a prod for a nascent research program, rather than as grounds for rejecting the whole thing out of hand" (Serchuk, 1989, p. 147). Developing such a program will take time, perhaps a good deal of time given the relatively small number of social psychologists interested in the task and the relatively ambitious causal generalization program previously outlined.

Second, basic laboratory experiments play a role in integrated programs of research that (1) use *sequences of experiments* to expand the warrant for causal generalization, and (2) use or refer to *multiple methods* such as case studies, relevant philosophical and sociological analysis, or quantitative studies of other sorts. Both these recommendations embed experiments in *programs of research* that might be conducted by those social psychologists who do have a persistent interest in science studies.

Programs of Research

Earlier we noted that the laboratory-based experimental version of SPS might be thought of as one of the basic research components of science studies. But we have also seen that any such laboratory-based experimental program needs to be coupled with research using other methodologies that can (1) answer questions other than those about the existence of cause–effect relationships, (2) provide other perspectives about those cause–effect relationships, and (3) explore the generalizability of experimental results. These other methodologies should include, for example, case studies and field studies of a correlational or quasi-experimental nature directed at social–psychological variables in science.

For example, one possible organization of such research programs suggested by Amabile's (this volume) work on creativity might be a 3 × 3 matrix in which three units of study (individuals, groups, organizations) are crossed with three investigatory styles differing in degree of leverage for drawing causal inference (randomized experiments, relational studies, descriptive studies):

1. Examples of experimental research:

   a. At the individual level include studies of the effects of self-expectations or coworker expectations or supervisor expectations on the individual's productivity, creativity, or effectiveness.
   b. At the group level include studies of the effects of any of the

above expectancy variables, or team-building types of interventions of the effectiveness of the group.

c. At the organizational level include studies of the effects of Amabile-like or Langer-like manipulations on creativity, or the effects of experimentally created microclimates on effectiveness might be studied or the effects of constituting work teams using sociometric and psychosocial variables.

2. Examples of relational research include studies of the prediction of productivity, creativity, or effectiveness from a knowledge of the means *and variances* of work group demographic, personality, and other measures.
3. Examples of descriptive research include studies of agreements and disagreements between the ways it is suggested that science ought to be done and the way in which it actually is done, perhaps in the style of research among the laboratory ethnographers, or using systematic interviewing and survey research.

Moreover, each of these styles of research can be applied to the different tasks in which scientists engage, including but not limited to hypothesis generation and problem formation, preparation for doing research (as when gathering background information), hypothesis testing, interpreting results, communication of results, and subsequent further refinement of ideas. Of course, no program of research is likely to be so heroic as to tackle all these tasks at once, but these tasks do suggest the range of possibilities.

An example of a research program follows (see Gorman, 1992b, for a more complete account). Schiaparelli described "channels" (canali) on the surface of Mars in 1877. Earlier observers had detected some vague, streak-like features, but Schiaparelli's "were finer, sharper and more systematically arranged on the surface" (Hoyt, 1976, p. 6). Percival Lowell, a wealthy American who developed a strong interest in astronomy, transformed these channels into canals and used them to make a strong argument for intelligent life on Mars. Lowell's hypothesis was that Mars was a dying planet, slowly drying out. The intelligent inhabitants, in order to survive, had built huge networks of canals to irrigate their world, using the water that melted from the polar caps. These canals were surrounded by belts of vegetation, which is why they were visible from the earth. Other observers disagreed, and a heated debate lasting over a quarter of a century was carried on in both scientific journals and the popular press. This scientific controversy raises interesting questions that SPS can help answer.

Throughout the controversy, Lowell argued persuasively for his point of view, and made no concessions to his opponents. This is what

Moscovici (1974) called a consistent behavioral style that ought to permit a minority to exert influence over a much larger majority. As Mitroff (1981) demonstrated in his study of Apollo moon scientists, this kind of stubbornness is often the hallmark of successful scientists:

> The three scientists most often perceived by their peers as most committed to their hypotheses . . . were also judged to be among the most outstanding scientists in the program. They were simultaneously judged to be the most creative and the most resistant to change. The aggregate judgment was that they were "the most creative" for their continual creation of "bold, provocative, stimulating, suggestive, speculative hypotheses" and "the most resistant to change" for "their pronounced ability to hang onto their ideas and defend them with all their might to theirs and everyone else's death." (p. 171)

Lowell certainly had a creative hypothesis, and he defended it until his death. He was also influential. His views were widely circulated and dominated popular thinking about Mars for over a decade. He also influenced a number of astronomers, who claimed to see the canals, including E.M. Antoniadi, director of the British Astronomical Association. As more powerful telescopes became available, more and more astronomers agreed that there were no canals. Still, as late as 1924 at least one astronomer claimed to see canals on Mars (Hoyt, 1976). Notice that all this social influence occurred despite the fact that Lowell was wrong about the existence of canals on Mars. Lowell himself died in 1916, confident that he had decisively confirmed the existence of canals and intelligent life on Mars.

The literature on social influence offers an explanation of Lowell's influence. Individual judgments of ambiguous stimuli are greatly influenced by others (Moscovici, 1974; Sherif, 1936). Even reports of unambiguous stimuli can be influenced by a unanimous majority (Asch, 1956). As long as Mars remained a fuzzy image in telescopes and photographs, Lowell could exert influence. He might have been more influential had he been a higher-status individual (Milgram, 1974) instead of an outsider to the scientific community, a serious amateur astronomer who built a major observatory but did not have the degrees or background of the professional astronomer.

Social psychologists should conduct more detailed analyses of case studies such as the "canals" controversy, perhaps in collaboration with historians. Social psychologists are sensitive to group dynamics and patterns of influence that others might miss. But the greatest strength of social psychology is its unique methods. Before considering other cases, let us sketch a program of social psychological research based on the canals controversy.

One of the reasons the canals controversy is uniquely suited to experimental simulation is that the participants themselves tried to resolve their differences experimentally. Douglass, one of Lowell's team of observers, tried to establish empirically whether Lowell's observations could be caused by some psychological phenomenon:

> I have made some experiments myself bearing on these questions by means of artificial planets which I have placed at a distance of nearly a mile from the telescope and observed as if they were really planets. I found at once that some well known planetary appearances could, in part at least, be regarded as very doubtful. (Hoyt, 1976, p. 124)

Douglass's loss of faith caused Lowell to fire him. Another astronomer, E. Walter Maunder, asked a group of English schoolboys to copy a picture of Mars; even though there were no lines on the original picture, the boys drew them in. Maunder concluded that the lines drawn by astronomers were no more real than those drawn by the boys: "It seems a thousand pities that all those magnificent theories of human habitation, canal construction, planetary crystallization and the like are based upon lines which our experiments compel us to declare non-existent" (Hoyt, 1976, p. 165). A similar set of tests were run on a group of French schoolboys, none of whom drew imaginary lines, but Maunder's critique caused the British Astronomical Association to reverse its position:

> The members . . . have themselves seen more than a hundred of the so-called canals during their observations since 1900, yet E. M. Antoniadi, their director and editor, apparently has their concurrence in holding that their eyes were probably deceived and that they really saw something very different from the straight lines they imagined they were looking at. (Hoyt, 1976, p. 165)

Details of these early experiments are hard to discover. A modern follow-up study might be able to address the issue whether the canals could have been a social–psychological phenomenon. Knowledge of the planets is much greater now than in Lowell's time. The Viking spacecraft demonstrated that there is almost certainly no life on Mars; there are features that resemble river beds, but these cannot be seen from earth. This knowledge would influence how subjects approached any stimulus resembling a planetary disk. But one could construct a situation in which people are shown blurred pictures of terrain features and asked to map them. In one condition, these subjects might be given some story about the presence of roads or rivers or other terrain features in this area and asked to try to include them in their maps. Another condition would be given no such story. If subjects in the first condition tended to draw the roads in, one could make the claim that a persuasive story could affect how people map terrain features.

We also noted that Lowell adopted an unwavering perspective on Mars for which he argued determinedly and persuasively. How effective was this argument in keeping the Mars debate alive? One could simulate its impact by using a Moscovici design in which a minority argue persuasively for the presence of roads on a map and attempt to persuade the rest of a group. One could couple this to an autokinetic effects design (Jacobs & Campbell, 1961) to explore factors affecting knowledge transmission. The autokinetic effect occurs when a point of light is shone at the front of an otherwise darkened room; individuals will see the light move differently, depending on their interpretation of their own eye movements. When subjects are run as a group, they influence each others' judgments, forming a kind of "group norm" regarding the distance the light appears to move (Sherif, 1936). Jacobs and Campbell established that this norm was at least partially passed along to new group members as the original group was gradually replaced, member by member (see Campbell, Chapter 2, this volume). An example is the passing on of norms against theorizing and for empirical research during the Devonian controversy; the norm continued over time, gradually easing as older members were replaced (Rudwick, 1985).

Similarly, one could take individuals who had already mapped roads on an ambiguous map and put them together as a group to compare their maps. One could arrange for a minority in this group to argue consistently for a particular system of features. One could gradually replace the original members with new members, providing additional opportunities to study similar ambiguous maps. Here, one could explore whether minority influence was transmitted across generations (see Rosenwein, Chapter 8, this volume). One could also investigate other mechanisms by which scientific "exemplars" are transmitted. For example, a debate could be arranged between groups that claimed to see roads and groups that did not, to study the way in which the groups sought and interpreted new evidence. One could even investigate which of two competing groups or "research programs" was most successful at attracting new converts, under a variety of circumstances.

The point is, designs of this sort permit us to study how social psychological variables affect judgments on tasks that simulate a scientific controversy. Independent variables could be derived from the philosophical literature, as was done in the early studies of confirmation and disconfirmation, or from case studies as in the canals controversy. Experimental studies could stimulate further case studies designed to investigate the external validity of the experimental results.

Another case eminently suited to experimental simulation is Rudwick's (1985), which we have mentioned extensively in this chapter. A group of "gentleman geologists" argued over the correct sequencing of geological strata in the 1830s and eventually reached consensus. However, the route to consensus was marked by disagreements over how particular areas should be sequenced and disputes concerning who was competent and who was in error. Again, this is exactly the sort of case study that cries out for a social psychological analysis, complemented by an experimental simulation. Unlike the Lowell case, we now need a better account of why the "right" view prevailed. It was by no means obvious at the time; the data did not fall into an unambiguous, easily resolved pattern. How do competing groups resolve anomalies in situations in which there is error and ambiguity?

One could set up a computerized task that would require subjects to resolve a complex sequence or pattern. As in the Devonian case, different groups could be given preferential access to computers containing information on different parts or areas within the pattern. This mirrors the situation faced by the geologists, who typically were experts in a particular geographical location and generalized from this area to others. Competing groups could be given brief opportunities to venture onto another group's "turf" and re-analyze their data. One could look at the effect of a consistent behavioral style, the amount of error in the data (cf. Gorman, 1986), or how often and under what circumstances groups shared data and hypotheses (cf. Laughlin, 1988).

Summary

SPS is likely to be characterized by the use of certain methods, such as the experiment, that have specific strengths and weaknesses. Nonetheless, such work will be most valuable when it is integrated into larger programs of research that include case studies, surveys, and a host of other methods. Such programs might be conducted by social psychologists themselves, or more likely in collaboration with colleagues in other science studies disciplines.

The Rest of This Book

In the remainder of this book, we continue to elaborate many of the ideas outlined in this extended introductory chapter. Chapters 2 (Campbell) and 3 (Fuller) are primarily conceptual and philosophical, exploring the warrant for SPS in past and present thinking in science studies. Campbell's description of SPS as straddling the divide between proepistemic and contraepistemic forces in sciences studies is perhaps the clearest program statement for SPS in this book; Fuller outlines a new strong program, the social psychology of scientific knowledge, as an agenda that those with general interests in SPS can pursue.

Chapters 4 through 12 present a host of applications of social psychology to science, all authored by social psychologists. Gorman (Chapter 4) describes what is probably the most laboratory-based, purely experimental SPS program in the book, with his studies on group confirmation and falsification strategies. Kruglanski (Chapter 5) extends his "lay epistemics" model to science, and especially to the question of motivated cognition in science. Rosenthal (Chapter 6) first provides us with a brief and highly personal reflection on his many years of work on interpersonal expectancy effects and then goes on to note some salutary effects that the application of meta-analysis might have in scientific work. Extending this latter discussion, Miller and Pollock (Chapter 7) turn reflexive, exploring the many ways that meta-analysis might be used to study, criticize, and improve the conduct of social psychology itself. Rosenwein (Chapter 8) then presents an extremely detailed examination of social influence processes in science, extending and broadening the work presented all too briefly in this chapter. Sieber (Chapter 9) turns to the moral and ethical context of science in her examination of the stages of scientists' reactions to raising ethical issues in their work. Wilkes (Chapter 10) summarizes his work on cognitive styles of thought in science, arguing that an emphasis on personality and creativity in SPS need not be inconsistent with an interest in sociological factors that affect creative scientific work. Amabile (Chapter 11) examines a host of social and contextual factors that contribute to or impede scientific creativity in research and development organizations. Finally, Westrum (Chapter 12) gives us a sociologist's view of SPS, describing his work on thinking by groups, organizations, and networks.

Chapters 13 through 15 present commentaries on the preceding chapters. We deliberately selected discussants with diverse views on the SPS enterprise and encouraged them to be critical. The result meets our every expectation (and then some). Turner (Chapter 13), a sociologist of science, cuts to the heart of the question of what it would take for SPS to attract much interest in science studies; his chapter should be required reading to keep us all humble. Tweney (Chapter 14), a cognitive psychologist of science, similarly challenges us to do SPS, not just to talk about it, and to show what SPS can do that cognitive science of science cannot do. Cook (Chapter 15), a social psychologist, reminds us of the many deficits that psychological social psychology has accumulated over the years, so that in some respects it is not the most ideal source of ideas for a truly social study of science. In Chapter 16, Shadish and Fuller respond to each of the three preceding commentaries.

These cogent criticisms give us all food for thought in developing the next generation of SPS. Toward that end, the book ends with three bibliographies of reading for those interested in further understanding and developing SPS. Chapter 17 (Fuller) presents a narrative guide to the philosophy and sociology of science for social psychology of science and Chapter 18 (Kruglanski) presents an annotated bibliography of social psychology readings that might be relevant to science studies. Chapter 19 (Lawless) discusses differences between psychological social psychology and sociological social psychology in topics covered and questions asked, with an extensive bibliography of examples. We trust that these three chapters, combined with the ideas and examples generated throughout this book, will provide the reader with sufficient grist to keep the mill busy for a long time to come.

Acknowledgments

The authors gratefully acknowledge permission to quote extensively from Martin J. S. Rudwick's *The Great Devonian Controversy*. Copyright 1985 by the University of Chicago. All rights reserved.

References

Aiken, L. S., West, S. G., Sechrest, L., & Reno, R. R. (1990). Graduate training in statistics, methodology, and measurement in psychology: A survey of Ph.D. programs in North America. *American Psychologist, 45,* 721–734.

Albaum, G. (1976). Selecting specialized creators: The independent inventor. *Psychological Reports, 39,* 175–179.

Allport, G. W. (1985). The historical background of social psychology. In G. Lindzey & E. Aronson (Eds.), *Handbook of social psychology* (3rd ed.) (Vol. 1, pp. 1–46). Hillsdale, NJ: Erlbaum.

Armstrong, J. S., & Soelberg, P. (1968). On the interpretation of factor analysis. *Psychological Bulletin, 70,* 361–364.

Aronson, E., Ellsworth, P. C., Carlsmith, J. M., & Gonzales, M. H. (1990). *Methods of research in social psychology* (2nd ed.). New York: McGraw-Hill.

Asch, S. E. (1951). Effects of group pressure on the modification and distortion of judgments. In H. Guetzkow (Ed.), *Groups, leadership, and men* (pp. 177–190). Pittsburgh, PA: Carnegie Press.

Asch, S. E. (1956). Studies of independence and conformity: A minority of one against a unanimous majority. *Psychological Monographs, 70* (9, Whole No. 416).

Ashmore, M. (1989). *The reflexive thesis: Wrighting the sociology of scientific knowledge.* Chicago: University of Chicago Press.

Ayer, A. J. (1936). *Language, truth and logic* (2nd ed.). New York: Dover Publications.

Bachelard, G. (1985). *The New Scientific Spirit*. Boston: Beacon Press (original work published 1934).

Barker, P. (1989). The reflexivity problem in the psychology of science. In B. Gholson, W. R. Shadish, R. A. Neimeyer, & A. C. Houts (Eds.), *Psychology of science: Contributions to metascience* (pp. 92–114). Cambridge, England: Cambridge University Press.

Barton, D. H. R. (1989). Foreword. In R. M. Roberts, *Serendipity: Accidental discoveries in science* (pp. vii–viii). New York: Wiley.

Bem, D. J. (1987). Writing the empirical journal article. In M.P. Zanna & J. M. Darley (Eds.), *The compleat academic: A practical guide for the beginning social scientist* (pp. 171–201). New York: Random House.

Berkowitz, L., & Donnerstein, E. (1982). External validity is more than skin deep: Some answers to criticisms of laboratory experiments. *American Psychologist, 37*, 245–257.

Berman, J. S., & Norton, N. C. (1985). Does professional training make a therapist more effective? *Psychological Bulletin, 98*, 401–407.

Blank, T. O. (1988). Reflections on Gergen's "Social psychology as history" in perspective. *Personality and Social Psychology Bulletin, 14*, 651–663.

Bloor, D. (1976). *Knowledge and social imagery*. London: Routledge.

Bollen, K. A. (1989). *Structural equations with latent variables*. New York: Wiley.

Boruch, R. F. (1975). On common contentions about randomized field experiments. In R. F. Boruch & H. W. Reicken (Eds.), *Experimental testing of public policy: Proceedings of the 1974 Social Science Research Council Conference on Social Experiments* (pp. 107–142). Boulder, CO: Westview Press.

Boruch, R. F. (1991). The President's mandate: Discovering what works and what works better. In M. W. McLaughlin & D. C. Phillips (Eds.), *Evaluation and education: At quarter-century* (pp. 147–167). Chicago: National Society for the Study of Education.

Boruch, R. F., McSweeney, A. J., & Soderstrom, E. J. (1978). Randomized field experiments for program planning, development, and evaluation: An illustrative bibliography. *Evaluation Quarterly, 2*, 655–695.

Brannigan, A., & Wanner, R. A. (1983a). Historical distributions of multiple discoveries and theories in scientific change. *Social Studies of Science, 13*, 417–435.

Brannigan, A., & Wanner, R. A. (1983b). Multiple discoveries in science: A test of communication theory. *Canadian Journal of Sociology, 3*, 135–151.

Brewer, M. B., & Kramer, R. M. (1985). The psychology of intergroup attitudes and behavior. *Annual Review of Psychology, 36*, 219–243.

Brown, H. (1989). Toward a cognitive psychology of what? *Social Epistemology, 3*, 129–138.

Brown, R. (1986). *Social psychology: The second edition*. New York: Free Press.

Buel, W. D. (1965). Biographical data and the identification of creative research personnel. *Journal of Applied Psychology, 49*, 318–321.

Campbell, D. T. (1957). Factors relevant to the validity of experiments in social settings. *Psychological Bulletin, 54*, 297–312.

Campbell, D. T. (1975). "Degrees of freedom" and the case study. *Comparative Political Studies, 8*, 178–193.

Campbell, D. T. (1982). Experiments as arguments. In E. R. House (Ed.), *Evaluation studies review annual* (Vol. 7, pp. 117–127). Newbury Park, CA: Sage.

Campbell, D. T. (1986). Relabeling internal and external validity for applied social scientists. In W. M. K. Trochim (Ed.), *Advances in quasi-experimental design and analysis* (pp. 67–77). San Francisco: Jossey-Bass.

Campbell, D. T. (1988). A tribal model of the social system vehicle carrying scientific knowledge. In E. S. Overman (Ed.), *Methodology and epistemology for social science: Selected papers* (pp. 489–503). Chicago: University of Chicago Press. (Original work published 1979).

Campbell, D. T. (1989). Fragments of the fragile history of psychological epistemology and theory of science. In B. Gholson, W. R. Shadish, R. A. Neimeyer, & A. C. Houts (Eds.), *Psychology of science: Contributions to metascience* (pp. 21–46). Cambridge, England: Cambridge University Press.

Campbell, D. T., & Fiske, D. W. (1959). Convergent and discriminant validation by the multitrait–multimethod matrix. *Psychological Bulletin, 56*, 81–105.

Campbell, D. T., & Stanley, J. C. (1963). *Experimental and quasi-experimental designs for research.* Chicago: Rand-McNally.

Chaiken, S. & Stangor, C. (1987). Attitudes and attitude change. *Annual Review of Psychology, 38*, 575–630.

Chambers, J. A. (1964). Relating personality and biographical factors to scientific creativity. *Psychological Monographs, 78* (7, Whole No. 584).

Chomsky, N. (1975). *Reflections on language.* New York: Pantheon.

Chubin, D. E., & Restivo, S. (1983). The "mooting" of science studies: Research programmes and science policy. In K. D. Knorr-Cetina & M. Mulkay (Eds.), *Science observed: Perspectives on the social study of science* (pp. 53–83). Newbury Park, CA: Sage.

Cicourel, A. (1973) *Cognitive sociology.* Harmondsworth, UK: Penguin.

Collins, H. M. (1983). An empirical relativist programme in the sociology of scientific knowledge. In K. D. Knorr-Cetina & M. Mulkay (Eds.), *Science observed: Perspectives on the social study of science* (pp. 85–113). Newbury Park, CA: Sage.

Collins, H. M. (1985). *Changing order: Replication and induction in scientific practice.* Newbury Park, CA: Sage.

Cook, T. D. (1990). The generalization of causal connections: Multiple theories in search of clear practice. In L. Sechrest, E. Perrin, & J. Bunker (Eds), *Research methodology: Strengthening causal interpretations of nonexperimental data* (pp. 9–31) (DHHS Publication No. PHS 90–3454). Rockville, MD: Department of Health and Human Services.

Cook, T. D. (1991). Clarifying the warrant for generalized causal inferences in quasi-experimentation. In M. W. McLaughlin & D. C. Phillips (Eds.), *Evaluation and education: At quarter-century* (pp. 115–144). Chicago: National Society for the Study of Education.

Cook, T. D., & Campbell, D. T. (1979). *Quasi–experimentation: Design and analysis issues for field settings.* Chicago: Rand–McNally.

Cook, T. D., & Shadish, W. R. (1986). Program evaluation: The worldly science. *Annual Review of Psychology, 37,* 193–232.

Cooper, H. M., & Lemke, K. M. (1991). On the role of meta-analysis in personality and social psychology. *Personality and Social Psychology Bulletin, 17,* 245–251.

Coulter, J. (1979) *Social construction of mind.* London: Macmillan.

Cowles, M. (1989). *Statistics in psychology: An historical perspective.* Hillsdale, NJ: Erlbaum.

Cronbach, L. J. (1982). *Designing evaluations of educational and social programs.* San Francisco: Jossey-Bass.

Cronbach, L. J., Ambron, S. R., Dornbusch, S. M., Hess, R. D., Hornick, R. C., Phillips, D. C., Walker, D. F., & Weiner, S. S. (1980). *Toward reform of program evaluation.* San Francisco: Jossey–Bass.

DeMey, M. (1992). *The cognitive paradigm* (2nd ed.). Chicago: University of Chicago Press.

Dennis, M. L. (1988). *Implementing randomized field experiments: An analysis of criminal and civil justice research.* Unpublished doctoral dissertation, Northwestern University, Evanston, IL.

Diesing, P. (1991). *How does social science work? Reflections on practice.* Pittsburgh, PA: University of Pittsburgh Press.

Doise, W. (1987) *Levels of explanation in social psychology.* Cambridge, England: Cambridge University Press.

Durlak, J. A. (1979). Comparative effectiveness of paraprofessional and professional helpers. *Psychological Bulletin, 86,* 80–92.

Eiduson, B. T., & Beckman, L. (Eds.). (1973). *Science as a career choice: Theoretical and empirical studies.* New York: Russell Sage.

Ellwood, C. A. (1925). *The psychology of human society.* New York: Appleton.

Ericsson, K. A., & Simon, H. A. (1984). *Protocol analysis: Verbal reports as behavior.* Cambridge, MA: MIT Press.

Faust, D. (1984). *The limits of scientific reasoning.* Minneapolis: University of Minnesota Press.

Festinger, L. (1954). A theory of social comparison processes. *Human Relations, 7,* 117–140.

Festinger, L. (1957). *A theory of cognitive dissonance.* Evanston, IL: Row, Peterson.

Festinger, L., & Hutte, H. A. (1954). An experimental investigation of the effect of unstable interpersonal relations in a group. *Journal of Abnormal and Social Psychology, 49,* 513–523.

Fienberg, S. E. (1971). Randomization and social affairs: The 1970 Draft Lottery. *Science, 171,* 255–261.

Fisch, R. (1977). Psychology of science. In J. Spiegel–Rosing & D. de S. Price (Eds.), *Science, technology, and society: A cross–disciplinary perspective* (pp. 277–318). Newbury Park, CA: Sage.

Fishbein, M., & Azjen, I. (1975). *Belief, attitude, intention, and behavior.* Reading, MA: Addison–Wesley.

Fisher, R. A. (1925). *Statistical methods for research workers.* Edinburgh, Scotland: Oliver & Boyd.
Fisher, R. A. (1926). The arrangement of field experiments. *Journal of the Ministry of Agriculture of Great Britain, 33,* 505–513.
Fisher, R. A. (1935). *The design of experiments.* Edinburgh, Scotland: Oliver & Boyd.
Fiske, S. T., Kenny, D. A., & Taylor, S. E. (1982). Structural models for the mediation of salience effects on attribution. *Journal of Experimental Social Psychology, 18,* 105–127.
Fodor, J. A. (1981). *Representation: Philosophical essays on the foundations of cognitive science.* Cambridge, MA: MIT Press.
Fox, M. F. (1983). Publication productivity among scientists: A critical review. *Social Studies of Science, 13,* 285–305.
Frank, M. L., & Rickard, K. (1988). Psychology of the scientist: LVIII: Anxiety about research: An initial examination of a multidimensional concept. *Psychological Reports, 62,* 455–463.
Fuller, S. (1991). Is history and philosophy of science withering on the vine? *Philosophy of the Social Sciences, 21,* 149–174.
Fuller, S. (1992). Epistemology radically naturalized. In R. Giere (Ed.), *Cognitive models of science* (pp. 427–459). Minneapolis: University of Minnesota Press.
Fuller, S. (1993). *Philosophy of science and its discontents.* (2nd ed.). New York: Guilford.
Fuller, S., DeMey, M., Shinn, T., & Woolgar, S. (Eds.). (1989). *The cognitive turn: Sociological and psychological perspectives on science.* Dordrecht, The Netherlands: Kluwer.
Geen, R. G. (1991). Social motivation. *Annual Review of Psychology, 42,* 377–400.
Gergen, K. J. (1973). Social psychology as history. *Journal of Personality and Social Psychology, 26,* 309–320.
Gholson, B., & Houts, A.C. (1989). Toward a cognitive psychology of science. *Social Epistemology, 3,* 107–127.
Gholson, B., Shadish, W. R., Neimeyer, R. A., & Houts, A. C. (Eds.). (1989). *Psychology of science: Contributions to metascience.* Cambridge, England: Cambridge University Press.
Giddens, A. (1984). *The constitution of society.* Berkeley: University of California Press.
Giere, R. N. (1989). The units of analysis of science studies. In S. Fuller, M. DeMey, T. Shinn, & S. Woolgar (Eds.), *The cognitive turn: Sociological and psychological perspectives on science* (pp. 3–11). Dordrecht, The Netherlands: Kluwer.
Glass, G. V. (1976). Primary, secondary, and meta-analysis of research. *Educational Researcher, 5,* 3–8.
Gorman, M. E. (1986). How the possibility of error affects falsification on a task that models scientific problem solving. *British Journal of Psychology, 77,* 85–96.
Gorman, M. E. (1992a). *Simulating science: Heuristics, mental models, and technoscientific thinking.* Bloomington: Indiana University Press.

Gorman, M. E. (1992b). Simulating social epistemology: Experimental and computational approaches. In R. N. Giere (Ed.), *Minnesota studies in the philosophy of science* (pp. 401–427). Minneapolis: University of Minnesota Press.

Gorman, M. E., & Carlson, B. (1989). Can experiments be used to study science? *Social Epistemology, 3*, 89–106.

Greenwood, J. D. (1989). *Explanation and experiment in social psychological science.* New York: Springer-Verlag.

Griggs, R., & Ransdell, S. (1986). Scientists and the selection task. *Social Studies of Science, 16*, 319–320.

Grover, S. C. (1981). *Toward a psychology of the scientist: Implications of psychological research for contemporary philosophy of science.* Lanham, MD: University Press of America.

Gruber, H. E. (1989). Networks of enterprise in creative scientific work. In B. Gholson, W. R. Shadish, R. A. Neimeyer, & A. C. Houts (Eds.), *Psychology of science: Contributions to metascience* (pp. 246–265). Cambridge, England: Cambridge University Press.

Hagstrom, W. (1967). *Competition and teamwork in science* (Final Report). Washington, DC: National Science Foundation.

Harris, M. J., & Rosenthal, R. (1985). Mediation of interpersonal expectancy effects: 31 Meta-analyses. *Psychological Bulletin, 97*, 363–386.

Hart, J. J. (1982). Psychology of the scientists: XLVI: Correlation between theoretical orientation in psychology and personality type. *Psychological Reports, 50*, 795–801.

Hastie, R., & Carlston, D. E. (1980). Theoretical issues in person memory. In R. Hastie, T. M. Ostrom, E. B. Ebbesen, R. S. Wyer, Jr., D. L. Hamilton, & D. E. Carlston (Eds.), *Person memory: The cognitive basis of social perception.* Hillsdale, NJ: Erlbaum.

Hattie, J. A., Sharpley, C. F., & Rogers, H. J. (1984). Comparative effectiveness of professional and paraprofessional helpers. *Psychological Bulletin, 95*, 534–541.

Haveman, R. H. (1986). *Social Experimentation* and social experimentation. *Journal of Human Resources, 21*, 586–605.

Hedges, L. V. (1983). A random effects model for effect sizes. *Psychological Bulletin, 93*, 388–395.

Hedges, L. V. (1987). How hard is hard science? How soft is soft science? The empirical cumulativeness of research. *American Psychologist, 42*, 443–455.

Hedges, L. V., & Olkin, I. (1985). *Statistical methods for meta-analysis.* Orlando, FL: Academic Press.

Heider, F. (1946). Attitude and cognitive organization. *Journal of Psychology, 21*, 107–112.

Helmreich, R. L., Spence, J. T., & Pred, R. S. (1988). Making it without losing it: Type A, achievement motivation, and scientific attainment revisited. *Personality and Social Psychology Bulletin, 14*, 495–504.

Heyes, C. M. (1989). Uneasy chapters in the relationship between psychology and epistemology. In B. Gholson, W. R. Shadish, R. A. Neimeyer, & A. C. Houts

(Eds.), *Psychology of science: Contributions to metascience* (pp. 115–137). Cambridge, England: Cambridge University Press.

Higgins, E. T., & Bargh, J. A. (1987). Social cognition and social perception. *Annual Review of Psychology, 38,* 369–426.

Hilgard, E. R. (1987). *Psychology in America: A historical survey.* San Diego, CA: Harcourt Brace Jovanovich.

Honorton, C. (1985). Meta-analysis of psi ganzfeld research: A response to Hyman. *Journal of Parapsychology, 49,* 51–91.

Houts, A. C. (1989). Contributions for the psychology of science to metascience: A call for explorers. In B. Gholson, W. R. Shadish, R. A. Neimeyer, & A. C. Houts (Eds.), *Psychology of science: Contributions to metascience* (pp. 47–88). Cambridge, England: Cambridge University Press.

Houts, A., & Gholson, B. (1989). Brownian notions: One historicist philosopher's resistance to psychology of science via three truisms and ecological validity. *Social Epistemology, 3,* 139–146.

Houts, A. C., & Haddock, C. K. (1992). Answers to philosophical and sociological uses of psychologism in science studies: A behavioral psychology of science. In R. N. Giere (Ed.), *Minnesota studies in philosophy of science.* (pp. 367–399) Minneapolis: University of Minnesota Press.

Hoyt, W. G. (1976). *Lowell and Mars.* Tucson: University of Arizona Press.

Hyman, R. (1985). The ganzfeld psi experiment: A critical appraisal. *Journal of Parapsychology, 49,* 3–49.

Hyman, R., & Honorton, C. (1986). A joint communique: The psi ganzfeld controversy. *Journal of Parapsychology, 50,* 351–364.

Jacobs, R. C., & Campbell, D. T. (1961). The perpetuation of an arbitrary tradition through several generations of a laboratory microculture. *Journal of Abnormal and Social Psychology, 62,* 649–658.

Johnson, J. A., Germer, C. K., Efran, J. S., & Overton, W. F. (1988). Personality as the basis for theoretical predictions. *Journal of Personality and Social Psychology, 55,* 824–835.

Jones, E. E. (1985). Major developments in social psychology during the past five decades. In G. Lindzey & E. Aronson (Eds.), *Handbook of social psychology* (3rd ed.) (Vol. 1, pp. 47–107). Hillsdale, NJ: Erlbaum.

Kahneman, D., Slovic, P., & Tversky, A. (Eds.). (1982). *Judgment under uncertainty: Heuristics and biases.* Cambridge, England: Cambridge University Press.

Kasperson, C. J. (1978). Psychology of the scientist: XXXVII. Scientific creativity: A relationship with information channels. *Psychological Reports, 42,* 691–694.

Katz, R. (1982). The effects of group longevity on project communication and performance. *Administrative Science Quarterly, 27,* 81–104.

Katz, R., & Tushman, M. (1979). Communication patterns, project performance, and task characteristics: An empirical evaluation and integration in an R&D setting. *Organizational Behavior and Performance, 23,* 139–162.

Kelley, H. H. (1967). Attribution theory in social psychology. In D.L. Vine (Ed.), *Nebraska symposium on motivation.* Lincoln, NB: University of Nebraska Press.

Knorr-Cetina, K. D., & Mulkay, M. (1983). Introduction: Emerging principles in social studies of science. In K. D. Knorr-Cetina & M. Mulkay (Eds.), *Science observed: Perspectives on the social study of science* (pp. 1–14). Newbury Park, CA: Sage.

Kraut, R. E., Galegher, J., & Egido, C. (1987–1988). Relationships and tasks in scientific research collaboration. *Human-Computer Interaction, 3*, 31–58.

Kruglanski, A. W., & Freund, T. (1983). The freezing and unfreezing of lay-inferences: Effects on impressional primacy, ethnic stereotyping, and numerical anchoring. *Journal of Experimental Social Psychology, 19*, 448–468.

Kuhn, T. S. (1970). *The structure of scientific revolutions* (2nd ed.). Chicago: University of Chicago Press.

Kuhn, T. S. (1977). *The essential tension* Chicago: University of Chicago Press.

Kulkarni, D., & Simon, H. A. (1988). The processes of scientific discovery: The strategy of experimentation. *Cognitive Science*, 139–175.

Latane, B., & Darley, J. M. (1970). *The unresponsive bystander: Why doesn't he help?* New York: Appleton-Century-Crofts.

Latane, B., & Nida, S. (1980). Social impact theory and group influence: A social engineering perspective. In P. B. Paulus (Ed.), *Psychology of group influence* (pp. 3–35). Hillsdale, NJ: Erlbaum.

Latour, B. (1987). *Science in action.* Cambridge, MA: Harvard University Press.

Latour, B. (1988). The politics of explanation: An alternative. In S. Woolgar (Ed.), *Knowledge and reflexivity: New frontiers in the sociology of knowledge* (pp. 155–176). London: Sage.

Latour, B., & Woolgar, S. (1979). *Laboratory life: The social construction of scientific facts.* Newbury Park, CA: Sage.

Laudan, L. (1984). *Science and values.* Berkeley: University of California Press.

Laudan, L. (1990). *Philosophy of science and relativism.* Chicago: University of Chicago Press.

Laughlin, P. R. (1988). Collective induction: Group performance, social combination processes, and mutual majority and minority influence. *Journal of Personality and Social Psychology, 54*, 254–267.

Lave, J., & Wenger, E. (1991). *Situated learning.* Cambridge, England: Cambridge University Press.

Law, J. (Ed.). (1986). *Power, action, and belief.* London: Routledge.

Locksley, A., Borgida, E., Brekke, N., & Hepburn, C. (1980). Sex stereotypes and social judgment. *Journal of Personality and Social Psychology, 39*, 821–831.

Mahoney, M. J. (1976). *Scientist as subject: The psychological imperative.* Cambridge, MA: Ballinger.

Maini, S. M., & Nordbeck, B. (1973). Critical moments, the creative process and research motivation. *International Social Science Journal, 25*, 190–201.

Markus, H., & Zajonc, R. B. (1985). The cognitive perspective in social psychology. In G. Lindzey & E. Aronson (Eds.), *Handbook of social psychology* (3rd ed.) (Vol. 1, pp. 137–230). Hillsdale, NJ: Erlbaum.

Maslow, A. H. (1966). *The psychology of science.* New York: Harper & Row.

McDougall, W. (1908). *Introduction to social psychology.* London: Methuen.

McGuire, W. J. (1985). Attitudes and attitude change. In G. Lindzey & E. Aronson (Eds.), *Handbook of social psychology* (3rd ed.) (Vol. 2, pp. 233–345). Hillsdale, NJ: Erlbaum.

McGuire, W. J., & McGuire, C. V. (1982). Significant others on self-space: Sex differences and developmental trends in the social self. In J. Suls (Ed.), *Psychological perspectives on the self* (Vol. 1, pp. 71–96). Hillsdale, NJ: Erlbaum.

Messeri, P. (1988). Age differences in the reception of new scientific theories: The case of plate tectonics theory. *Social Studies of Science, 18*, 91–112.

Messick, D. M., & Mackie, D. M. (1989). Intergroup relations. *Annual Review of Psychology, 40*, 45–81.

Milgram, S. (1963). Behavioral study of obedience. *Journal of Abnormal and Social Psychology, 67*, 371–378.

Milgram, S. (1974). *Obedience to authority*. New York: Harper & Row.

Miller, D. T., & Turnbull, W. (1986). Expectancies and interpersonal processes. *Annual Review of Psychology, 37*, 233–256.

Mitroff, I. I. (1974). *The subjective side of science*. Amsterdam: Elsevier.

Mitroff, I. I. (1981). Scientists and confirmation bias. In R. D. Tweney, M. E. Doherty, & C. R. Mynatt (Eds.), *On scientific thinking* (pp. 170–175). New York: Columbia University Press.

Mitroff, I. I., & Fitzgerald, I. (1977). On the psychology of the Apollo Moon Scientists: A chapter in the psychology of science. *Human Relations, 30*, 657–674.

Moscovici, S. (1974). Social influence I: Conformity and social control. In C. Nemeth (Ed.), *Social psychology: Classic and contemporary integrations* (pp. 217–250). Chicago: Rand-McNally.

Moscovici, S. (1985). Social influence and conformity. In G. Lindzey & E. Aronson (Eds.), *Handbook of social psychology* (Vol. 2, pp. 347–412). Hillsdale, NJ: Erlbaum.

Moscovici, S., Lage, E., & Naffrechoux, M. (1969). Influence of a consistent minority on the responses of a majority in a color perception task. *Sociometry, 32*, 365–379.

Moscovici, S., & Personnaz, B. (1980). Studies in social influence V: Minority influence and conversion behavior in a perceptual task. *Journal of Social and Experimental Psychology, 10*, 270–282.

Mulkay, M. (1988). Don Quixote's double: A self-exemplifying text. In S. Woolgar (Ed.), *Knowledge and reflexivity: New frontiers in the sociology of knowledge* (pp. 1–11). London: Sage.

Mullins, N. C. (1973). *Theories and theory groups in contemporary American sociology*. New York: Harper & Row.

Neimeyer, R. A., Shadish, W. R., Freedman, E. G., Gholson, B., & Houts, A. C. (1989). A preliminary agenda for the psychology of science. In B. Gholson, W. R. Shadish, R. A. Neimeyer, & A. C. Houts (Eds.), *Psychology of science: Contributions to metascience* (pp. 429–448). Cambridge, England: Cambridge University Press.

Nemeth, C. J., & Staw, B. M. (1989). The trade-offs of social control and innovation in groups and organizations. In L. Berkowitz (Ed.), *Advances in experimental social psychology* (Vol. 22, pp. 175–210). San Diego, CA: Academic Press.

Nersessian, N. (Ed.). (1987). *The process of science.* The Hague: Martinus Nijhoff.

Nickles, T. (Ed.). (1980). *Scientific discovery: Logic, and rationality.* Dordrecht, The Netherlands: Reidel.

Nickles, T. (1989). Integrating the science studies disciplines. In S. Fuller, M. DeMey, T. Shinn, & S. Woolgar (Eds.), *The cognitive turn: Sociological and psychological perspectives on science* (pp. 225–256). Dordrecht, The Netherlands: Kluwer.

Nisbett, R., & Ross, L. (1980). *Human inference: Strategies and shortcomings of social judgment.* Englewood Cliffs, NJ: Prentice-Hall.

Pearson, K. (1904). Report on certain enteric fever inoculation statistics. *British Medical Journal, 2,* 1243–1246.

Pelz, D. C., & Andrews, F. M. (1976). *Scientists in organizations: Productive climates for research and development.* Ann Arbor: University of Michigan Press.

Peters, D. P., & Ceci, S. J. (1982). Peer-review practices of psychological journals: The fate of published articles, submitted again. *Behavioral and Brain Sciences, 5,* 187–195.

Pettigrew, T. F. (1979). The ultimate attribution error: Extending Allport's cognitive analysis of prejudice. *Personality and Social Psychology Bulletin, 5,* 461–476.

Pittman, T. S., & Heller, J. F. (1987). Social motivation. *Annual Review of Psychology, 38,* 461–490.

Popper, K. R. (1970). Normal science and its dangers. In I. Lakatos & A. Musgrave (Eds.), *Criticism and the growth of knowledge* (pp. 51–58). Cambridge, England: Cambridge University Press.

Radnitzky, G. (1987). *Economic imperialism.* New York: Paragon House.

Reicken, H. W., Boruch, R. F., Campbell, D. T., Caplan, N., Glennan, T. K., Pratt, J. W., Rees, A., & Williams, W. (Eds.). (1974). *Social experimentation: A method for planning and evaluating social interventions.* New York: Academic Press.

Roberts, R. M. (1989). *Serendipity: Accidental discoveries in science.* New York: Wiley.

Rorty, R. (1979). *Philosophy and the mirror of the nature.* Princeton: Princeton University Press.

Rosenthal, R. (1966). *Experimenter effects in behavioral research.* New York: Appleton-Century-Crofts.

Rosenthal, R. (1973a). The mediation of Pygmalion effects: A four-factor theory. *Papua New Guinea Journal of Education, 9,* 1–12.

Rosenthal, R. (1973b). *On the social psychology of the self-fulfilling prophecy: Further evidence for Pygmalion effects and their mediating mechanisms.* New York: MSS Modular Publications, Module 53.

Rosenthal, R. (1984). *Meta-analytic procedures for social research.* Newbury Park, CA: Sage.

Rosenthal, R. (1986). Meta-analytic procedures and the nature of replication: The ganzfeld debate. *Journal of Parapsychology, 50,* 315–336.

Rosenthal, R. (1990). Replication in behavioral research. In J. W. Neuliep (Ed.), *Replication research in the social sciences* (pp. 1–30). Newbury Park, CA: Sage.
Rosenthal, R., & Rosnow, R. L. (Eds.). (1969). *Artifact in behavioral research.* New York: Academic Press.
Rosenthal, R., & Rosnow, R. L. (1991). *Essentials of behavioral research: Methods and data analysis* (2nd ed.). New York: McGraw-Hill.
Rosenthal, R., & Rubin, D. B. (1978a). Interpersonal expectancy effects: The first 345 studies. *Behavioral and Brain Sciences, 3,* 377–386.
Rosenthal, R., & Rubin, D. B. (1978b). Issues in summarizing the first 345 studies of interpersonal expectancy effects. *Behavioral and Brain Sciences, 3,* 410–415.
Ross, E. A. (1908). *Social psychology.* New York: Macmillan.
Ross, M., & Fletcher, G. J. O. (1985). Attribution and social perception. In G. Lindzey and E. Aronson (Eds.), *Handbook of social psychology* (3rd ed.) (Vol. 2, pp. 73–122). Hillsdale, NJ: Erlbaum.
Rossi, P. H., & Wright, J. D. (1984). Evaluation research: An assessment. *Annual Review of Sociology, 10,* 331–352.
Rudwick, M. J. S. (1985). *The great Devonian controversy.* Chicago: University of Chicago Press.
Schneider, D. J. (1991). Social cognition. *Annual Review of Psychology, 42,* 527–561.
Schutz, A. (1962). *Collected papers* (Vol. 1). The Hague: Martinus Nijhoff.
Sedelow, W. A. (1976). Some implications of computer networks for psychology. *Behavior Research Methods and Instrumentation, 8,* 281–282.
Serchuk, A. (1989). What can the cognitive psychology of science bring to science and technology studies? *Social Epistemology, 3,* 147–152.
Sewell, W. H. (1989). Some reflections on the golden age of interdisciplinary social psychology. *Annual Review of Sociology, 15,* 1–16.
Shadish, W. R. (1989a). The perception and evaluation of quality in science. In B. Gholson, W. R. Shadish, R. A. Neimeyer, & A. C. Houts (Eds.), *Psychology of science: Contributions to metascience* (pp. 383–426). Cambridge, England: Cambridge University Press.
Shadish, W. R. (1989b). Science evaluation: A glossary of possible contents. *Social Epistemology, 3,* 189–204.
Shadish, W. R., Cook, T. D., & Leviton, L. C. (1991). *Foundations of program evaluation: Theories of practice.* Newbury Park, CA: Sage.
Shadish, W. R., & Haddock, C. K. (in press). Combining estimates of effect size. In H. M. Cooper & L. V. Hedges (Eds.), *Handbook of research synthesis.* New York: Russell Sage.
Shadish, W. R., Houts, A. C., Gholson, B., & Neimeyer, R. A. (1989). The psychology of science: An introduction. In B. Gholson, W. R. Shadish, R. A. Neimeyer, & A. C. Houts (Eds.), *Psychology of science: Contributions to metascience* (pp. 1–16). Cambridge, England: Cambridge University Press.
Shadish, W. R., & Neimeyer, R. A. (1989). Contributions of psychology to an integrative science studies: The shape of things to come. In S. Fuller, M. De Mey, T. Shinn & S. Woolgar (Eds.), *The cognitive turn: Sociological and*

psychological perspectives on science (pp. 13–38). Dordrecht, The Netherlands: Kluwer.

Shaycoft, M. F. (1970, March). *The eigenvalue myth and the data reduction fallacy.* Paper presented at the meeting of the American Educational Research Association, Minneapolis, MN.

Sherif, M. (1936). *The psychology of social norms.* New York: Harper.

Sherman, S. J., Judd, C. M., & Park, B. (1989). Social cognition. *Annual Review of Psychology, 40,* 281–326.

Showers, C., & Cantor, N. (1985). Social cognition: A look at motivated strategies. *Annual Review of Psychology, 36,* 275–305.

Shrager, J., & Langley, P. (1990). *Computational models of scientific discovery and theory formation.* San Mateo, CA: Morgan Kaufmann.

Simon, H. A. (1976). *Administrative behavior.* New York: Free Press.

Simonton, D. K. (1989). Chance–configuration theory of scientific creativity. In B. Gholson, W. R. Shadish, R. A. Neimeyer, & A. C. Houts (Eds.), *Psychology of science: Contributions to metascience* (pp. 170–213). Cambridge, England: Cambridge University Press.

Singer, B. F. (1971). Toward a psychology of science. *American Psychologist, 26,* 1010–1016.

Slezak, P. (1989). Scientific discovery by computer as empirical refutation of the strong programme. *Social Studies of Science, 19,* 563–600.

Stephan, W. G. (1985). Intergroup relations. In G. Lindzey & E. Aronson (Eds.), *Handbook of social psychology* (3rd ed.) (Vol. 2, pp. 599–658). Hillsdale, NJ: Erlbaum.

Suls, J., & Fletcher, B. (1983). Social comparison in the social and physical sciences: An archival study. *Journal of Personality and Social Psychology, 44,* 575–580.

Susser, M. (1977). Judgment and causal inference: Criteria in epidemiologic studies. *American Journal of Epidemiology, 105,* 1–15.

Tesser, A., & Shaffer, D. (1990). Attitudes and attitude change. *Annual Review of Psychology, 41,* 479–524.

Thagard, P. (1989). Scientific cognition: Hot or cold? In S. Fuller, M. DeMey, T. Shinn, & S. Woolgar (Eds.), *The cognitive turn: Sociological and psychological perspectives on science* (pp. 71–82). Dordrecht, The Netherlands: Kluwer.

Thomas, W. I., & Znaniecki, F. (1918). *The Polish peasant in Europe and America.* Chicago: University of Chicago Press.

Trochim, W. M. K. (Ed.). (1986). *Advances in quasi-experimental design and analysis.* San Francisco: Jossey-Bass.

Tversky, A., & Kahneman, D. (1971). Belief in the law of small numbers. *Psychological Bulletin, 76,* 105–110.

Tversky, A., & Kahneman, D. (1974). Judgment under uncertainty: Heuristics and biases. *Science, 185,* 1124–1131.

Tweney, R. D., Doherty, M. E., & Mynatt, C. R. (Eds.). (1981). *On scientific thinking.* New York: Columbia University Press.

Wachter, K. W., & Straf, M. L. (Eds.). (1990). *The future of meta-analysis.* New York: Russell Sage.

Wagner, R. K. (1987). Tacit knowledge and everyday intelligent behavior. *Journal of Personality and Social Psychology, 52,* 1236–1247.

Wason, P. C. (1960). On the failure to eliminate hypotheses in a conceptual task. *Quarterly Journal of Experimental Psychology, 12,* 129–140.

Watson, J. B. (1925). *Behaviorism.* New York: W. W. Norton.

Watson, J. D. (1968). *The double helix.* New York: Atheneum.

Wilson, E. B. (1952). *An introduction to scientific research.* New York: McGraw-Hill.

Woolgar, S. (1983). Irony in the social study of science. In K. D. Knorr-Cetina & M. Mulkay (Eds.), *Science observed: Perspectives on the social study of science* (pp. 237–266). Newbury Park, CA: Sage.

Woolgar, S. (Ed.). (1988). *Knowledge and reflexivity: New frontiers in the sociology of knowledge.* London: Sage.

Woolgar, S. (1989). Representation, cognition, and self: What hope for an integration of psychology and sociology? In S. Fuller, M. DeMey, T. Shinn, & S. Woolgar (Eds.), *The cognitive turn: Sociological and psychological perspectives on science* (pp. 201–223). Dordrecht, The Netherlands: Kluwer.

Woolgar, S., & Ashmore, M. (1988). The next step: An introduction to the reflexive project. In S. Woolgar (Ed.), *Knowledge and reflexivity: New frontiers in the sociology of knowledge* (pp. 1–11). London: Sage.

Wrong, D. (1961) The oversocialized conception of man, *American Sociological Review, 26,* 184–193.

Ziman, J. (1984). *An introduction to science studies: The philosophical and social aspects of science and technology.* Cambridge, England: Cambridge University Press.

CHAPTER 2

The Social Psychology of Scientific Validity: An Epistemological Perspective and a Personalized History

DONALD T. CAMPBELL

The Structure of Scientific Revolutions (Kuhn, 1962) was the major singly identifiable catalyst for the current flowering of science-of-science studies. At the time, Kuhn's area of specialty was history of science. But in a prefatory description of his agenda, Kuhn (1962) says: "Many of my generalizations are about the sociology or social psychology of scientists, yet at least a few of my conclusions belong traditionally to logic and epistemology" (p. 8). Speaking of crisis periods in the history of science, he says "The questions to which this leads demand the competence of the psychologist even more than that of the historian" (Kuhn, 1962, p. 85). Kuhn also cited the work of social psychologists much more than he cited the work of the sociologists of science.

The catalyzed response of both the philosophers of science and the sociologists of science to Kuhn have been enormous, but we social psychologists, although at least equally relevant, did not similarly respond. Even though we all read Kuhn with excitement, and even though it can be seen in retrospect that we had been making minor contributions to the social psychology of science long before 1962, we were not stimulated to generate a new specialty. The present book more than any other is both a symptom of and, I hope, a stimulus to the belated development of a vigorous social psychology of science (SPS).

My special emphasis in this chapter is signaled by the word *validity* in the title. Not only do I want our SPS to be epistemologically relevant, I feel that social psychologists are particularly suited to *combining a social psychology of validity with a social psychology of invalidity*, or, rather, to providing an integrated perspective in which both are understood as aspects of the same processes, from which contingent norms for optimizing validity may be generated.

I first outline the epistemological predicament as a would-be knower, attempting to speak for the emerging postpositivist, post-Kuhnian philosophy of science. I include in this section a review of the sociology of scientific knowledge (SSK). Following that is a sketch of social psychology's past contributions and aborted initiatives. While I attend to the work of others, this sketch has an autobiographical emphasis. Inserted midway is a discussion of the "Comptean divide," and an appeal that social psychologists not only contribute to an individual psychology of cognition but also emphasize interpersonal and social processes.

Our Epistemological Predicament

If there is an emerging postpositivist, post-Kuhnian, antifoundationalist consensus among philosophers of science, it can perhaps in part be epitomized by Figure 2.1, and by this useful if "invalid" syllogism:

> If Newton's theory A is true, then it should be observed that the tides have period B, the path of Mars shape C, the trajectory of a cannonball form D.
> Observation confirms B, C, & D (as judged by the scientific consensus of the day, Quine–Duhem cop-outs notwithstanding).
> Therefore, Newton's theory A is "true."

The invalidity lies in the cross-hatched areas representing the other potential explanations for observations B, C, and D. The conclusion, like all scientific "conclusions," can be termed an *incomplete* induction.

Where "observations" are visual perceptions, we see (as in Figure 2.2), that these too are incomplete inductions, as Rene Descartes, the greatest physiological psychologist of his day, clearly understood. But the observations or facts, B, C, and D of Figure 2.1 (and the syllogism above), are socially achieved scientific consensuses, heavily dependent on trust in verbal communications. Figure 2.3 illustrates the additional equivocalities that verbal transmission introduces, equivocalities that are particularly relevant to the social psychology of knowing. For each of the facts of Figure 2.1, similar diagrams could be drawn. (The Quine–Duhem

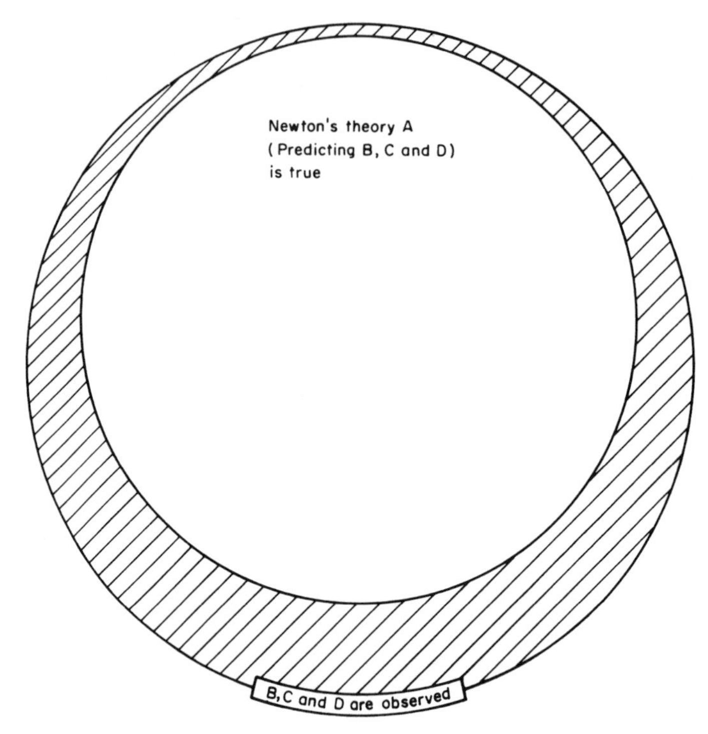

FIGURE 2.1. Newton's Gravitational Theory as an "Incomplete Induction." From Campbell (1990a, p. 4). Copyright 1990 by University Press of America. Reprinted by permission.

copouts referred to in the syllogism are uses of this equivocality to explain away failed theoretical predictions.)

These three diagrams illustrate the possibility that beliefs (theories, facts) may be valid and also that they may be invalid. Technically, they support the skeptic's point that the truth of any belief is unprovable; that induction, even when seemingly "successful," is logically "scandalous." The antifoundationalism of most current epistemology is an acceptance of this much of the skeptic's message. While the cross-hatched areas of the diagrams are technically "infinite" in size, in practice science can only deal with those rival hypotheses that are explicitly articulated within the scientific community of the day. If there are none that do as well, scientists will accept a theory as true even if the inner circle overlaps the outer one a little bit. Thus, a century before Einstein it was known that Newton's theory failed to predict the annual shift in the location of Mercury's closest point to the sun. As Kuhn (1962), Krech and

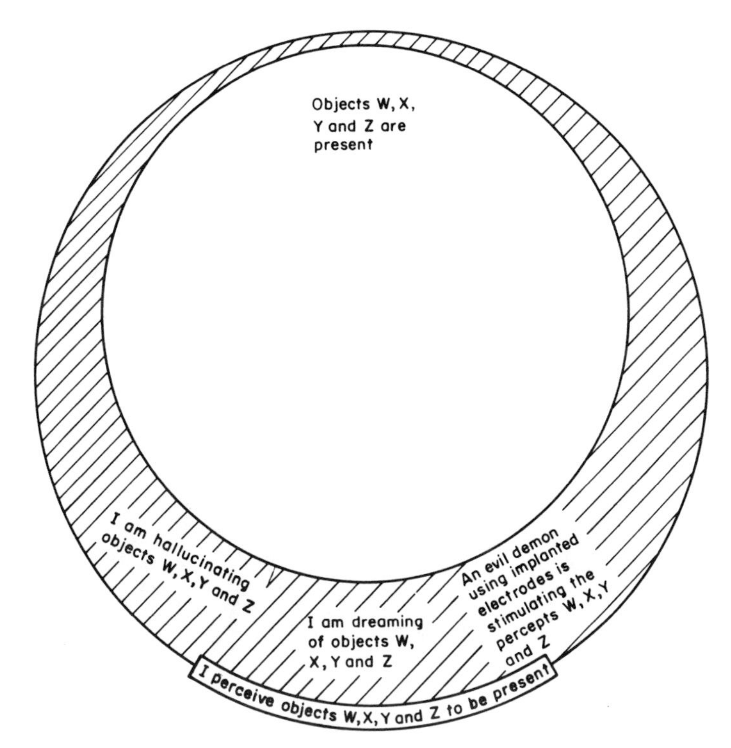

FIGURE 2.2. Visual Perception as "incomplete induction," with apologies to Descartes. (The relative size of the spaces is to illustrate the scarcity of available plausible hypotheses rival to the belief that the objects are present.) From Campbell (1990a, p. 4). Copyright 1990 by University Press of America. Reprinted by permission.

Crutchfield (1948), and no doubt many others have noted, we tolerate anomalies until a better theory comes along. In this postpositivist, post-Kuhnian consensus, belief in either a theory or a fact is recognized to be radically underjustified.

Quine (1951) accurately anticipated this postpositivist consensus. (For me, the last 15 Gestaltish paragraphs are the core.) Quine recommends what Cook and Campbell (1986) have called "omnifallibilist trust." In this view, we have no choice but to trust the great bulk of our beliefs while revising but a few. Peripheral perceptual beliefs and core beliefs in order and logic are those that are properly most resistant to revision, but even these are to be revisable. Beliefs in "theories" lie in between and are most revisable. According to Quine, we do (and should) revise our beliefs both to increase their coherence with the rest of the whole web of belief and to minimize the changes required in other beliefs

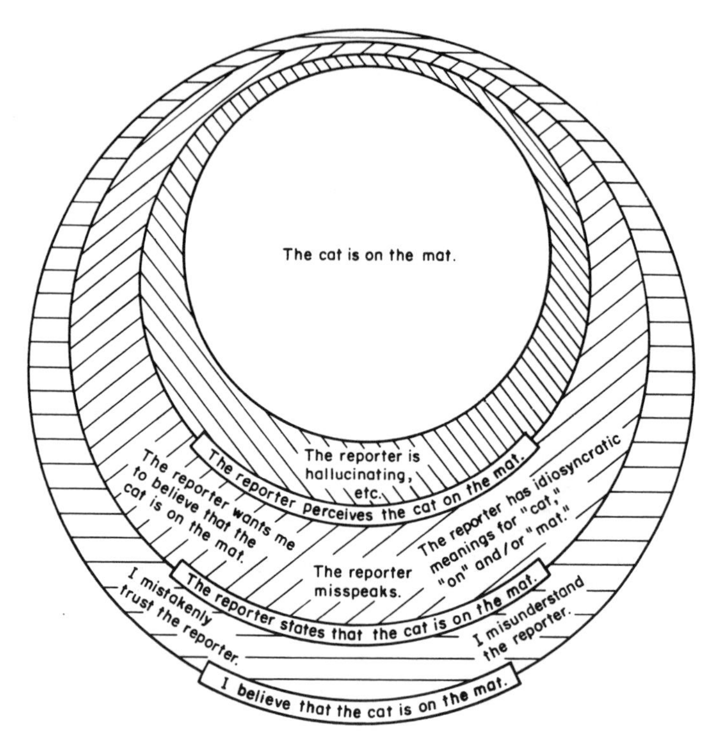

FIGURE 2.3. Verbally transmitted belief as an "incomplete induction." From Campbell (1993, p. 93). Copyright 1993 by the University of Chicago Press. Reprinted by permission.

already held. Recommendation of such a coherence strategy of belief revision is now widespread in current philosophy of science (e.g., Lehrer, 1974, 1990; Campbell, 1991). This is accompanied by a recognition that one cannot use the fit of a belief to "reality" as a truth test. Instead, we must make do with comparing belief with belief. But this is not incompatible with retaining a correspondence goal and meaning for "truth."

Study Figures 2.1, 2.2, and 2.3 closely. While they come from the skeptic's tool kit, note that they are also quite compatible with valid scientific (or perceptually based, or verbally transmitted) belief. What they rule out is any kind of *certain proof* (logical or inductive) of a belief's validity. Reactions to this epistemological predicament have been of two kinds, proepistemic, emphasizing the likelihood of validity, and contraepistemic, emphasizing invalidity. Much too rarely are these polar possibilities encompassed within a unified viewpoint. Most current philosophers of science and epistemologists stress the possibility of validity through concepts such as "coherentism" and "inference to the best explanation" (although many of them are at the same time debunking beliefs in the reality of successful scientific constructs).

All the exciting recent Western European SSK has the contraepistemic emphasis. My sustained interaction with this SSK group during the last 15 years has been fully as important to me as has my increased contact with philosophers during the same period. I recommend a similar involvement to my fellow social psychologists moving into the theory of science, and offer the next few paragraphs as a reading guide.

Scientific Knowledge and Sciological Theory (Barnes, 1974) was my first major introduction to the SSK movement. Kuhn was a major stimulus and model for Barnes (1974; see also 1982). This can be crudely measured by the fact that Barnes cites Kuhn 24 times, in contrast to 6 citations for Robert Merton, the sociologist of science, and 8 for Mary Hesse, the historian and philosopher of science, who is the next most frequently cited. Barnes (1974) and his Edinburgh colleague Bloor (1976) created what has become known as the "Strong Programme" in the sociology of science. That term is not mnemonic enough to point to their special emphasis, however. Much better is their later term *symmetry* (Barnes & Bloor, 1982). As stressed in Barnes's (1974) first chapter, sociological explanations of scientific beliefs should *not* be limited to beliefs we now regard as mistaken (as has been the case, Barnes asserts, in past sociology of science). Instead, the adoption of scientific beliefs now regarded as valid also requires sociological and psychological explanations, in which ideologies, cultural beliefs, individual interests, and social persuasion processes also operate. That is, believed-to-be-valid and believed-to-be-false scientific beliefs are to be treated symmetrically by sociologists of science.

From the epistemological perspective just presented, this is a valid and important point. The alternative is a "direct" or "clairvoyant," realism for valid beliefs, which the current antifoundationalist epistemological consensus rejects. My own version of the symmetry principle does not preclude hypothetically differentiating the social persuasion and interest negotiation processes of various social institutions. This allows us to speculate on which of these (e.g., astronomy vs. astrology or religion vs. science) might maximize the possible role of a hypothesized real world in the process of selection among alternative beliefs to achieve a new consensus. I anticipate that the social negotiation system of the physical and biological sciences in the last two centuries will turn out to be plausibly superior. Nonetheless, these social processes, even at their most validity optimizing, sometimes produce erroneous consensus and never provide proof in any logical sense (Campbell, 1986/1988b).

Barnes (1974, 1982) does not rule out such a reading, but his research emphasis has primarily been on the contraepistemic. In the much cited MacKenzie and Barnes (1975, 1979) study of the victory of Mendelianism (under William Bateson's leadership) over the continuous variation biometry of Pearson and Weldon, intrascientific social processes are neglected in favor of tangential extrascientific social class and clique interests.(Kyung-Man Kim's [in press] restudy of this episode fills in this gap, and, I would argue, is equally sociological.) The University of Edinburgh Science Studies Unit has used history of science examples as its main source of evidence. A famous review by one of its members (Shapin, 1982) provides a good entry into this literature, and is triumphantly contraepistemic in its emphasis. The most famous of its recent products, *Constructing Quarks: A Sociological History of Particle Physics* (Pickering, 1984), while contraepistemic in its overall emphasis, can be read as also allowing proepistemic outcomes from the social processes described, at least to the extent of acknowledging an indirect role for physical restraints on the social constructions produced (see also Pickering, 1989).

The SSK movement consists of several subschools or emphases, divergent enough to have vigorous disagreements among themselves, coming from a variety of intellectual streams, but also with interlocking memberships worthy of study in their own right. An important figure is Michael Mulkay. While Barnes (1974) identified the new sociology of science with the old sociology of knowledge, it was Mulkay (1979) who made this point most effectively and, at the same time, identified both with issues in the philosophy of science. The sociology of knowledge tradition is one of which the new social psychologists of science should be aware. (Bar-Tal and Kruglanski's [1988] collection, *The Social Psychology of Knowledge*, is mistitled in that it makes no such contact.) As exemplified by Mannheim (1946) (and still more clearly in the work of his translators Wirth and Shils), *Ideology and Utopia* is a critique of ideological and social class biases with the goal of improving the validity of social science, rather than abandoning the goal of validity. Mannheim's use of analogues to visual perception (perspectivism, relationism) makes an attractive integrative epistemological position. This can be asserted while still accepting Mulkay's and Barnes's criticism of Mannheim for exempting the physical sciences.

Mulkay (1979) does not rule out a proepistemic aspect, but his later contributions, particularly his current "discourse analysis" (Gilbert & Mulkay, 1984), are so contraepistemic as to allow no role for validity-producing aspects of the social processes of science. (While I treat Mulkay as a separate school, or schools, he was an early collaborator with David Edge, founding head of the Edinburgh Science Studies Unit, and teacher of Steve Woolgar and Malcolm Ashmore, mentioned later.)

A third intellectual stream (or streams) contributing to SSK has come to be epitomized by the notion of the "social construction" of scientific beliefs. Latour and Woolgar (1979) and Knorr-Cetina (1981) exemplify this movement. Both also exemplify the ethnographic study of social interaction in scientific laboratories. Here again, I heartily endorse science's status as a social product, with transient consensuses socially negotiated at every level from agreement on laboratory "facts" to theory construction and theory choice. What I reject is the implication that "socially constructed" must entail "invalid," or that *all* socially constructed beliefs are equally invalid, or that the issue of validity is irrelevant for socially constructed beliefs. I would like to argue that the sorts of processes of social construction that characterize science at its best make it plausible that "the way the world is" has been a "coselector" of new socially constructed consensuses (Campbell, 1986/1988b, 1993).

The laboratory ethnographies of Latour and Woolgar (1979) and Knorr-Cetina (1981) (and also Pickering [1984]) are replete with instances in which scientific conjectures are offered and abandoned. These are all instances of "selection," grist for the mill of "selection theory" (my recent relabeling of "evolutionary epistemology"), with its epistemological focus on "plausible selection by referent." What needs to be done is to tally such episodes by type. Some will be due to contraepistemic selectors: The hunch or hypothesis or explanation was dropped because it would offend the funding agency, the laboratory head, the scientific community, political powers, coreligionists, and so on, or would give comfort to the laboratory's scientific rivals. But from my casual reading, I am sure that an equally large number will not be explicable this way. For these other cases, there are reports that it "could not be made to work," or that it was inconsistent with other trusted facts, and so forth. Such episodes, of course, do *not* show that the resulting consensus would be iconic with *Ding an sich* reality. But they may often plausibly be regarded as instances in which such reality, or the "referent," has played a role in belief selection. The culture and norms of science may be such as to optimize such "selection by referent," compared to other belief-conveying social systems (e.g., more likely in today's astronomy than astrology). (In my full model, there is a third or hybrid category of "vehicular" selectors, belief selection related to maintaining the social system that carries scientific "knowledge," resulting in choices often different from those that would maximize validity per se [Campbell, 1979/1988c, 1987].)

Of the social constructionists, Karin Knorr-Cetina has moved toward a position I find compatible with this view (Knorr-Cetina & Mulkay, 1983; Knorr-Cetina, 1987, 1990). Latour's (1987) *Science in Action* is recognized as a very important statement. In its explicit rejection of social constructivism, it disrupts the oversimple subcategorizing of the general SSK movement I offer here. Attending only to its explicit propositions, it is enthusiastically contraepistemic, going beyond mere agnosticism about validity enhancement in science to a militant ontological nihilism. Social psychologists of science should welcome radical Gestalt switches in the view of the social system of science, and Latour makes a dramatic effort to provide one. Victories of points of view in science are political/military victories. Rather than being produced in any part by the hidden hand of reality (as my own model might be fairly caricatured), these victories create "scientific reality," which is nothing more than past victories no longer contested. However, one of the editors of this volume (WRS) has convinced me that, implicitly at least, there is lots of room in his description for selection by referent. For example, the political actors described include microbes and coral reefs, which competing scientists must coopt as allies, and which implicitly are not merely the products of scientific inscription processes, rhetoric, and alliance formation. Two reviews in key sociology of science journals help in this exploration: Amsterdamska's (1990) and Shapin's (1988). (The latter, by a key SSK leader, portrays the divisions within SSK much more complexly than I have here.) Woolgar (e.g., 1988) has remained the able advocate of a totally skeptical position, allying himself with the sociological school of ethnomethodology. The latter position is also exemplified by Lynch (1985), a student of ethnomethodology's founder, Harold Garfinkel.

A fourth school within SSK is the "empirical programme of relativism," led by Harry Collins at the University of Bath. In a triumphant summary of the research of this school and related SSK positions, Collins (1981) asserts that they have demonstrated that "the natural world has a small or nonexistent role in the construction of scientific knowledge" (p. 3). In an exciting overall statement, Collins (1985) uses Luis Borges to illustrate an extreme Sapir/Whorf type of linguistic relativism, and includes in his footnotes criticism of Barnes and Bloor for being too realistic! I have answered by arguing the evolutionary and psychological fact that underjustified perceptual reification of ordinary objects and movements is a widespread *precursor* of language, shared by many animals, and that our similarity in such reifications makes useful language possible (Campbell, 1989). I also assert (with Barnes, 1974) an epistemological relativism: "Cousin to the amoeba: How could we know for certain?" and quote Popper's "We don't know: We can only guess." Collins's relativism often seems to go beyond this to a conjectural denial of any reality to be referred to (however imperfectly), that is, to an ontological relativism or ontological nihilism.

The SSK movement has recently turned to technology (see, e.g., Bijker, Hughes, & Pinch, 1987). This setting permits dramatizing the social constraints on those technologies that become dominant without denying the physical–feasibility constraints simultaneously at work. In this setting, physical reality (however incompletely known) need not be portrayed as socially constructed to make the indisputable point of the social shaping of the bicycle, automobile, or jet engine. Here the scientific realist and SSK can agree. We can now take the SSK point as saying that so-called pure science is not different from technology in this regard, that it too has been shaped by social forces as well as by physical reality, and is no different than technology in this regard. Latour (1987) can be construed as making this point in denying any science–technology dichotomy. I find it compatible with my fallibilist selection theory of incomplete inductions.

As a final SSK reading assignment to new participants in the social psychology of science (or even as a first entry), I suggest Ashmore (1989). Under the guise of a semifictional spoof, this is a serious review of the whole field, turning SSK's deconstructionist tools on its own enterprise. In the process, it provides a gossipy roll call of major and minor participants. It also has a very extensive bibliography and much interview material that I take to be authentic. In the process, Ashmore introduces one social psychologist reflexively doing discourse analysis on social psychology, Jonathan Potter (Potter & Wetherell, 1987). Ashmore's emphasis seems entirely contraepistemic.

Whereas many philosophers of science who have paid attention to the SSK movement have been offended by the overwhelming agnosticism or outright denial of the validity of science, the Philosophy of Science Association, during recent meetings has shown increasing acceptance. I myself find the SSK movement legitimized by the philosophical analyses of science's epistemological predicament (i.e., by present-day antifoundationalism and underjustificationism), but I regard SSK as itself a conjectural enterprise, dependent on coherentist plausibility over and above skeptical logic. I reject only those forms of SSK that eschew supplementing the contraepistemic with a coherentist, presumptive sociology of scientific validity. A dialectical dynamics is involved. Those overtrained in an exaggerated faith in science's discovery and "proof" of absolute truths may well react to the discovery that no truth is eternal with a total rejection of science's claim to validity, finding excitement, novelty, and discovery in the contraepistemic emphasis. However, this antithesis should later generate a balanced thesis in which proepistemic possibilities are also included.

Historians of science seem the farthest along in this dialectic. Embarrassed by a long tradition of celebratory hagiography of episodes in scientific discovery of new truth, they have reacted with a phase of debunking and/or a self-conscious agnosticism regarding the truth of beliefs in the scientific communities they study. The contributions of this phase must not be abandoned as one goes on to the more balanced synthesis I recommend. Such agnosticism has usually improved the validity of the history and provides better raw material for the social psychology of scientific validity than did earlier celebratory histories. Thus, England's leading physicist of 1870, Lord Kelvin, being religiously offended by Darwin's evolutionary history of humankind (Wilson, 1974), turned his efforts to estimating the age of the earth from its rate of cooling, "proving" by application of the best physics of his day that the earth was too young for Darwinian biological evolution to have taken place. At that time, physicists had no knowledge of radioactivity as a continuing source of the earth's heat. The coherence theory of belief revision makes such a mistake quite comprehensible as part of a process that, overall, can produce improved validity and has subsequently produced the presumably more valid physics and biology of today.

Historians of psychology are now in the contraepistemic stage (e.g., Morawski, 1988; Ash & Woodward, 1987). Vidal (1987) provides what may eventually be seen as an episode parallel to Kelvin's. To our image of the young Piaget as a 13-year-old author of a published scientific paper on a fresh water mollusk, Vidal now adds the charismatic 18-year-old Christian Youth Group leader, whose later scientific theories had to be consistent with his religious convictions, and who rejected "Mendelism," "mutationism," and neo-Darwinian evolutionary theory, perhaps for that reason.

Much of current exciting history of science still has a preponderantly contraepistemic emphasis (an example is the widely acclaimed Shapin and Schaffer [1985], although allowing proepistemic possibilities), but we now have a growing responsive literature that may be a step on the way to a synthesis that stresses the proepistemic (e.g., Rudwick, 1985; Galison, 1987; Kim, in press). These three studies take cognizance of the new SSK, attempt to give full attention to social and individual interests and prejudices that are involved, but nonetheless end up with arguments for the improved validity of the new scientific consensuses that emerged. They are far from the earlier purely adulatory histories of science, but they fail, nonetheless, to be quite the balanced synthesis that I feel is needed. They overly reject the sociologists of scientific knowledge.

Let me end this epistemological section by calling attention to *the hypothetically normative agenda*. The studies of these sociologists of scientific knowledge are at present descriptive and antinormative. I have ineffectively exhorted them to move to the second phase of the symmetrical social–constructivist, relativist sociology of scientific knowledge in which they would invade the territory of the philosophers of science and make speculative recommendations as to the social–organizational forms, norms, and sanctioning mechanisms that would optimize the achievement of new scientific consensuses of increased validity (were there to be a physical world independent of the beliefs of the physicists, and also contingent upon the validity of our current knowledge of social humanity and physical reality) (see Campbell, 1979/1988c, 1986, 1989, 1993).

Many philosophers of science now believe that *any* epistemology will depend on unproven ontological hypotheses; that any normative theory for science will be contingent, presumptive, and fallible, just as is science itself. This "naturalistic epistemology," founded by Quine (1969; see also Kornblith, 1985) has been extended to science by Giere (1985) and others (listed in Campbell, 1986). It is characteristic of our epistemological predicament that even were these hypothetical norms to be perfectly correct, they would not guarantee the validity of every scientific belief that was produced in accordance to the norms (any more than every belief produced by vision is valid, even though most probably are).

Sociologists of scientific knowledge will probably remain reluctant to contribute to this normative agenda, even in its conjectural, contingent, science-of-science form. We social psychologists, for many reasons, have the tradition and orientation to undertake the normative agenda.

The Social Psychology of Individual Cognition

Most of social psychology's pre-Kuhnian and post-Kuhnian contributions also turn out to have been contraepistemic in that they have focused on processes leading to invalidity. However, this emphasis has not been accompanied by the dialectical swing to debunking science's claim to producing validity found among the SSK. Perhaps the reasons for this place us in the best position of all to achieve an integrated formulation in which the contra- and proepistemic effects are seen as aspects of a single process. I return to this theme at many points in the survey that follows, but first a perspective from general psychology.

Both perception and learning are epistemic processes, producing belief or knowledge. Both have been studied with the issue of validity in the forefront. Research on visual perception has from the beginning included both the conditions affecting valid acuity and those producing illusions. Psychologists specializing in illusions have nonetheless implicitly or explicitly assumed that most perceptions were *not* illusory (in contrast to the uses of illusion by philosophers as early as Locke and Berkeley). In the Helmholtz, Brunswik tradition in which I was trained, a neurological "constructivism" (combined with an evolutionary selection of constructions) best unifies the theories of validity and illusion (Campbell, 1987). Our conscious perceptions are based on neurological processing of the physical stimulus array that is characterized by presumptive static-reducing, contrast-enhancing, form–constancy-maximizing processes. All of the illusion examples used by philosophers (except for the stick bent at the water's surface) are the product of neurological data processing that *usually* enhances the validity of perceptions. Illusions occur in those ecologically atypical situations in which the implicit "presumptions" built into the input transformations are mistaken. While studies of learning have in general neglected "illusory" outcomes, they are nonetheless there, as in Skinner's superstitious pigeons, and in the studies of popular errors in the Berkeley 30-unit multiple T rat maze of the 1930s (for details, see Campbell & Gruen, 1958). Systematic biases (as well as haphazard error) are characteristic of early or intermediate (Campbell, 1956) stages of learning, and tend to be eliminated as learning progresses toward asymptote. It is perhaps this background that led social psychologists to fail to join the sociologists of scientific knowledge in the full antithesis of contraepistemic emphasis. (Our greater ideological commitment to being "scientific" may have played an equally strong role.)

In the years immediately following World War II, social psychology was a very exciting field. Among the several flourishing movements was the "New Look" social psychology of perception led by Jerome Bruner and Leo Postman at Harvard. Their contribution to the new field of the social psychology of science can be documented by citing Kuhn's use of it at some length:

> To a greater or lesser extent (corresponding to the continuum from the shocking to the anticipated result), the characteristics common to the three examples above are characteristic of all discoveries from which new sorts of phenomena emerge. Those characteristics include: the previous awareness of anomaly, the gradual and simultaneous emergence of both observational and conceptual recognition, and the consequent change of paradigm categories and procedures often accompanied by resistance. There is even evidence that these same characteristics are built into the nature of the perceptual process itself. In a psychological experiment that deserves to be far better known outside the trade, Bruner and Postman [1949] asked experimental subjects to identify on short and controlled exposure a series of playing cards. Many of the cards were normal, but some were made anomalous, e.g., a red six of spades and a black four of hearts. Each experimental run was constituted by the display of a single card to a single subject in a series of gradually increased exposures. After each exposure the subject was asked what he had seen, and the run was terminated by two successive correct identifications.
>
> Even on the shortest exposures many subjects identified most of the cards, and after a small increase all the subjects identified them all. For the normal cards these identifications were usually correct, but the anomalous cards

were almost always identified, without apparent hesitation or puzzlement, as normal. The black four of hearts might, for example, be identified as the four of either spades or hearts. Without any awareness of trouble, it was immediately fitted to one of the conceptual categories prepared by prior experience. One would not even like to say that the subjects had seen something different from what they identified. With a further increase of exposure to the anomalous cards, subjects did begin to hesitate and to display awareness of anomaly. Exposed, for example, to the red six of spades, some would say: That's the six of spades, but there's something wrong with it—the black has a red border. Further increase of exposure resulted in still more hesitation and confusion until finally, and sometimes quite suddenly, most subjects would produce the correct identification without hesitation. Moreover, after doing this with two or three of the anomalous cards, they would have little further difficulty with the others. A few subjects, however, were never able to make the requisite adjustment of their categories. Even at forty times the average exposure required to recognize normal cards for what they were, more than 10 percent of the anomalous cards were not correctly identified. And the subjects who then failed often experienced acute personal distress. One of them exclaimed: "I can't make the suit out, whatever it is. It didn't even look like a card that time. I don't know what color it is now or whether it's a spade or a heart. I'm not even sure now what a spade looks like. My God!"[13][*] In the next section we shall occasionally see scientists behaving this way too . . . . (Kuhn, 1962, pp. 62–64)

The subjects of the anomalous playing-card experiment discussed in Section VI experienced a quite similar transformation. Until taught by prolonged exposure that the universe contained anomalous cards, they saw only the types of cards for which previous experience had equipped them. Yet once experience had provided the requisite additional categories, they were able to see all anomalous cards on the first inspection long enough to permit any identification at all. Still other experiments demonstrate that the perceived size, color, and so on, of experimentally displayed objects also varies with the subject's previous training and experience [Hastorf, 1950; Bruner, Postman, & Rodrigues, 1951]. Surveying the rich experimental literature from which these examples are drawn makes one suspect that something like a paradigm is prerequisite to perception itself. What a man sees depends both upon what he looks at and also upon what his previous visual–conceptual experience has taught him to see. (Kuhn, 1962, pp. 111–112)

Note Kuhn's balance in that last sentence: "What a man sees depends both upon what he looks at" (proepistemic) "and also upon . . . " (contraepistemic). To judge from his later writings, Kuhn feels he has

*Footnote 13 from Kuhn (1962): Bruner & Postman, 1949, p. 218. My colleague Postman tells me that, though knowing all about the apparatus and display in advance, he nevertheless found looking at the incongruous cards acutely uncomfortable.

always wanted to retain that balance, even though his work was often read as predominantly contraepistemic. He nonetheless still rejects any simple or direct scientific realism.

In their overall summaries of the "New Look" research program, Bruner (1951) and Postman (1953) provided a point of view that integrated the pro- and contraepistemic. However, they failed to note that their proepistemic principle of "perceptual sensitization" and their contraepistemic "value-resonant errors" were aspects of the same process, as now seen from the point of view of signal detection theory (see also Campbell, 1963, pp. 112–117). Since that time, Bruner (1962, 1973) has gone on to contribute directly to the theory of knowledge. (Postman, after shifting first to associative learning, has gone on to history of psychology. It would be worth examining his historical work to see if his own early research, and Kuhn's use of it, has been influenced in the theory-of-science direction.)

Many other social psychologists have contributed directly or indirectly to our field. Leon Festinger's theory of cognitive dissonance was immediately picked up by E. G. Boring (1964) in his last essay on the theory of science. (I return to Festinger later.) The social psychologists Nisbett and Ross (1980), through their book *Human Inference* and through personal contacts and collaborations, are having a very strong impact on the philosophy of science and on epistemology. While their emphasis is almost entirely contraepistemic, overall they show a strong commitment to the possibility of valid science, as well as a desire to improve the human discretionary processes involved in valid science. Their work has convinced their philosopher friend, Stephen Stich (1985), to conclude that humans are not rational animals. Stich combines this conclusion with a proepistemic call for an educated rationality that would overcome this weakness. Kahneman, Slovic, and Tversky (1982) are having a similar impact on philosophy.

While until the last decade I did not regard the theory of science as a major scholarly commitment, a back-burner interest has been present all along (see, e.g., Wyatt & Campbell, 1951). It is very much a product of my Berkeley undergraduate and graduate school environment. My teacher Tolman's lecture presentations of Krechevsky's work on hypotheses in rats (Krechevsky, 1932; Tolman & Krechevsky, 1933) made the point that the same processes that lead to valid learning in learnable mazes will produce stimulus-response consistencies in an unlearnable mazes in which the blind alleys are continually randomly rearranged so that there is no environmental consistency to be learned. This constitutes a fine unification of the contra- and proepistemic viewpoints. In my 1936–1947 period at Berkeley there was a great enthusiasm for projective techniques, and I took the relevant clinical courses, including two

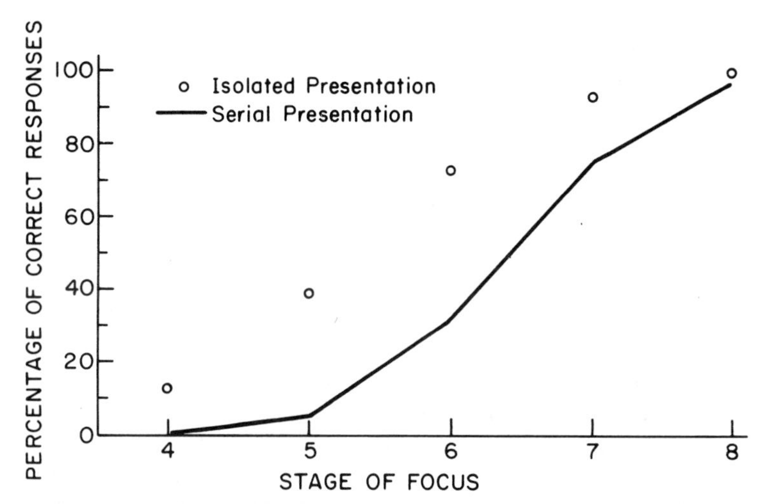

FIGURE 2.4. Results from Galloway's experiment. (Percentage of correct responses for judgments of six slides at each stage of focus. Eighty- eight subjects and 292 judgments are involved in each stage for serial persentations, which began at the least well-focused stage. For the isolated presentations, the figure varies between 15 and 292.) From Wyatt and Campbell, (1951, p. 497). Copyright 1951 by The American Psychological Association. Reprinted by permission.

seminars (on psychoanalytical theory and children's psychopathology) from Erik Erikson. I worked as a teaching assistant in perception with Egon Brunswik, was a research assistant to Else Frenkel-Brunswik, and was introduced to social psychology through Robert Tryon, who enthusiastically assigned *The Psychology of Social Norms* (Sherif, 1936). Out of this came a perspective quite compatible with Bruner's and Postman's "New Look" and the four-page Wyatt and Campbell paper (which Kuhn could have equally well used to make the same points). I quote from Wyatt and Campbell's (1951, p. 496) opening paragraphs, and the first of its two graphs is shown as Figure 2.4.

> Common sense observation, social psychology, and the sociology of knowledge emphasize the effect of stereotypes, snap judgments, preconceptions, or sets in reducing the adequacy of observations. Usually it is the objectivity of the man-in-the- street that is challenged, but frequently enough the problem has been stated for the theory-builder. Thus Tolman has said:
>
>> And once set up, a system probably does as much harm as it does good. It serves as a sort of sacred grating behind which each novice is commanded to kneel in order that he may never see the real world, save through its interstices. (1932, page 394)

Krech and Crutchfield comment in a similar vein:

> This principle also helps us to understand the tenacity with which people hold on to "disproved" scientific theories or economic and political dogmas. No matter how much evidence one can bring to bear that a scientific theory does not fit the known facts, scientists are reluctant to give it up until one can give them another integration in place of the old. In the absence of some other way of organizing facts, people will frequently hold onto the old, for no other reason than that. (1948, page 87)

Tolman, of course, was the teacher of both Krech (formerly Krechevsky) and Crutchfield, and they dedicated their great social psychology text to him. Wyatt and Campbell (1951) confirmed the prior results obtained by Galloway (1946/1948), a clinical psychology student exploring a new projective test in a master's thesis supervised by Egon Brunswik. Galloway had asked students to guess at the content of slides projected out of focus. His primary interest was in the content of the wrong guesses, but, probably due to Brunswik's influence, he had asked on some slides for repeated guessing at a series of stages of clarity beginning with the most unfocused and steadily improving the focus. In a poorly counterbalanced way, other students guessed at a given slide only once, at specified degrees of focus. Galloway had also scored the guesses for accuracy. Those with previous experience in guessing at the slide when less well focused performed much worse at a given stage than did those who were seeing the slide at that stage of focus for the first time. Presumably the interpretations they developed early then biased later guesses. Wyatt had replicated the effect with much better counterbalancing. Our final paragraph follows:

> Whatever the theoretical explanation chosen, the experiment provides an excellent paradigm for social-attitude formation. In the cognition of the physical environment, the process herein isolated offers little practical difficulty. In the cognition of social reality, the slide is usually left at stage 4 or 5, with reality testing so difficult and with the facts so little coercive that the disparate hypotheses, stereotypes, or prejudices are rarely if ever brought into agreement. In social problem solving, "experience" per se without reality testing or verification may often be a liability, bringing reduced objectivity. For those seeking to perceive the causal structure of their social environment (be it the man-on-the-street or the social scientist) the slide is indeed blurred and will not soon be brought into focus. (Wyatt & Campbell, 1951, p. 499)

Figure 2.4 points toward a balance of the contra- and proepistemic. In the end, or when environmental clarity or stability is great enough, even the serial-presentation participants get it right. The theory of the conditions of validity includes both the "clarity" of the environment (something akin to what I later denote as "entitativity," Campbell, 1973, and "ostensionability," Campbell & Paller, 1989), and the preservation of erroneous old beliefs. Greenwald (Greenwald, Pratkanis, Leippe, & Baumgarder, 1986) has made use of Galloway's paradigm in a great recent contribution to SPS (the social psychology of social psychology, to be specific).

From the beginning of my teaching in 1947, my lectures in social psychology were contraepistemic in emphasis (within a framework always assuming that most perception, learning, and science were valid), drawing from experimental psychology bias tendencies in psychophysics, perception, and learning and applying them to understanding social perceptions, attitudes, and stereotypes (1959a for a review, see Campbell). I tried to use both perceptual and learning theory in a parallel explanation of the same social perceptual phenomenon. When, in 1959, Robert Rosenthal began our friendship by recruiting me to a pioneering symposium on SPS, "The Problem of Experimenter Bias," I used this set of biases to describe the predicament of social scientists (Campbell, 1959/in press). Most of this was social psychology of individual cognition, explicitly related to Francis Bacon's "Idols of the Tribe."

Social Psychology on the Sociological Side of Compte's Divide

It is said that in his hierarchical developmental division of the sciences, Auguste Compte, writing around 1840, provided no room for psychology, dividing its contents between biology and sociology. [For social psychologists, Allport (1954) provides a convenient recounting.] Herbert Spencer (1855–1896) disagreed, and put 1,000 pages of psychology between biology and sociology in his systematic volumes. But Compte's division remains with us, a constant source of division within psychology departments, and surprisingly enough, within social psychology itself, and indeed within the career of such major refounders of the field as Leon Festinger. Festinger's contributions in his early work on social adjacency and communications (Festinger, Schachter, & Back, 1950), on social comparison processes (Festinger, 1950/1989), and on patterns of communication toward dissenting minorities (Festinger, 1950/1989; see also Rosenwein, Chapter 8, this volume) are all on the sociological side; all depend on recognition of the social environment of other persons. On the other hand, Festinger's dissonance theory has no such dependency. It would characterize a solitary problem-solving animal and might be due to innate brain mechanisms. Indeed, Festinger extended it to research on rats (Schachter & Gazzaniga, 1989).

As we in this volume seek to carve out a specialty called SPS, we need to recognize this Comptean division among our own efforts in the social psychology of science and epistemology. It turns out that the great bulk of our contributions to date have been on the biological, individual, cognitive–science side of Compte's divide, as in the examples of the previous section. While not all of social psychology's contributions to the theory of knowledge have been asocial, there is danger that our genuinely social contributions will decrease in the future.

As I see the last quarter century in psychology, the long behaviorist domination of experimental psychology left a lacuna in individual cognitive psychology which social psychology filled, having become cognitive much earlier (Sherif, 1936; Klineberg, 1940; Krech & Crutchfield, 1948; Asch, 1952/1987), even when the problems worked on were not particularly social. At the same time, led by Festinger's post-World War II Stanford University career, high-prestige social psychology as done in psychology departments became increasingly a pure laboratory science. While this does not preclude social variables, dissonance theory research led to their neglect, and most social psychology training programs dropped the sociological and anthropological components that had characterized them in the 1930s and early 1940s. Now that experimental cognitive psychology dominates the departmental slots and space once given over to learning and perception, the prestige environment of social psychology fosters continued neglect of the social sciences in our training and research programs. Partly because of my old-fashionedness, I oppose this neglect both in general and in future research on the social psychology of knowing and science in particular. In my judgment, social psychologists have an important contribution to make on both sides of the Comptean Divide, perhaps by creatively straddling the divide. I hope in this section to make it clear that we can make contributions that the sociologists of knowledge and science are not now making, nor are they apt to make. For this contribution, however, we must do more than borrow their terms. We must be familiar with their contributions and locate our essays in a domain that includes both philosophy and sociology of science.

Social psychologists have all along made a few contributions on the social side of the Comptean divide that are applicable to SPS. Festinger's (1954) early work on social comparison processes is ripe for such application. It immediately fits in with SPS to be found in philosopher of science David Hull's (1988) *Science as a Process*. What Hull calls the drive for "conceptual inclusive fitness" is better described in Festinger's analysis of whose esteem we covet. That scientists value the esteem of fellow scientists much more than that of the larger society is certainly an important factor, plausibly contributing to validity enhancement in the social system of science.

Michael Polanyi (1958) was a great social psychologist of science (see in particular Polanyi, 1958, chap. 7) His model of mutual monitoring among overlapping competence areas in the self-governing republic of science needs elaboration (Polanyi, 1966; Campbell, 1969). Somewhere else, Polanyi uses this analogy: For the collective solution of a jigsaw puzzle, Polanyi says, it is optimal for each participant to see the whole board and the pieces each of the others is holding. Polanyi uses this statement to justify the norm of open information sharing in science, but he did no research on it. However, social psychologists Alex Bavelas (1950) and Harold Guetzkow and Herbert Simon (1955) have. They connected six-person teams in the different patterns of communication links shown in Figure 2.5. They distributed fragments of a poker deck among the participants and asked them to select the strongest possible poker hand from the totality of cards they held. The center pattern, spokes, was best (even when the least able person occupied the central coordinator role). The circle was worst, and the fully connected pattern evolved by neglect of channels into the spoke pattern as the same participants repeatedly played the game. As a model for science, which is correct, Polanyi's or Bavelas's? Which is the better analogy to the collective task of science? Both models are thoroughly social, and both have normative content as competing models for optimizing collective validity.

The sketches above are but analogical suggestions for SPS. In that eventual textbook on SPS, there should be a full chapter on experimenting with the process of science making directly. The value of the present book in this regard is weakened by the fact that Robert Rosenthal has chosen to write on meta-analysis rather than to recount his heroic decade or so of experimenting with experimenter effects (for a summary, see Rosenthal, 1976). Note that this profuse demonstration of self-deceptive bias in our own science is a part of a vigorous proepistemic commitment to making our science *more* valid, rather than giving up on the goal of

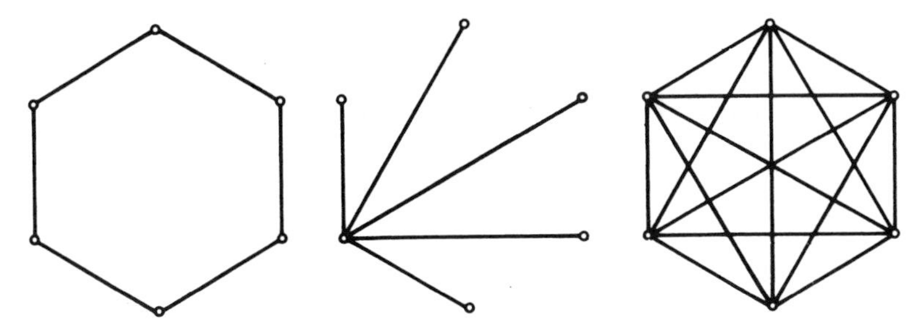

FIGURE 2.5. Bavelas-type communication patterns in six-person teams. (From left to right, circle, spokes, and fully connected.)

validity. Others have shared this perspective. The subtitle of Adair (1973) says it well: *The Social Psychology of the Psychological Experiment*. Faust (1984) also belongs. So, too, do experimental studies of the manuscript review process, in which the same manuscripts are submitted with male or female authorship, high- or low-prestige institutional addresses, results confirming or refuting established views, and so on (e.g., Abramowitz, Gomes, & Abramowitz, 1975; Mahoney, 1977; Mahoney, Kazdin, & Kenigsberg, 1978; Peters & Ceci, 1982). Once again, such studies could be taken as proving the invalidity of the process of science, with the renunciation of the goal of validity, or (as in these examples) as a part of an effort to design review processes that minimize such biases (see also extensive review with critiques by Cicchetti, 1991). These are all contributions to a sociology of scientific validity that the sociologists of scientific knowledge have not made, and are unlikely to make.

In the remainder of this chapter, I return to the autobiographical mode, on the sociological side of the Comptean divide, and to my earliest, still unpublished contribution.

Through my undergraduate course in sociology taught by Robert Nisbett, as well as his service on the qualifying exams for my doctorate, I had heard of the sociology of knowledge by the time I finished my degree, but I had not had time to explore the field. My real education on this topic began in 1947 with my first appointment as assistant professor of psychology at Ohio State University. It was my good fortune, during my 3 years there, to spend 2 hours each Friday afternoon at a "philosopher's table" in the Faculty Club. In addition to the philosopher leaders Eliseo Vivas and Virgil Hinshaw, Kurt Wolff was a major participant. Wolff was a sociologist of knowledge, a student of Karl Mannheim and the Frankfurt School, and the translator of Georg Simmel. From Wolff and his reading suggestions, I learned about the sociology of knowledge and its then tenuous relationship to the sociology of science. From that model, I formed my image of what a "psychology of knowledge" would be. During the summer session of 1950, in my final teaching assignment at Ohio State, I offered a course on the psychology of knowledge. Later that year, I delivered to the University of Chicago Psychology Club what I regarded as my "inaugural lecture" for my new assistant professorship there. I have since judged this to be a publishable paper but somehow never submitted it. (I still keep copies available, useful at least as a historical document.) My beginning paragraphs were as follows:

> Under this very general heading, I have found it possible to bring together a number of my preoccupations, ranging from such menial endeavors as the indirect measurement of social attitudes, to more generalized interests in learning theory, and the science of science. I hope in the process to provide

the beginnings of a rationale under which the attention of psychologists can be turned to that most human of human products—"Knowledge." May I start with a concrete example from my own work.

As in personality and clinical psychology, so also in social attitude research there has been in recent years increasing effort to free our research from dependence upon "voluntary self-description" in attitude measurement (Campbell, 1950a). In the service of this goal there have been a number of attempts to substitute objective, projective, disguised or indirect attitude measurements. If we search for a common feature among these, we can find that they all involve this one aspect of indirection: *They ask the respondent* (client or subject) *to report on the world around him rather than to report upon himself.* In this sense, they provide an objective, object oriented, task; an assignment dealing with external reality, with *knowledge*.

The information test is a promising example of such approaches. As exemplified in research on attitudes toward the Negro [forgive the dated language] started at the Ohio State University, and being continued here under the support of the University of Chicago Committee on Education, Training, and Research in Race Relations, the following approach is used: Facts are collected relevant to the controversial topics, stereotypes, etc. Around these facts are built multiple choice alternatives which differ from each other in degree or directionality. There is an attempt made to state these at a level of detail beyond common knowledge, and to state them simply enough so that the implications of each alternative are clear. Here is an example:

According to Army surveys made during World War II of the attitudes of White soldiers toward Negroes:

1. The attitudes were most hostile among White troops that had been in combat with Negro troops.
2. The attitudes were most hostile among White troops that had worked in port battalions and other service units with Negro troops.
3. Experience with Negro troops made no difference in the attitudes of White soldiers.
4. The attitudes were most favorable among White troops that had been in combat with Negro troops.

We expect individual differences in response to such items to be diagnostic of attitudes on the basis of such sociological and psychological mechanisms as the following: The kinds of correct information that a person collects are a function of his attitudes, and of the selective sources of information which are available in his particular social milieu. Thus if we had a test embodying both "pro-Negro" facts and "anti-Negro" facts, we might expect a person's attitude to be revealed by which kind of items that he got right. Similarly, if there has been systematic distortion in memory, or in sources of information, such distortion is expected to reflect itself in the direction of the errors made. And if guessing is involved, we would expect that guessing as to the detailed facts would be done on the basis of more generalized attitudes.

Now I would like to propose that over and beyond the usefulness of such a test as a tool in other research [for a later rejection of applied use, see

Campbell, 1988d, pp. 8-9] such studies have an intrinsic value in themselves. They can be justified, I believe, as explorations in the psychology of "sociology of knowledge" of the man-on-the-street. Popular notions of the nature of reality can be compared with those of the social scientist. Systematic deviations can be noted, and can be related to the particular social locus of the individual. (Campbell, 1950b, pp. 1–2)

We have hypothesized a general "knowledge" process, superordinate to the concepts of perception and learning. We would suggest that it is an appropriate activity of psychologists to build theory at this level, and that much of our theory from perception and learning can be used as such and is congruent once the difficulties of "inside" and "outside" language have been removed. We would like to further suggest that such a theory of the knowledge process is appropriately extendable to encompass much of the behavior we call "science."

Sciences can be regarded as essentially an extension of the reality-testing process, extending the number of observations, the points of observation, and the kinds of manipulation involved, and thus developing a model of reality of still more generalized utility, of greater stability, than that achievable by the single mind. The human mind has certain limitations: Its storage capacity is limited, the number of variables it can observe at one time is still more limited, the complexity of the integrative and predictive formulas it can use is also finite. In these and other ways, science—in some degree—extends the experimental basis of our image of reality. Science is a refinement of, but essentially partakes of the same nature as our general knowledge processes. And individual scientists are finite biological organisms whose activities must certainly be subsumed under the same general laws as those of other problem-solving organisms.

Currently the social sciences are keenly aware of the fact that objectivity is not easily achieved, and out of these problems a minor discipline—the sociology of knowledge—has emerged, specifying the social conditioners of what man and science take for knowledge. A psychology of the knowledge processes can support and extend such understanding, and throw additional light, I believe, upon the special problems of the social sciences. (Campbell, 1950b, pp. 6–7)

Rather than further expand the discussion along this line, however, I would like to pass on to a purely social principle of knowledge behavior. It is this: Individuals solving problems in a group will tend to come out with similar solutions, regardless of the accuracy of such solutions. This pervasive law of learning has recurred under many different labels in psychological and sociological observations. The experiments of Sherif (1936) on group judgments of the autokinetic phenomena are classic in this regard. Older research on "majority prestige" or "suggestibility to group opinion" is also relevant. Festinger (1950), in summarizing a series of group-dynamic experiments on communication within the group, emphasizes this under the general rubric of "pressures toward uniformity within the group." . . . While such a finding is certainly not characteristic of all group efforts it serves to illustrate the point that among scientists as among other persons, groups create a tendency toward consensus of belief, and that such increased consensus may be independent of any increase in the validity of that "knowledge."

As seen in the light of these considerations, social science faces some problems of objectivity in a more acute form than do the physical sciences. Paramountly, the social sciences may be thought to be operating in a much more ambiguous field, working possibly on more complex problems, and in a poorer situation for reality testing. This increases susceptibility to the effects of all of the various nonobjective tendencies, and makes more necessary that extended form of reality testing known as the scientific method. To take a primitive example, the unassisted observer can be better trusted to tell whether or not a lever has moved a rock than to tell whether or not a speaker has moved an audience. And while both the physical scientist and the social scientist in the selection of their problems reflect value judgments, the social scientist may be more apt to be overmotivated. The problems of the social scientist are more immediately related to the problems of his everyday living and his social locus. Wirth (1946, page xxiv) has stated it thus: "The fact that in the realm of the social the observer is part of the observed and hence has a personal stake in the subject of observation is one of the chief factors in the acuteness of the problem of objectivity in the social sciences . . . . When we advocate something, we do not do so as complete outsiders to what is and what will happen." This higher motivation coupled with a lack of reality testing may reduce the "accuracy" of our solutions.

A third special problem involves competition with the content of culture. At the present time, at least, the important problems which the social scientist undertakes are those about which culture already has answers. Both at the social level, of censorship and pressure, and at the individual level of habit interference and set, this competition is important in determining the end form of knowledge, under conditions of high ambiguity and low opportunity for reality testing. (Campbell, 1950b, pp. 9–10)

If our motivational analysis is correct, there are many areas of inquiry in the social sciences which the well-adjusted tribe member would never undertake. As Robert Park (1928) has pointed out, the culturally uprooted have expanded greatly the scope of problems and perspectives of social science. Indeed, some such maladjustment or personality divergence from the norm may be a prerequisite to any interest in these areas. We should expect then to find rigid people studying rigidity, minority group members studying prejudice, normless sociologists studying norms, rebellious sons studying institutional authority, etc. We should expect to find personality differences between those who propose theories and those who criticize, check, or reject them. And this interaction of the variety of human adjustment patterns and the topics of inquiry should advance rather than hinder science, where reality testing is done. No better example of this interaction can be found than in the area of social attitudes, in which minority group members are

> financing, in the interest of their own survival, research upon the characteristics of prejudiced majority group members. While some of this research has resulted in wishful thinking, autistic projection, and the like, in those instances in which scientific methods have been used, and more extensive testing of theories has been made possible, the results represent genuine advances over previous speculative treatments of the same problems. And without such motivation, these particular contributions would never have been made. (Campbell, 1950b, p. 11)

While I did not publish the 1950 address, my science-of-science interests continued as a minor part of my scholarly effort. In 1958, I published a paper that remains a favorite of mine. It is a contribution to the "groups are real" literature, using the Gestalt principles of perceptual organization to relativize the concept of thing or entity even for its physical–object exemplifications. This relativization allows, and provides imperfect criteria for, positing "real" ontological status for some social organizations. In 1959 I initiated my naturalistic, evolutionary epistemology series (Campbell, 1959b). Still more germane to the agenda I now recommend for SPS is Jacobs and Campbell's (1961) paper, which has no explicit metascientific content.

Conformity research provides evidence of social processes leading to the cultural determination of beliefs, including social beliefs. The trust required of other scientists makes these processes relevant in science as well. How is this to be reconciled with the reality-testing emphasis of the 1950 paper just quoted? I wish I had an ideal experimental exemplar for "cultural transmission with directional selection" (Boyd & Richerson, 1985) that showed movement toward increased validity, that would show how to combine the SSK contraepistemic emphasis with a proepistemic one. Lacking that, Jacobs and Campbell (1961) come closest.

Figure 2.6 shows the transmission of an "arbitrary tradition" over eight "generations" in a laboratory microculture. Using the autokinetic phenomenon as introduced by Sherif, subjects judged the number of inches of movement of an actually stationary point of light exposed for 5 seconds in an otherwise dark room. Of the first generation of four persons, the three giving their judgments first were confederates whose judgments consistently averaged around 16 inches. After every 30 judgments, the "oldest" of the four persons was removed and a new, naive, participant introduced, who was then the last of the four to call out his or her judgments for the next round of 30, and so on. By the fourth generation, the arbitrary norm is being inadvertently transmitted by the "elders" of the now all naive subjects. This fact serves as an analogue to the "social construction of reality" (Berger & Luckmann, 1967) emphasized by the older sociologists of knowledge, and reemphasized by the new SSK.

In Figure 2.6, we see this socially constructed reality both being naively perpetuated in reports on objective reality and being steadily undermined by what the people are "seeing with their own eyes," just as the ideology of science says should happen (see Campbell, 1986/1988b; 1988, Chapter 19, especially, p. 515). Jacobs and Campbell (1961) provide ample evidence of a natural illusory perception of movement, under their conditions averaging around 3.8 inches to begin with and decreasing to 2.9 inches over three sets of 30 judgments. (In addition to a group series run with no confederates, they also ran 24 persons in isolation, confirming this natural range of perception.) It might have been better for my present illustrative purposes if the naive, "culture-free" perceptions had been completely veridical. That the 3.9–2.8 inches of perceived movement were illusory reminds us of psychology's arguments against perceptual epistemological foundationalism. The social psychology of scientific validity must never claim to have compellingly answered the skeptics. Nor must it claim direct realism for visual processes. The fact that interpersonal perceptions tend strongly to agree is what makes language and science possible (Campbell & Paller, 1989), not the foundational accuracy or unbiasedness of these perceptions. So this illustration of the potential power of individual perceptions to undermine a "social construction out of whole cloth" is, I believe, relevant. This is not to

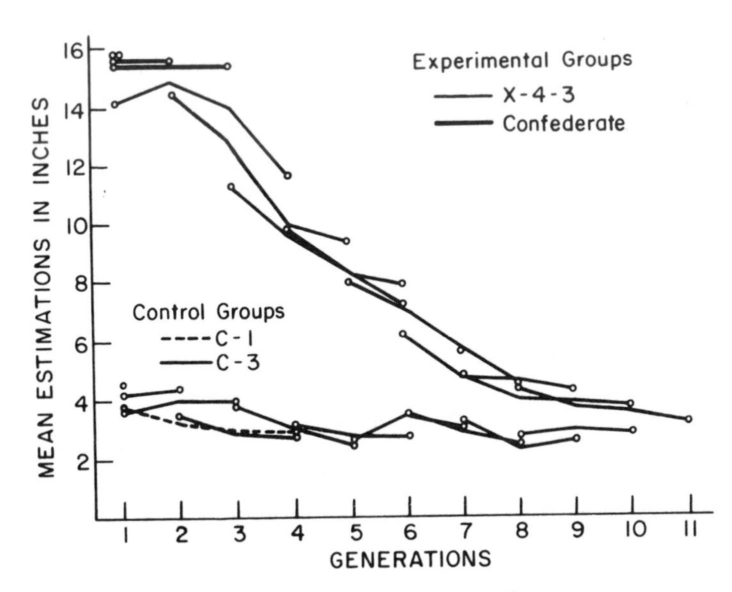

FIGURE 2.6. Transmission of an arbitrary norm in four-person groups, including judgments of two types of control group. From Figures 1 and 4 of Jacobs and Campbell (1961, pp. 651, 653). Copyright 1961 by the American Psychological Association. Adapted by permission.

discourage rerunning Jacobs and Campbell's (1961) experiment, however, with a highly ambiguous task in which socially untutored individual judgments have, on the average, a "correct" value rather than an "illusory" one.

The boundaries of SPS as distinguished from the psychology of science on the one side and the sociology of science on the other, are admittedly vague. Metascience done by career social psychologists is one crude criterion. Epistemological essays triggered by clearly social–psychological research programs thus deserve to be included. As a by-product of *The Influence of Culture on Visual Perception* (Segall, Campbell, & Herskovits, 1966), I published a favorite along these lines (Campbell, 1964), which confidently reports cultural differences in susceptibility to the Muller–Lyer and other optical illusions, with details as to direction and magnitude. But from a foundationalist perspective, our study is built on quicksand. We raised the possibility that earlier study results due to respondents' failure to understand the task, and we built in many features to avoid such errors. One such feature was to begin the set of test items with four items as comprehension checks. The respondent had to "get right" (i.e., to choose the way the investigator would have chosen) these very simple items or we would not use the data. The fourth of these comprehension checks was a very extreme Muller–Lyer figure. If other-culture respondents to the question, "Which red line is longer?" answered differently than would the researcher on this item, it was taken as *evidence of failure of communication*, of misunderstanding the task. The illusion-supported choice on that exemplar was 700% longer than the correct line. In contrast, the 12 Muller–Lyer test items had discrepancies ranging from -5% to +50%. If, on these items, the respondents answered differently than would the researcher, this was taken as *evidence of perceiving differently*. As it turned out, the 50% item could also have been used as a comprehension check.

The epistemological moral is that one can only establish cultural differences in perception if they are relatively small. The coherence-based determination of a cross-cultural difference requires one to assume that most of the other's perceptions are the same as one's own. This illustrates very well the merging of the contraepistemic and the proepistemic that I have repeatedly called for in this chapter. (This collaboration with the anthropologist Herskovits also led me to a metascientific introduction to a posthumous collection of his essays on cultural relativism, emphasizing the role of that doctrine in improving the validity of anthropology as an alternative to using it to argue giving up the goal of validity [Campbell, 1972].)

Were I to have been asked to submit already published essays of mine to a modern anthology on SPS, I would have suggested only two.

The first of these would have been "A Tribal Model of the Social System Vehicle Carrying Scientific Knowledge" (1979/1988c). My topics were the social norms of science and scientific leadership. I used data that David Krantz (Krantz & Wiggins, 1973), a historian and social psychologist of science, collected in a study that he and I jointly planned. My puzzle was why Spence's students had been so much more loyal to Spence's (and Hull's) theory than Tolman's had been, especially since I believe we can now judge that Tolman had the much better learning theory. Using the responses of more than 20 former students and colleagues, Krantz obtained the results shown in Tables 2.1 and 2.2. In loyalty to the theory (even though showing my own great love for Tolman), I charged Tolman with a "default in tribal leadership," and speculated on younger-brother, baby-of-the-family personality dynamics.

I argued that it is desirable in science that each competing theory be thoroughly explored, and that it may take an irrational degree of faith in the theory by a thought collective of believing scientists to achieve that thorough exploration and elaboration. Tolman's whimsical, self-deprecating humor combined with the traits shown in Table 2.2, failed to instill such faith, and once away from Berkeley, his students did no further research on the theory. This was in sharp contrast to the behavior of Spence's students. From the standpoint of a social psychology of scientific leadership, it may also have helped that Hull was the founder of the theory, Spence merely the preaching disciple. (Tolman's earlier student, Calvin Hall, once told me I had Tolman's personality just right. But almost everyone who knew Tolman loved him, and both they and many who knew Spence well don't want me to make any criticisms of Tolman.)

The final example I offer of SPS on the social side of the Comptean divide (and the other paper I would offer to that anthology) is my exegesis on a theme from Asch (1952/1987) rather than my own work. The nor-

TABLE 2.1. Traits Shared by Spence and Tolman

	Spence	Tolman
Inspiring	91%	93%
Creative	85	96
In teaching, presented primarily his own position	64	96
Critical of others' systematic positions	96	61
Overly influenced by others' views of his approach	10	18

Note. Percentage of the raters attributing the characteristic to the theorist, condensed from the top three categories of a nine-point rating scale in Krantz's study. From Campbell (1979/1988c).

TABLE 2.2. Traits on Which Spence and Tolman Differed in Krantz's Study

	Spence	Tolman
Convinced of value of own systematic position	98%	46%
Took himself seriously	98	32
Expected strong commitment to his approach from me	98	4
Authoritarian	88	0
Aggressive	84	4
Strong-willed	96	36
Accepted criticism well	13	86
Allowed me autonomy in choice of research problems	26	96
Open-minded	6	100
Humorous	28	96

Note. From Campbell (1979/1988c).

mative component is stressed in my title "Asch's Moral Epistemology for Socially Shared Knowledge" (Campbell, 1990b). My introductory abstract for that paper is as follows:

> From 1938 on, Asch has pointed out our ubiquitous dependence upon the reports of others for our knowledge of the world. Most conformity studies should be reinterpreted as illustrating this process (rather than showing the character defect of suggestibility). The Aschian moral norms for socially achieved knowledge are:
>
> **Trust:** It is our duty to respect the reports of others and be willing to base our beliefs and actions on them.
> **Honesty:** It is our duty to report what we perceive honestly, so that others may use our observations incoming to valid beliefs.
> **Self respect:** It is our duty to respect our own perceptions and beliefs, seeking to integrate them with the reports of others without deprecating them or ourselves.
>
> Social consensuses based on this morality have a much better chance of being valid than do consensuses where agreement per se has been the goal. (Campbell, 1990b, p. 39)

Asch says that my paper touches on the aspect of his work most important to him. He likes my abstract. But in the full article, I do not get it quite right. So, in Rock's (1990) *Festschrift* for Asch, Asch (1990) has included a commentary.

There are several famous "Asch experiments." For the purpose of my exegesis, most appropriate is his study on the intelligence required for politics (Asch, 1940). Students told that last year's class had ranked politics first among the 12 professions, gave politics a high rank. When later asked what they meant by "politicians," they gave "statesmen" as examples and synonyms. But those students who were told that fellow students had previously ranked "politics" at the bottom of the professional hierarchy gave politics low rankings, and later gave the term "politician" the connotation of "ward heeler." They had illustrated trust and self-respect, and had searched out a resolution that made compatible both their own judgments and those reported for their fellow students.

Asch and I differ, however, on my next step. I say that were only the three Aschian norms at stake (as in the most famous Asch conformity experiments with the seven or eight confederates and only one real subject [Asch, 1952/1987, 1956]), the participants under "trust" should have said, "you fellows' 8 pairs of eyes are no doubt better than my one pair, so I have to believe line A is longer. But it is my duty, under honesty, to report what I see, and I see line B as longer." That is, private reports (focused on net belief rather than own perception) should have shown greater conformity than did public (these latter being the oral reports on which others had to depend). Asch's results, and those of many other experimenters, show exactly the reverse. So, I conclude that even such newly organized minimal groups value consensus over truth (though valuing both). I note, too, that this preference for consensus also produces a problem for honesty in science, even though in scientific groups truth is officially the paramount goal.

By inserting the term *validity* into the title of this chapter, I have expressed my hope that some of the contributions of SPS will be hypothetically *normative*, rather than purely *descriptive*. In this we will want to make contact with the normative agenda in the philosophy of science. For this same reason, I ended my Asch exegesis with a section "Introducing Asch to Philosophers." There I identify Asch as antifoundationalist with regard to visual perception (rather than a direct realist à la James Gibson), a coherence theorist of belief revision who has not made coherence per se the definition of truth, an epistemological relativist, and a naturalistic epistemologist. I close with a discussion of the epistemological aspects of the sociology of science (Campbell, 1990b):

> *Sociology of knowledge and sociology of science traditions* are nowadays merging with naturalistic epistemology, and a major point of view is the "social constructionist." I see Asch, as well as myself, as also social constructionists, since the beliefs one has about the world are collective social products. The social constructionists call attention to the relativity of belief thus resulting. At the level of *epistemological* relativism, I see Asch as joining them: One cannot know for sure that one's beliefs are true, as they are all dependent on the assumed completeness of the available frames of references, most of them socially provided.

> The bulk of the social constructionists go on from there to ontological relativism, agnosticism, or nihilism. The goal of *truth* or *objectivity*, they assert, must be given up as naive and incoherent since consensus is all we can be sure science achieves. I see Asch as departing from them at this point. If we are to be rationally authentic to our experienced purposes, we must retain the goal of knowing how the world is in a way that is independent of our particular vantage point and limited frame of reference. In this regard, all consensus processes are not equivalent. Some of them, as found in science when working at its best, can be seen to be rationally more likely to improve the validity of the resulting beliefs. These are the ones that follow the Aschian epistemological morality. Consensuses achieved where consensus itself has been the ultimate goal must be sharply distinguished from consensuses discovered in the collaborative search for truth. (p. 50)

Summary

SPS has long existed, but it is only since Kuhn that it has started to flourish. Most of this social psychology of knowing has been individual and cognitive, including the Bruner and Postman work that influenced Kuhn. Social psychology, both of ordinary knowing and of science, has always been hypothetically normative, relating social–cognitive processes to validity. This is quite in contrast to the sociologists of scientific knowledge, with whom we should nonetheless become much more thoroughly acquainted, and whose emphases are also legitimized by the logical analysis of our epistemological predicament.

Social psychologists of knowing can also make (and have made) contributions on the social side of the Comptean divide. These—or some of these—should also be hypothetically normative, as illustrated by Asch and others.

Acknowledgment

The author gratefully acknowledges permission to quote extensively from Thomas S. Kuhn's *The Structure of Scientific Revolutions*. Copyright 1962 by the University of Chicago Press. All rights reserved.

References

Abramowitz, S. I., Gomes, B., & Abramowitz, C. V. (1975). Publish or politic: Referee bias in manuscript review. *Journal of Applied Social Psychology, 5*, 187–200.

Adair, J. G. (1973). *The human subject: The social psychology of the psychological experiment*. Boston: Little, Brown.
Allport, G. W. (1954). The historical background of modern social psychology. In G. Lindzey (Ed.), *Handbook of social psychology* (Vol. 1, pp. 3–56). Reading, MA: Addison-Wesley.
Amsterdamska, O. (1990). Surely you are joking, Monsieur Latour! [Review of *Science in action*]. *Science, Technology, & Human Values, 15*(4), 495–504.
Asch, S. E. (1940). Studies in the principles of judgments and attitudes: II. Determination of judgments by group and by ego standards. *Journal of Social Psychology, 12*, 433–465.
Asch, S. E. (1956). Studies of independence and conformity: I. A minority of one against a unanimous majority. *Psychological Monographs, 70* (9, Whole No. 416).
Asch, S. E. (1987). *Social psychology*. Cambridge, England: Cambridge University Press. (Original work published 1952)
Asch, S. E. (1990). Comments on D. T. Campbell's chapter. In I. Rock (Ed.), *The legacy of Solomon Asch: Essays in cognition and social psychology* (pp. 53–55). Hillsdale, NJ: Erlbaum.
Ash, M. G., & Woodward, W. B. (Eds.). (1987). *Psychology in twentieth-century thought and society*. Cambridge, England: Cambridge University Press.
Ashmore, M. (1989). *The reflexive thesis: Wrighting the sociology of scientific knowledge*. Chicago: University of Chicago Press.
Barnes, B. (1974). *Scientific knowledge and sociological theory*. London: Routledge.
Barnes, B. (1982). *T. S. Kuhn and social science*. New York: Columbia University Press.
Barnes, B., & Bloor, D. (1982). Relativism, rationality, and the sociology of knowledge. In M. Hollis & L. Lukes (Eds.), *Rationality and relativism* (pp. 21–47). Oxford, England: Blackwell.
Bar-Tal, D., & Kruglanski, A. W. (Eds.). (1988). *The social psychology of knowlege*. New York: Cambridge University Press.
Bavelas, A. (1950). Communication patterns in task-oriented groups. *Journal of the Acoustical Society of America, 22*, 725–730.
Berger, P. L., & Luckmann, T. (1967). *The social construction of reality*. Garden City, NY: Doubleday.
Bijker, W. E., Hughes, T. P., & Pinch, T. J. (Eds.). (1987). *The social construction of technological systems*. Cambridge, MA: MIT Press.
Bloor, D. (1976). *Knowledge and social imagery*. London: Routledge.
Boring, E. G. (1964). Cognitive dissonance: Its use in science. *Science, 145*, 680–685.
Boyd, R., & Richerson, P. J. (1985). *Culture and the evolutionary process*. Chicago: University of Chicago Press.
Bruner, J. S. (1951). Personality dynamics and the process of perceiving. In R. R. Blake & G. V. Ramsey (Eds.), *Perception: An approach to personality* (pp. 121–147). New York: Ronald Press.
Bruner, J. S. (1962). *On knowing: Essays for the left hand*. Cambridge, MA: Harvard University Press.

Bruner, J. S. (1973). *Beyond the information given: Studies in the psychology of knowing*. New York: W. W. Norton.

Bruner, J. S., & Postman, L. (1949). On the perceptions of incongruity: A paradigm. *Journal of Personality, 18*, 206–223.

Bruner, J. S., Postman, L., & Rodrigues, J. (1951). Expectations and the perception of color. *American Journal of Psychology, 64*, 216–227.

Campbell, D. T. (1950a). The indirect assessment of social attitudes. *Psychological Bulletin, 47*, 15–38.

Campbell, D. T. (1950b, November 9). *On the psychological study of knowledge*. Paper presented before the Psychology Club, University of Chicago, November 9, 1950.

Campbell, D. T. (1956). Enhancement of contrast as composite habit. *Journal of Abnormal and Social Psychology, 53*(3), 350–355.

Campbell, D. T. (1958). Common fate, similarity, and other indices of the status of aggregates of persons as social entities. *Behavioral Science, 3*(1), 14–25.

Campbell, D. T. (1959a). Systematic error on the part of human links in communication systems. *Information and Control, 1*, 334–369.

Campbell, D. T. (1959b). Methodological suggestions from a comparative psychology of knowledge processes. *Inquiry, 2*, 152–182.

Campbell, D. T. (1963). Social attitudes and other acquired behavioral dispositions. In S. Koch (Ed.), *Psychology: A study of science: Vol. 6. Investigations of man as socius* (pp. 94–172). New York: McGraw-Hill.

Campbell, D. T. (1964). Distinguishing differences of perception from failures of communication in cross-cultural studies. In F. S. C. Northrop & H. H. Livingston (Eds.), *Cross-cultural understanding: Epistemology in anthropology* (pp. 308–336). New York: Harper & Row.

Campbell, D. T. (1969). Ethnocentrism of disciplines and the fish-scale model of omniscience. In M. Sherif & C. W. Sherif (Eds.), *Interdisciplinary relationships in the social sciences* (pp. 328–348). Chicago: Aldine.

Campbell, D. T. (1972). Herskovits, cultural relativism, and metascience. In M. J. Herskovits, *Cultural relativism* (pp. v–xxiii). New York: Random House.

Campbell, D. T. (1973). Ostensive instances and entitativity in language learning. In W. Gray & N. D. Rizzo (Eds.), *Unity through diversity, Part 2* (pp. 1043–1057). New York: Gordon & Breach.

Campbell, D. T. (1986). Science policy from a naturalistic sociological epistemology. In P. D. Asquith & P. Kitcher (Eds.), *PSA 1984* (Vol. 2, pp. 14–29). East Lansing, MI: Philosophy of Science Association.

Campbell, D. T. (1987). Neurological embodiments of belief and the gaps in the fit of phenomena to noumena. In A. Shimony & D. Nails (Eds.), *Naturalistic epistemology* (pp. 165–192). Boston: Reidel.

Campbell, D. T. (1988a). Prospective: Artifact and control. E. S. Overman (Ed.), *Methodology and epistemology for social science: Selected papers* (pp. 167–190). Chicago: University of Chicago Press. (Original work published 1969).

Campbell, D. T. (1988b). Science's social system of validity-enhancing collective belief change and the problems of the social sciences. E. S. Overman (Ed.), *Methodology and epistemology for social science: Selected papers* (pp. 504–523). Chicago: University of Chicago Press. (Original work published 1986).

Campbell, D. T. (1988c). A tribal model of the social system vehicle carrying scientific knowledge. E. S. Overman (Ed.), *Methodology and epistemology for social science: Selected papers* (pp. 489–503). Chicago: University of Chicago Press. (Original work published 1979).

Campbell, D. T. (1988d). Perspective on a scientific career. E. S. Overman (Ed.), *Methodology and epistemology for social science: Selected papers* (pp. 1–26). Chicago: University of Chicago Press.

Campbell, D. T. (1989). Models of language learning and their implications for social constructionist analyses of scientific belief. In S. L. Fuller, M. De-Mey, T. Shinn, & S. Woolgar (Eds.), *The cognitive turn: Sociological and psychological perspectives on science* (pp. 153–158). Dordrecht, The Netherlands: Kluwer.

Campbell, D. T. (1990a). Epistemological roles for selection theory. In N. Rescher (Ed.), *Evolution, cognition, realism* (pp. 1–19). Lanham, MD: University Press of America.

Campbell, D. T. (1990b). Asch's moral epistemology for socially shared knowledge. In I. Rock (Ed.), *The legacy of Solomon Asch: Essays in cognition and social psychology* (pp. 39–52). Hillsdale, NJ: Erlbaum.

Campbell, D. T. (1991). Coherentist empiricism, hermeneutics, and the commensurability of paradigms. *International Journal of Educational Research, 15*(6), 587–597.

Campbell, D. T. (1993). Plausible coselection of belief by referent: All the "objectivity" that is possible. *Perspectives on Science: Historical, philosophical, social, 1,* 85–105.

Campbell, D. T. (in press). Systematic errors to be expected of the social scientist on the basis of a general psychology of cognitive bias. In P.D. Blanck (ed.), *Interpersonal expectations: Theory, research and application.* New York: Cambridge University Press. (Original work written 1959)

Campbell, D. T., & Gruen, W. (1958). Progression from simple to complex as a molar law of learning. *Journal of General Psychology, 59,* 237–244.

Campbell, D. T., & Paller, B. T. (1989). Extending evolutionary epistemology to 'justifying' scientific beliefs. (A sociological rapprochement with a fallibilist perceptual foundationalism?) In K. Hahlweg & C. A. Hooker (Eds.), *Issues in evolutionary epistemology* (pp. 231–257). Albany: State University of New York Press.

Cicchetti, D. V. (1991). The reliability of peer review for manuscript and grant submissions: A cross-disciplinary investigation. *Behavioral and Brain Sciences, 14,* 119–186.

Collins, H. M. (1981). Stages in the empirical programme of relativism. *Social Studies of Science, 11,* 3–10.

Collins, H. M. (1985). *Changing order: Replication and induction in scientific practice.* Newbury Park, CA: Sage.

Cook, T. D., & Campbell, D. T. (1986). The causal assumptions of quasi-experimental practice. *Synthese, 68*, 141–180.

Faust, D. (1984). *The limits of scientific reasoning.* Minneapolis: University of Minnesota Press.

Festinger, L. (1954). A theory of social comparison processes. *Human Relations, 7*, 117–140.

Festinger, L. (1989). Informal social communication. In S. Schachter & M. Gazzaniga (Eds.), *Extending psychological frontiers: Selected works of Leon Festinger.* New York: Russell Sage. (Original work published 1950)

Festinger, L., Schachter, S., & Back, K. (1950). *Social pressures in informal groups: A study of human factors in housing.* New York: Harper.

Galison, P. (1987). *How experiments end.* Chicago: University of Chicago Press.

Galloway, D. W. (1948). An experimental investigation of structural lag in perception. *American Psychologist, I*, 450. (Abstract)

Giere, R. N. (1985). Philosophy of science naturalized. *Philosophy of Science, 52*, 331–356.

Gilbert, G. N., & Mulkay, M. (1984). *Opening Pandora's box: A sociological analysis of scientists' discourse.* New York: Cambridge University Press.

Greenwald, A. G., Pratkanis, A. R., Leippe, M. R., & Baumgardner, M. H. (1986). Under what conditions does theory obstruct research progress? *Psychological Review, 93*, 216–229.

Guetzkow, H., & Simon, H. A. (1955). The impact of certain communication nets upon organization and performance in task-oriented groups. *Management Science, 1*, 233–250.

Hastorf, A. H. (1950). The influence of suggestion on the relationship between stimulus size and perceived distance. *Journal of Psychology, 29*, 195–217.

Hull, D. L. (1988). *Science as a process.* Chicago: University of Chicago Press.

Jacobs, R. C., & Campbell, D. T. (1961). The perpetuation of an arbitrary tradition through several generations of a laboratory microculture. *Journal of Abnormal and Social Psychology, 62*, 649–658.

Kahneman, D., Slovic, P., & Tversky, A. (Eds.). (1982). *Judgment under uncertainty: Heuristics and biases.* Cambridge, England: Cambridge University Press.

Kim, Kyung-Man. (in press). *Explaining scientific consensus. A sociological analysis of the reception of Mendelian genetics.* New York: Guilford Press.

Klineberg, O. (1940). *Social psychology.* New York: Holt.

Knorr-Cetina, K. D. (1981). *The manufacture of knowledge: An essay on the constructivist and contextual nature of science.* Oxford, England: Pergamon Press.

Knorr-Cetina, K. D. (1987). Evolutionary epistemology and the sociology of science. In W. Callebaut & R. Pinxten (Eds.), *Evolutionary epistemology: A multiparadigm program* (pp. 179–201). Dordrecht, The Netherlands: Reidel.

Knorr-Cetina, K. D. (1990, October 19). The nature of things: On the organization of openness in two epistemic cultures. Paper presented at the annual meeting of the Society for the Social Studies of Science, Minneapolis, MN.

Knorr-Cetina, K. D., & Mulkay , M. (1983). Introduction: Emerging principles in social studies of science. In K. D. Knorr-Cetina & M. Mulkay (Eds.), *Science observed* (pp. 1–17). Newbury Park, CA: Sage.

Kornblith, H. (Ed.). (1985). *Naturalizing epistemology*. Cambridge, MA: MIT Press.

Krantz, D. L., & Wiggins, L. (1973). Personal and impersonal channels of recruitment in the growth of theory. *Human Development, 16*, 133–156.

Krech, D., & Crutchfield, R. S. (1948). *Theory and problems of social psychology*. New York: McGraw-Hill.

Krechevsky, I. (1932). *The genesis of "hypotheses" in rats: Vol. 6, University of California Publications in Psychology*. Berkeley: University of California Press.

Kuhn, T. S. (1962). *The structure of scientific revolutions*. Chicago: University of Chicago Press.

Latour, B. (1987). *Science in action*. Cambridge, MA: Harvard University Press.

Latour, B., & Woolgar, S. (1979). *Laboratory life: The social construction of scientific facts*. Newbury Park, CA: Sage.

Lehrer, K. (1974). *Knowledge*. Oxford, England: Clarendon Press.

Lehrer, K. (1990). *Theory of knowledge*. Boulder, CO: Westview Press.

Lynch, M. (1985). *Art and artifact in laboratory science*. London: Routledge.

MacKenzie, D. A., & Barnes, B. (1975). Biometrician versus Mendelian: A controversy and its explanation [in German]. *Kolner Zeitschrift für Soziologie und Sozialpsychologie, 18*, 165–196.

MacKenzie, D. A., & Barnes, B. (1979). Scientific judgment: The biometry–Mendelism controversy. In B. Barnes & S. Shapin (Eds.), *The natural order: Historical studies of scientific culture* (pp. 191–210). Newbury Park, CA: Sage.

Mahoney, M. J. (1977). Publication prejudices: An experimental study of confirmatory bias in the peer review system. *Cognitive Therapy Research, 1*, 161–175.

Mahoney, M. J., Kazdin, A. E., & Kenigsberg, M. (1978). Getting published. *Cognitive Therapy and Research, 2*, 69–70.

Mannheim, K. (1946). *Ideology and utopia: An introduction to the sociology of knowledge*. New York: Harcourt Brace.

Morawski, J. (1988). *The rise of experimentation in American psychology*. New Haven, CT: Yale University Press.

Mulkay, M. (1979). *Science and the sociology of knowledge*. London: Allen & Unwin.

Nisbett, R., & Ross, L. (1980). *Human inference: Strategies and shortcomings*. Englewood Cliffs, NJ: Prentice-Hall.

Park, R. (1928). Human migration and the marginal man. *American Journal of Sociology, 33*, 881–893.

Peters, D. P., & Ceci, S. J. (1982). Peer-review practices of psychological journals: The fate of published articles submitted again. *Behavioral and Brain Sciences, 5*, 187–255.

Pickering, A. (1984). *Constructing quarks: A sociological history of particle physics*. Chicago: University of Chicago Press.

Pickering, A. (1989). Living in the material world: On realism and experimental practice. In D. Gooding, T. J. Pinch, & S. Schaffer (Eds.), *The uses of experiment: Studies of experimentation in the natural sciences* (pp. 275–297). Cambridge, England: Cambridge University Press.

Polanyi, M. (1958). *Personal knowledge*. London: Routledge.

Polanyi, M. (Ed.). (1966). A society of explorers. *The tacit dimension* (pp. 55–92). Garden City, NY: Doubleday.

Postman, L. (1953) The experimental analysis of motivational factors in perception. In J. S. Brown, H. Harlow, L. Postman, V. Nowlis, T. Newcomb, & O. Mowrer (Eds.), *Current theory and research in motivation* (pp. 59–108). Lincoln: University of Nebraska Press.

Potter, J. & Wetherell, M. (1987). *Discourse and social psychology*. Newbury Park, CA: Sage.

Quine, W. V. (1951). Two dogmas of empiricism. *Philosophical Review, 60*, 20–43.

Quine, W. V. (1969). *Ontological relativity*. New York: Columbia University Press.

Rock, I. (Ed.). (1990). *The legacy of Solomon Asch: Essays in cognition and social psychology*. Hillsdale, NJ: Erlbaum.

Rosenthal, R. (1976). *Experimenter effects in behavioral research* (enl. ed.). New York: Irvington.

Rudwick, M. J. S. (1985). *The great Devonian controversy: The shaping of scientific knowledge among gentlemanly specialists*. Chicago: University of Chicago Press.

Schachter, S., & Gazzaniga, M. (Eds.). (1989). *Extending psychological frontiers: Selected works of Leon Festinger*. New York: Russell Sage.

Segall, M. H., Campbell, D. T., & Herskovits, M. J. (1966). *The influence of culture on visual perception*. Indianapolis, IN: Bobbs-Merrill.

Shapin, S. (1982). History of science and its sociological reconstructions. *History of Science, 20*, 157–211.

Shapin, S. (1988). Following scientists around [Review of Latour, Science in action]. *Social Studies of Science, 18*, 533–550.

Shapin, S., & Schaffer, S. (1985). *Leviathan and the air pump: Hobbes, Boyle, and the experimental life*. Princeton, NJ: Princeton University Press.

Sherif, M. (1936). *The psychology of social norms*. New York: Harper & Row.

Stich, S. P. (1985). Could man be an irrational animal? Some notes on the epistemology of rationality. In H. Kornblith (Ed.), *Naturalizing epistemology*. Cambridge, MA: MIT Press.

Tolman, E. C. (1932). *Purposive behavior in animals and men*. New York: Century.

Tolman, E. C., & Krechevsky, I. (1933). Means–end–readiness and hypothesis: A contribution to comparative psychology. *Psychological Review, 40*, 60–70.

Vidal, F. (1987). Jean Piaget and the liberal protestant tradition. In M. G. Ash & W. B. Woodward (Eds.), *Psychology in twentieth-century thought and society*. (pp. 271–294). Cambridge, England: Cambridge University Press.

Wilson, D. B. (1974). Kelvin's scientific realism: The theological context. *Philosophical Journal* (Transactions of the Royal Philosophical Society of Glasgow), 11, 41–60.

Wirth, L. (1946). Preface. In K. Mannheim (Ed.), *Ideology and utopia* (pp. xiii–xxxi). New York: Harcourt Brace.

Woolgar, S. (1988). *Science: The very idea.* London: Tavistock.

Wyatt, D. F., & Campbell, D. T. (1951). On the liability of stereotype or hypothesis. *Journal of Abnormal and Social Psychology, 46,* 496–500.

CHAPTER 3

The Social Psychology of Scientific Knowledge: Another Strong Programme

STEVE FULLER

Science: A Theory about Society and a Constitution for Society

Over the past 5 years, I have been developing a way to accommodate social scientific and philosophical approaches to the knowledge enterprise under a single research program, called social epistemology. Social scientists tend to propose empirical descriptions and hypotheses, whereas philosophers tend to issue prescriptions or speak in more generally normative terms. Discussions between the two sides frequently end in stalemate. Either the social scientists claim to find no evidence for philosophical norms or the philosophers accuse the social scientists of methodological naivete for coming up dry. In my attempts to mediate this dispute, one conclusion that critics have urged on me is that the very existence of a knowledge enterprise—"science," if you will—should not be taken for granted. Rather, "knowledge" should be treated as shorthand for a system of hypotheses, a theory that normally remains implicit but can nevertheless be subject to explicit empirical test. In this chapter I defend the centrality of this task to the social psychology of science (SPS).

In both academic and lay discourse, behavior is frequently explained in terms of one's possession of knowledge-related things, or "epistemic entities" (cf. Fuller, 1991b, on "intellectual property"). Among the things possessed include beliefs, reasons, intelligence, cognitive processes, (access to) information, disciplinary or technical training, theoretical assumptions, world views, and lines of thought. Although it is easy to imagine contexts in which one or more of these entities might be used to account for what someone has done, it is difficult to see how all of them could be accommodated within a single conceptual framework. That would seem to demonstrate the extent to which appeals to epistemic entities are typically made without much reflection. Social psychologists have been in forefront of challenging the power of beliefs and reasons to explain people's behavior, including one's own behavior (Nisbett & Wilson, 1977). I see SPS pushing the challenge still further, encompassing tests of the other entities mentioned above. To stress the continuity between this program and another radical empiricist study of knowledge of recent vintage, the sociology of scientific knowledge (SSK) (Bloor, 1976), I call this area of research the social psychology of scientific knowledge (SPSK).

What is at stake here? A useful way to think about the stakes is to consider what would follow from a demonstration that all our appeals to epistemic entities commit a version of what social psychologists call the fundamental attribution error (Ross, 1977). In other words, suppose it were shown that human behavior is more a function of one's situation than of whatever knowledge one brings to the situation. Consider the sorts of studies that would be relevant to such a finding. For example, social psychologists may fail to find significant differences between the decision-making patterns of economists and noneconomists; in fact, manipulating the decision task itself may be a more reliable way of changing the behavior of both groups than providing additional training in economics. Would it any longer make sense to defer to someone's judgment because of the economic knowledge he/she possesses? What would become of education in economics? What would "credentials" and "expertise" in economics mean under these circumstances? More generally, how can someone be held accountable for his/her actions if claims to knowledge do not seem to make an empirical difference to his/her behavior? Indeed, what would be the status of the social psychological research that had established this finding?

Consider another example. When people who have been antecedently identified as intelligent fail to excel on an intelligence test, the test is understandably blamed. However, if no test can be designed such that all these people outperform those who have been antecedently identified as unintelligent, it makes sense to ask whether those who are said to possess intelligence truly have something in common. To say, as has been the fashion in psychology recently, that there are "multiple intelligences" (e.g., Sternberg, 1985; Gardner, 1987) is only to push the problem to another level, to wit: Why should all these admittedly different abilities be subsumed under the same general category of intelligence? These are the sorts of probes that a truly empirical SPSK needs to make of most epistemic entities (cf. Danziger, 1990). Many such probes have already been conducted by selfstyled "constructivists"—ethnomethodological sociologists (e.g., Knorr-Cetina & Cicourel, 1981; Coulter, 1979), social psychologists (e.g., Gergen & Gergen, 1982; Gergen, 1985; Billig, 1987), and sociologists of science (Gilbert & Mulkay, 1984; Woolgar, 1988; Mulkay, 1990)—who have studied the situational variability of knowledge–talk in ordinary and scientific settings.

At this point, let me confront one obvious objection. I have assumed that if epistemic entities explain anything—or, for that matter, if there are such entities at all—they must stand in some systematic relation to some class of social phenomena. Yet, this assumption seems to beg the question of the significance of our knowledge–talk, the self-contained nature of which is supposed to show that the sorts of pursuits alternatively deemed rational, cognitive, epistemic, or scientific are governed by standards internally defined by such talk, standards sustainable under a variety of social arrangements. Thus, Popper's (1963) adaptation of Clausewitz's famous saying—that modern science is Socratic dialectic continued by other means—reflects the widespread belief among philosophers and scientists that if one is willing and able to rise to the level of critical scrutiny demanded of scientific inquiry, it does not matter whether one is tackling problems in the lab or in the marketplace. Such a view poses a direct threat to the possibility of SPSK because it envisages no "in principle" reason why knowledge could not be reliably produced by individuals as well as groups, by large groups as well as small, by groups whose members are concentrated or dispersed in time and space.

The general SPSK response to this kind of objection should be to treat as empirically open the issue of exactly which social arrangements foster (or impede) the production of which kinds of knowledge (Fuller, 1992b). But a more specific answer can also be given to the alleged autonomy of genuine knowledge production from social factors. Consider as an example the establishment of "neutral current" in high-energy physics, a case widely discussed among sociologists of science, whereby the existence of a highly counterintuitive entity was confirmed by several competing laboratories and ultimately accepted by virtually the entire physics community (Pickering, 1984; Roth & Barrett, 1990). This episode epitomizes the dual sense in which scientific knowledge has been traditionally regarded as "autonomous." On the one hand, the physicists appeared capable of rising above their local interests to recognize the existence of neutral current; on the other hand, hypotheses concerning neutral current managed to survive a variety of challenges so as to count presumptively as knowledge for all members of the physics community.

This dual sense of autonomy may itself be met by a twofold strategy.

First, one may question the extent to which the laboratory settings of the competing physicists were really different or independent from one another (e.g., potential contact between research teams and sufficient similarity in scientists' training to condition similar responses). Second, if the perceived differences among laboratories hold up under such scrutiny, one can engage in the reverse tactic of questioning whether these differences themselves might be responsible for the physicists' agreeing on the existence of neutral current. In other words, the different laboratories may be shown to have behaved in ways that could have been predicted on the basis of their particular social arrangements alone, even if the existence of neutral currents were not at issue. To put the point more provocatively: Just because the laboratories end up agreeing on the existence of a particular entity, it does not follow that their agreement is due to the existence of that entity. One laboratory may come to believe in neutral currents because it always follows whatever the research director thinks, whereas another laboratory may come to the very same belief as a result of a weighted average of what the entire research team thinks. If each laboratory is operating in its customary fashion, the convergence in beliefs could have been predicted simply on the basis of knowing the laboratory's decision-making procedures, without knowing anything about the content of the belief on which they converged. The natural conclusion to this line of thought is that the convergent belief in neutral currents is "epiphenomenal" to the diverse social processes that issued in assertions of that belief. Thus, as philosophers of mind are wont to say about such situations, to assert a belief in neutral currents is to make "opaque reference" to certain social practices. The reference is opaque in that if one takes knowledge–talk at face value, especially as it is expressed in such official communications as journal articles, one would easily get the impression that the various laboratories reached the same conclusions for largely the same reasons. The diversity of the underlying social practices would thus be masked, and hence rendered opaque, to the analyst.

We see, then, that knowledge production appears autonomous only when the corresponding sociological story remains incomplete, which makes both knowledge products and producers seem underdetermined. However, I do not mean to imply that each laboratory can be understood only in terms of its own local practices. As we have seen, the decision-making procedures that distinguished our hypothetical laboratories—deference to a superior and the weighted averaging of peers—are ones that can be found in other, nonscientific sectors of society (cf. Fuller, 1992a). These considerations go to the very heart of how modern society is organized and maintained. They bring out the extent to which our systematic talk about the possession and transformation of knowledge is itself a kind of social theory, that is, an attempt to explain the orderly flow of people, goods, and services in society (Bell, 1973; Boehme & Stehr, 1986; Stehr & Ericsson, 1992). Specifically, science, or "the scientific method," purports to be a means of reaching consensus that is more publicly accountable and less emotionally charged than other decision-making processes (for a history and critique of this view, see Ezrahi, 1990). Not surprisingly, the framers of the U.S. Constitution, Mill, Dewey, and Popper have all seen in scientific inquiry the deep structure of democratic politics. Whatever one makes of their observation, it does draw attention to two points:

> (1) Unlike other social theories, the scientific method is designed to justify what it explains: it performs normative as well as descriptive functions. For nearly three centuries, social theorists have invoked class, status, or power as entities that enabled the people who had them to get what they want. However, it is only when people are said to possess "knowledge" that the theorist tacitly accepts the legitimacy of their actions. But if the referentiality of knowledge–talk is as opaque as my hypothetical analysis of the neutral current episode suggests, then these legitimatory moves may be called into question.
>
> (2) A likely outcome of SPSK research, especially in light of the considerations that I will raise below about the transmission of knowledge, is that certain traditionally valued forms of knowledge—such as those that result from a rigorous enforcement of the method of conjectures and refutations—may be possible only under certain social arrangements which, for all practical purposes, today's transnational knowledge enterprises have outgrown. This conclusion would be all too familiar to political theorists, who must struggle with what "democracy" can mean in contemporary nation-states, whose size and heterogeneity have rendered obsolete the classical model of the democratic state, the Athenian polis (Dahl, 1989). An SPSK-informed philosophy of science may likewise be forced to contend with what "science," especially "autonomous science," can mean in a world where the amount of labor and capital bound up in the knowledge enterprise exceeds virtually all industrial production. (Fuller, 1993a, chap. 8)

The full significance of the second point would take us far afield from our presentation of the SPSK research agenda. Nevertheless, this much is worth stressing. To the eye and ear untutored in SPSK, the knowledge production system has not changed. People called scientists or scholars are still teaching courses, doing research, writing articles, and having conferences. They are even talking about these activities by making the traditional appeals to epistemic entities. At this naive level, all that has changed since the Scientific Revolution is that more people in more places are doing these things. However, to the tutored eyes and ears of the SPSK researcher, some radical changes have occurred. The activities in question are prompted, maintained, and transformed under rather different sets of conditions, which means that they may no longer serve the same functions they once did. Consider just one vivid case in point. Historians and philosophers of science of all denominations presume that the division of cognitive labor into special disciplines is a "natural" development in the growth of knowledge, comparable to the differentiation of organs in a maturing young animal. It is easy to cast this story in a progressive light by talking about the more finely grained theoretical languages and the more precise and accurate data-gathering techniques that specialization brings. Yet, it has been recently argued that the pervasive, accelerated, and seemingly irreversible quality of specialization has less to do with the internal imperatives of the knowledge system and more with transnational publishing houses discovering that they can turn a bigger profit by subsidizing journals, whose entire raison d'être is specialization, than by publishing books, which would typically appeal to a less lucrative but intellectually broader-based market (Horowitz, 1986). Such a thesis would tell us to look not at what the scientists have reason to say, but rather at what they have incentive to do.

The Social Psychology of Scientific Knowledge Transmission

A Theoretical Model

A striking feature of the way the sciences are organized is that the perceived "hardness" of a science varies directly with the acceptance rate of articles for journal publication: The harder the science, the easier it would seem to get into print. The surface paradox of this claim quickly disappears, however, once it is recalled that in graduate school, natural scientists learn to formulate issues according to the writing canons of their discipline. Moreover, natural scientists learn to associate quality work more with publication in specific journals than with the impact that their work actually has on the readers of those journals. While this practice consigns the natural scientist to a relatively small audience, it is an audience on which he/she can depend for an appropriate level of attention and response. In marked contrast, what a natural scientist learns in graduate school, a social scientist will only pick up haphazardly from the gatekeeping practices of the journals to which he/she submits articles. I say "haphazardly" because social scientists constantly get mixed signals in their efforts to communicate with a potentially wide audience. As Cicchetti (1991) has observed, while rejection rates in the social sciences (and, interestingly, in interdisciplinary forms of natural science) are very high, it is also true that eventually most articles find a place in some journal or other. Is this form of institutionalization good or bad?

Much depends here on whether one sees science primarily as a vehicle for each scientist's self-expression or as a collective enterprise issuing in knowledge products, the significance of which transcends any particular scientist's concerns. The expressive latitude permitted by the proliferation of social science journals clearly speaks to the former conception of science. But like most people who write about peer review, Cicchetti seems to believe that the fairer the system is to the individual scientist, the higher the quality of science that is likely to result. The key bridging concept is innovation, which he sees the natural sciences as systematically stifling. Indeed, the nearest Cicchetti comes to drawing prescriptive conclusions from his research is to say that good scholarship should not be lost to the world because of standards that are more stringent than they are reliable. Thus, he seems to prefer the social scientific mode of publication. But is this preference well grounded? Some considerations from classical political theory and what more recent theorists would call the social ecology of science offer a different slant that may inform SPSK research.

The classical political theorists of the Enlightenment tended to agree that two conditions preclude the existence of society: Complete abundance renders society unnecessary, while extreme scarcity renders it impossible. The resource of concern to us here is access to publication. Given Cicchetti's analysis, the social sciences fail to optimize knowledge production because publication access is too abundant. Thus, each social scientist can diverge into his/her own separate niche (i.e., a unique discourse and audience) without fearing marginalization—mainly because there is no center, no forum in which all knowledge claims must be exchanged. Following the model developed by the social ecology school of American sociology (e.g., Hawley, 1950), we can project the consequences of resources becoming progressively scarcer, as measured in terms of a decline in publication possibilities. In particular, let us distinguish between *bounded abundance* and *moderate scarcity* as two intermediate states between abundance and scarcity in their pure forms.

In terms of a community of inquirers, pure abundance would be represented by an anything-goes approach to knowledge, whereas pure scarcity would take the form of radical skepticism. In the state of bounded abundance, however, possibilities have been cut sufficiently that each inquirer must take the behavior of other inquirers into account when judging how he/she should proceed. Simply repeating or reinventing what other inquirers have done will fail to merit publication. The best strategy under these circumstances is to carve out a niche within what others have done, using one's own work to complement theirs. It is a strategy that characterizes the natural sciences, one that instills a distinctly "market" mentality among scientific inquirers, especially if one imagines the market as a classical free trader like Ricardo would have, namely, as spontaneously generating an efficient division of labor on the basis of the "comparative cost" of alternative pursuits (White, 1981).

Yet, if Popper is right that the method of "conjectures and refutations" is the key to knowledge growth, a state of moderate scarcity is to be preferred to that of bounded abundance. Here publication possibilities are tightened some more, to the point that some inquirers must be displaced in order for others to thrive. In that case, inquirers must adopt a more mutually critical attitude, which, while leading to the elimination of most inquirers in the long run, will also likely improve the quality of knowledge claims proposed by those who remain. Thus, every individual scientist's loss is potentially an overall gain for science.

An Historical Controversy

Conceptualizing scientific inquiry as a social ecology enables us to see the blind spots in a recent American debate, between political scientist Karl Deutsch and sociologist Daniel Bell, over the progressiveness of the social sciences. The debate was occasioned by a list of "advances" that Deutsch and his associates compiled which purported to show that the social sciences were just as much given to cumulative growth as are the natural sciences (Deutsch, Markovits, & Platt, 1986). Deutsch then submitted this list to the most sophisticated statistical methods that the social sciences had to offer (of course) in order to discern the social conditions that fostered social scientific creativity. Deutsch (1986) saw many similarities between developments in the natural and social sciences, especially with regard to the delay between the introduction of an innovation and its widespread acceptance in one or more disciplines. Bell (1986) countered that Deutsch would start seeing some sharp dissimilarities if he turned to the long-term survival rates of natural vis-à-vis social scientific innovations. To end the story, as Deutsch does, with the widespread acceptance of an innovation is to take for granted that the innovation has become a permanent part of the corpus of knowledge—at least as a basis for making subsequent innovations. While this probably holds for the natural sciences, Bell argued, it is hardly the case in the social sciences. From a long-term perspective, social scientific innovations appear to be little more than fads and fashions.

Part of the problem with assessing the merits of Deutsch's and Bell's claims is that they are concerned almost exclusively with what philosophers of science have called the discovery and justification contexts of the scientific enterprise, which capture only the rhetorical surface of knowledge production. This is not to deny that the rhetoric has its uses. For example, any scientific innovation can be described as either a discovery or an invention. To "invent" is simply to discover an application of real principles, whereas to "discover" is none other than to invent a new way of accessing real principles. Nevertheless, the rhetorical contexts in which these two descriptions work are quite different. Discovery–talk functions well when making a case before a university committee or a scientific forum, whereas invention–talk is essential to making claims to corporate funding and legal rights. Similarly, the rhetoric of justification, with its appeal to the logical unfolding of inquiry, is persuasive in contexts where one needs to muster up the past in a selectively memorable fashion so as to highlight a preferred research trajectory. Classrooms and grant agencies come to mind as sites for such rhetoric. However, all this talk obscures the central questions of SPSK, answers to which are needed to resolve the Deutsch–Bell debate. In particular, we need to learn what governs the ways in which scientists direct attention to each other's work.

What is it about the work of their colleagues that scientists find sufficiently compelling that they end up spending their time either building on it (i.e., bounded abundance) or criticizing it (i.e., moderate scarcity)? As I (Fuller, 1991a, 1993b) have argued, recent historiographical models of the scientific inquirer drawn from cognitive psychology have failed miserably in coming to grips with this question, as historians simply take for granted that if a scientific community is working "rationally," its members reproduce in their own minds the innovation of a Faraday, Einstein, or some other "genius." Social psychological research on the appropriation of ideas shows that the "contagion" model of knowledge transmission presupposed by these historians grossly oversimplifies the actual cognitive processes, even in children (Wicklund, 1989). To propose a hypothesis for consideration—even one that has enough evidential support to merit publication in a mainstream forum—does not guarantee that it will be given any attention whatsoever. Deutsch's mistake in drawing parallels between the delays that, say, Newtonianism or Darwinism had to endure before being fully accepted by the community of natural scientists and the delays faced by today's social scientific innovators is that whereas there was continuous debate for the 50-odd years from the time Newton and Darwin first proposed their views to their general acceptance, often social scientific innovations died stillborn, only to be reinvented years later in a more hospitable intellectual environment. It was only when the reinventors needed to distance themselves from other social scientific fashion-mongers that they started looking for "precursors." (In this respect, Mendel's postmortem fame as the "father of genetics" is the relevant analogue from the natural sciences;

see Brannigan, 1981.) Thus, Bell's skepticism is probably warranted, insofar as Deutsch has confused the rhetorical contexts of discovery and justification with the actual transmission and institutionalization of ideas. But how might these processes be experimentally studied?

An Experimental Research Program

Jacobs and Campbell (1961) developed an influential experimental paradigm for studying the transmission and institutionalization of knowledge. They examined the extent to which a false viewpoint (a so-called arbitrary tradition) can be transmitted intact as members of a group are gradually replaced. Do new members simply conform to the reigning viewpoint, or does an independent assessment of the situation cause them to change their mind? Jacobs and Campbell found that the false viewpoint was perpetuated in an eroded form, which suggested to them that reality slowly makes some headway into collective illusion. They argued that the false viewpoint was not more robustly transmitted because there was no social function served by holding it, unlike the superstitions that persist in real societies. In a series of follow-up experiments, MacNeil and Sherif (1976) largely confirmed Jacobs and Campbell's hypotheses, from which they argued the general claim that the stability of a tradition across subject generations is inversely related to the arbitrariness of the tradition. This argument seems to imply that truth is naturally transmitted, while special effort must be made to maintain error. However, one of the lessons of SSK is that such a conclusion is methodologically suspect. It caters to the philosophical tendency to provide "asymmetrical" explanations for "good" and "bad" science, as if our present-day evaluations would naturally correspond with distinct causal processes (Bloor, 1976).

However, the results of these original experiments were reanalyzed and contested from an ethnomethodological standpoint by Zucker (1977), who showed that regardless of its level of arbitrariness, a tradition would be transmitted intact if subjects perceived themselves and each other as acting in an official capacity rather than as issuing personal judgments. Thus, Zucker diagnosed the deterioration of the arbitrary tradition in the earlier experiments as resulting from the subjects simply being asked to make a personal judgment pertaining to the tradition, rather than to adopt a role that might lead them to discount or, in some way to revise, their personal judgment. For the ethnomethodologist, the mark of a highly institutionalized tradition is the ease with which it is transmitted, usually just by telling someone that it is the case; hence, the "inertial" character of institutions. Indeed, when subjects occupying an official capacity supplement their remarks with personal judgments, the tradition they are attempting to transmit loses some of its efficacy. Zucker related these findings to ones in cognitive dissonance that suggest that subjects tend to lower their degree of belief if they perceive that their agreement had to be secured on grounds independent of the belief's merits (e.g., by special pleading, coercing, or financial incentive). In such situations, subjects seem to presume that the explicit personal appeal is needed to compensate for the belief's own uncertain status. Zucker's "new look" at the original transmission experiments raises three points of interest here.

1. The social function performed by an apparently arbitrary tradition generally does not enter into the reasons that social actors give for maintaining the tradition. Why? Presumably, to raise such an issue would be to invite discussion of alternative practices that may serve the same function more effectively, and hence "less arbitrarily." Indeed, this is how such functions stay "latent," in Robert Merton's sense. Thus, a scientist typically justifies the way he/she writes a journal article by referring to its content rather than to the literary conventions that his/her discipline upholds.
2. The original experimenters implicitly begged the question of knowledge having a significant social component when they presumed that a subject's personal account of some phenomenon would be truer than the official account that he/she was taught. Kuhn-inspired talk about "theory-laden observation" has led both philosophers and sociologists to underestimate the extent to which scientists often use theories, not to inform experience but to counteract or compensate for it. Indeed, in the case of theories that revolutionized scientific inquiry, such as Copernicanism, Newtonianism and Darwinism, scientists were converted by learning to distrust their intuitions about "natural motions" and "natural kinds." If these theories constituted scientific progress, there would seem to be nothing inherently wrong with subjects discounting their personal experience when making judgments.
3. It is nevertheless also clear that scientists should not *always* prefer official accounts to personal ones. The question of how much and how often one should "go official" or "go personal" is the reverse side of our original question of how scientists decide to pay attention to each other. For, to attend to another's officially presented observations is to implicitly discount one's own personal ones. And as we saw above, if the natural sciences err toward discounting the personal too often, the social sciences surely err toward discounting the official. In conclusion, I will

outline a procedure for studying this phenomenon experimentally.

One way of conceptualizing the difference between "humanistic" (i.e., social scientific) and "scientific" (i.e., natural scientific) approaches to knowledge is in terms of how scholars manage their time and energy. I have argued that the rate of scientific progress accelerated when scholars no longer needed to understand fully the intentions of a precursor's research before using it for their own purposes. This freed up time for the scholar's own research that would have otherwise been spent in recapitulating his/her precursors. The legacy of this approach may be seen today in the lack of commentary about the texts cited in the literature reviews of science articles. To show this point in the laboratory, subjects may be given one of two sets of instructions on how they should write up their results. The humanistic group is told that their work will be judged primarily on how well they have understood the sources on which their solution explicitly relies, whereas the scientific group is told that theirs will be judged primarily on how well their solution addresses the problem at hand. The interesting thing to look for here is the extent to which humanistic reports tend to solve their problems less incisively than do scientific ones, and scientific reports tend to be more opportunistic and perfunctory in their use of sources than are humanistic ones.

Experimental groups may also be divided into those who are told that the results of their research may be published in a professional journal and those who are led to believe that only the experimenter will be interested in reading their results. The key dependent variable here is the extent to which this difference in expectations affects the time and energy the group allots for writing up its results: How is the labor divided between making progress and representing that progress? If several groups are studied in competition, hypotheses can be tested concerning the differences between expository and persuasive texts. On the basis of a wide range of experiments, McGuire (1985) has argued that a text that conforms to the canons of clear expository writing—that is, states the thesis up front and then defends it with arguments ranging from the most to the least important—will seem persuasive only to those who are already inclined to accept the thesis. To persuade the unpersuaded, however, one would do better to lead up to the thesis after raising considerations in order of likely importance to the reader. In competitive group situations, where there is potentially an advantage to have others join one's group, to what extent can this artful inversion be detected in the communication between members of different groups—and how does this affect the group's conceptualization of its task? For example, do the members draw a sharp distinction between the expository and persuasive ways of presenting their activities?

Libraries: The Neglected Ecology of Science

A useful way to summarize the research agenda that has been sketched for SPSK is in terms of the sequence of ecologies that constitute the scientific enterprise. Each ecology consists of a selection mechanism that operates in a particular human environment. More plainly, it is a field of competition for the attention of an audience that is crucial for expediting the spread of science at a particular juncture. Four ecologies have been implicated in this chapter, three of which have already been discussed explicitly:

1. *The ecology of the self or the workplace*. This was previously described as the self-editing of ideas and expressions, a process at which natural scientists become especially adept in the course of their professional training. It accounts for the high publication acceptance rates in natural science journals. This process is continued in the workplace through the mutual shaping of research team members, which includes brainstorming and transactive remembering, whereby team members remind each other of information or arguments that need to be taken into account before an acceptable scientific paper is produced (Clark & Stephenson, 1989). Depending on whether one adopts a cognitivist or a behaviorist orientation, this mutual shaping may be seen as externalizing the self-editing process or self-editing may be seen as internally anticipating the response of colleagues.
2. *The ecology of the editorial office*. This consists of the often controversial "gatekeeping" practices of the peer-review system, which covers both journal publication and such mechanisms of resource allocation as grant agencies (Chubin & Hackett, 1990). As mentioned earlier, such practices perform a much more pronounced selection function in the social sciences than in the natural sciences.
3. *The ecology of the university library*. This crucial ecology has been largely ignored, although it will be discussed below. Here the expansive tendencies of academic professionalism most intimately interface with the commercial interests of transnational publishing: Who will buy the specialty journals, and how available will they be to potential readers? While the neglect of librarians may be partly explained by their lack of involvement in producing

the scholarship of science, most of the neglect is probably traceable to a general failure to take seriously the *scale* of the scientific enterprise as a significant social psychological variable. Consequently, it is still common to find students of science writing as if a scientific author communicates with a readership, if not directly, then only with the assistance (or impedence) of an editor. Few take seriously the fact that books and journals first need to be found on library shelves.

4. *The ecology of the reading room.* The literature on the reception and spread of scientific ideas has markedly grown and improved with the emergence of the historical sociology of science (Shapin, 1982). Nevertheless, as suggested earlier, the presumptive psychology of readers leaves much to be desired. There remain two unresolved issues of major import: (a) How much of an author does one need to understand in order to put that work to use in one's own research? (b) Is the goal of scholarly reading really something like persuasion and information, or is it something more self-serving and strategic? Answers to these questions will no doubt force a new conception of the sense in which scientists can be said to contribute to a cumulative body of knowledge.

To highlight the importance of the university library to the continued production of knowledge, consider the fate of interdisciplinary research. To talk the language of markets, how does the scientific "knowledge consumer" decide which of the variety of interdisciplinary knowledge products is best suited to his/her cognitive needs? The interdisciplinary markets are flooded with more articles in more obscure journals than the specialty markets are. Since the cognitive limitations of the consumer remain fixed as the number of markets and products grow, *physical access* is becoming an increasingly important determinant of which research turns out to be influential. Is it published in a journal that I routinely peruse? Is the journal copy readily available in the library? Does the article appear indexed in many databanks? The answers to these questions depend on issues quite incidental to the intellectual merits of a given article: Can I afford the journal, and does it publish other articles that I normally find interesting? Is the current periodicals section properly policed and updated? Does the title of the article contain words that make the right associations with other words in the databank? All the best laid plans to reform peer review will have been for naught if the high-quality journal that publishes the high-quality scholarship turns out to be low on physical access.

My point, then, is that interdisciplinary research may give an illusory sense of preserving good scholarship simply because of its more liberal publication policies. This illusion is fostered by focusing on the editorial office as the only clearinghouse for knowledge products. It is not that interdisciplinarians do bad work, but rather that their work is so diffusely placed that access to such work, and hence its ultimate impact, is limited. Given the inaccessibility of some journals, the work might as well have never made it into print. This suggests some policy implications.

1. Editors should forge closer links to the library and information systems that will determine the access that potential consumers have to journals and books.
2. The goals of peer review should be oriented more to the interests of a given journal's readership. At the moment, when there is a conflict of goals, peer review aims more at publishing papers that exemplify the methodological standards of the journal's discipline than papers that are likely to be taken up by the readership in their own research.
3. Regardless of whether one thinks that more scientists make for better science, the growing number of submissions may, at some point in the future, have to be checked by requiring that authors restrict the number of papers they publish over a given period. (In fact, Campbell [1987] has suggested that the use authors put to such self-restraint could be weighed in tenure and promotion decisions.)
4. Tighter control of the knowledge markets, including increased self-selection of paper submissions, could encourage specialization and rigidify disciplinary boundaries. Much depends here on whether journal editorial policies are more dictated by the character of the field of study or the field of study comes to have its character in virtue of the journal editorial policies.

References

Bell, D. (1973). *The coming of post industrial society*. New York: Basic Books.

Bell, D. (1986). The limits of the social sciences. In K. Deutsch, A. Markovits, & J. Platt (Eds.), *Advances in the social sciences, 1900–1980* (pp. 312–324). Lanham, MD: University Press of America.

Billig, M. (1987). *Arguing and thinking*. Cambridge, England: Cambridge University Press.

Bloor, D. (1976). *Knowledge and social imagery*. London: Routledge.

Boehme, G., & Stehr, N. (Eds.). (1986). *The knowledge society*. Dordrecht, The Netherlands: Reidel.

Brannigan, A. (1981). *The social basis of scientific discoveries*. Cambridge, England: Cambridge University Press.

Campbell, D. (1987). Guidelines for monitoring the scientific competence of preventive intervention research centers. *Knowledge, 8*, 389–430.

Chubin, D., & Hackett, E. (1990) *Peerless science: Peer review and American science policy*. Albany: State University of New York Press.

Cicchetti, D. (1991). The reliability of peer review for manuscript and grant submissions. *Behavioral and Brain Sciences, 14*, 119–150.

Clark, N., & Stephenson, G. (1989). Group remembering. In P. Paulus (Ed.), *Psychology of group influence* (2nd ed.) (pp. 357–392). Hillsdale, NJ: Erlbaum.

Coulter, J. (1979). *Social Construction of Mind*. London: Macmillan.

Dahl, R. (1989). *Democracy and its critics*. New Haven, CT: Yale University Press.

Danziger, K. (1990). *Constructing the subject*. Cambridge, England: Cambridge University Press.

Deutsch, K. (1986). Substantial advances: real but elusive. In K. Deutsch, A. Markovits, & J. Platt (Eds.), *Advances in the social sciences, 1900–1980* (pp. 361–372). Lanham, MD: University Press of America.

Deutsch, K., Markovits, A., & Platt, J. (Eds.). (1986). *Advances in the social sciences, 1900–1980*. Lanham, MD: University Press of America.

Ezrahi, Y. (1990). *The descent of Icarus*. Cambridge, MA: Harvard University Press.

Fuller, S. (1988). *Social epistemology*. Bloomington: Indiana University Press.

Fuller, S. (1991a). Is history and philosophy of science withering on the vine? *Philosophy of the Social Sciences, 21*, 149–174.

Fuller, S. (1991b). Studying the proprietary grounds of knowledge. *Journal of Social Behavior and Personality* *6*(6), 105–128.

Fuller, S. (1992a). Social epistemology and the research agenda of science studies. In Pickering, *Science as practice and culture* (pp. 390–428). Chicago: University of Chicago Press.

Fuller, S. (1992b). Epistemology radically naturalized. In R. Giere (Ed.), *Minnesota studies in the philosophy of science. Vol. 15. Cognitive models of science*. Minneapolis: University of Minnesota Press.

Fuller, S. (1993a). *Philosophy, rhetoric, and the end of knowledge: The coming of science and technology studies*. Madison: University of Wisconsin Press.

Fuller, S. (1993b). *Philosophy of science and its discontents* (2nd ed.). New York: Guilford Press.

Gardner, H. (1987). *The mind's new science*. New York: Basic Books.

Gergen, K. (1985). The social constructionist movement in modern psychology. *American Psychologist, 40*, 266–275

Gergen, K., & Gergen, M. (1982). Explaining human conduct. In P. Secord (Ed.), *Explaining human behavior* (pp. 127–154). London: Sage.

Gilbert, N., & Mulkay, M. (1984). *Opening pandora's box: A sociological analysis of scientists' discourse*. New York: Cambridge University Press.

Hawley, A. (1950). *Human ecology*. New York: Ronald Press.

Horowitz, I. (1986). *Communicating ideas*. Oxford, England: Oxford University Press.

Jacobs, R., & Campbell, D. (1961). The perpetuation of an arbitrary tradition through several generations of a laboratory microculture. *Journal of Abnormal and Social Psychology, 62*, 649–658.

Knorr-Cetina, K., & Cicourel, A. (Eds.). (1981) *Advances in Social Theory*. London: Routledge.

MacNeil, M., & Sherif, M. (1976). Norm change over subject generations as a function of arbitrariness of prescribed norms. *Journal of Personality and Social Psychology, 34*, 762–773.

McGuire, W. (1985). Attitudes and attitude change. In G. Lindzey & E. Aronson (Eds.), *Handbook of social psychology* (3rd ed.) (pp. 233–346). Hillsdale, NJ: Erlbaum.

Mulkay, M. (1990). *The sociology of science*. Bloomington: Indiana University Press.

Nisbett, R., & Wilson, T. (1977). Telling more than we can know. *Psychological Review, 84*, 231–259.

Pickering, A. (1984) *Constructing quarks: A sociological history of particle physics*. Chicago: University of Chicago Press.

Popper, K. (1963). *Conjectures and refutations*. New York: Harper & Row.

Ross, L. (1977). The intuitive psychologist and his shortcomings. In L. Berkowitz (Ed.), *Advances in experimental social psychology* (Vol. 10). New York: Academic Press.

Roth, P., & Barrett, R. (1990). Deconstructing quarks. *Social Studies of Science, 20*, 579–632.

Shapin, S. (1982). History of science and its sociological reconstructions. *Isis, 20*, 157–219.

Stehr, N., & Ericson, R. (Eds.). (1992). *The culture and power of knowledge*. Berlin, Germany: Walter de Gruyter.

Sternberg, R. (1985). *Beyond IQ: A triarchic theory of human intelligence*. London: Cambridge University Press.

White, H. (1981). Where do markets come from? *American Journal of Sociology, 87*, 517–547.

Wicklund, R. (1989). The appropriation of ideas. In K. Paulus (Ed.), *Psychology of group influence* (2nd ed.) (pp. 393–424). Hillsdale, NJ: Erlbaum.

Woolgar, S. (1988). *Science: The very idea*. London: Tavistock.

Zucker, L. (1977). The role of institutionalization in cultural persistence. *American Sociological Review, 42*, 726–743.

Part II

Social Psychological Contributions to Science Studies

CHAPTER 4

Toward an Experimental Social Psychology of Science: Preliminary Results and Reflexive Observations

MICHAEL E. GORMAN

In this chapter, I describe a series of experiments I designed to explore group processes in scientific reasoning. These studies are used as a springboard for discussing the potential role of experiments in a social psychology of science.

A Majority–Minority Confrontation

In graduate school, I became interested in how the literature on social influence might relate to scientific debate and consensus. From Moscovici (1974), I derived four predictions concerning the outcome of a confrontation between a scientific majority supporting one theory and a minority supporting a rival theory: (1) conformity to the majority when its theory was confirmed and the minority's disconfirmed; (2) innovation, or minority influence, when this pattern was reversed; (3) polarization when both subgroups' theories were confirmed; (4) normalization or convergence when both were disconfirmed. Moscovici emphasized the importance of a "consistent behavioral style" in influencing others; I made the assumption that recent confirmation or disconfirmation would affect a group's confidence and therefore its "behavioral style."

How might one test these hypotheses? One could look at actual scientific controversies, trying to see whether majorities and minorities acted in the way I predicted. But suppose one found a counterexample. One could always point to specific aspects of the case and say, "Well, if it hadn't been for thus-and-such, my prediction would have been confirmed." There are so many variables interacting in a real scientific controversy that it is hard to assess the effect of confirmation or disconfirmation. Indeed, that is part of the reason why philosophers, sociologists, and historians can draw widely different conclusions from the same case.

My answer to this problem was to design an experiment to test my predictions. I would like to say that I had thought through all the philosophical ramifications of conducting experimental simulations of science. In actuality, I was taking a method I had been trained to use and applying it to a problem I was interested in. Before considering the validity of experimental simulations of science, I will briefly review my first one.

I established two subgroups, a four-person majority and a two-person minority, each of which worked separately at first, using two theories to make predictions about the level of drug use in a series of (fictitious) societies. The majority found that one theory worked well on almost every trial; the minority found that the other theory worked better. Therefore, when the two subgroups were brought together to discuss two more societies, they had to resolve a conflict (see Gorman, Lind, & Williams, 1977).

To create Moscovici's four conditions, I manipulated whether each subgroup obtained confirmation or disconfirmation of its approach on the last trial before meeting the other sub-group. If both subgroups were successful, I expected them to polarize; if both failed, I expected normalization; if the majority succeeded and the minority failed, I expected the latter to conform to the former; if the majority failed and the minority succeeded, I expected the former to conform to the latter.

None of these predictions was confirmed. Subgroups stuck publicly to their positions, although they tended to show more sympathy for the others' point of view on anonymous questionnaires. Had I falsified the Gorman–Moscovici hypothesis?

Generalizing Results

Another way of putting this question is to ask, How far can I generalize my negative result? If the hypothesis failed in a single experimental test, could I infer that it would fail in other tests and in actual scientific controversies?

This question forces us to consider three types of validity (see Shadish, Fuller, & Gorman, Chapter 1, this volume, for a more detailed discussion).

1. *Internal validity* has to do with whether the study was properly conducted. A flawed experiment cannot be generalized. In my case, I could see no obvious methodological flaws: Subjects were randomly assigned to conditions, appropriate statistics were used, and so on.
2. *External validity* has to do with whether the conclusions of an experiment can be generalized to other places, times, or settings (Cook & Campbell, 1979). Given the fact that I used college students as subjects and had them work under highly artificial circumstances that bore only a distant resemblance to scientific practice, my ability to generalize beyond the laboratory would be very limited. But as Berkowitz and Donnerstein (1982) argue, "artificiality is the strength and not the weakness of experiments" (p. 256). To discover why they make this provocative claim, we need to consider a third type of validity.
3. *Ecological validity* has to do with whether the pool of subjects and the experimental situation are representative of some real-world situation. To have ecological validity, experimental studies of scientific reasoning would have to use scientists working on tasks that bear a close resemblance to actual scientific problems and set up an environment that resembles an actual laboratory, or laboratories, in competition (for a discussion, see Houts & Gholson, 1989).

But if one's experimental designs closely approximate a real scientific situation, one runs into another kind of limit on generalizability. Consider an example. Suppose I had used a task based closely on a particular scientific controversy in my majority-minority confrontation. Then how could I be sure my results would generalize beyond that controversy?

Thus, artificiality can impose limitations on the generalizability of experiments, but so can ecological realism. However, the limitations imposed by each are different, and whether one wants to err in one direction or the other depends on the purpose of one's experiment.

Consider my majority–minority study again. I wanted to test a hypothesis. This meant that I wanted to see whether the theory would work *under ideal conditions*: what the philosophers call *ceteris paribus* (all other things being equal). I knew that even a strong positive result would say nothing about whether actual scientific controversies would be resolved in the same way; to answer that question, I would have to conduct ecologically valid follow-up studies. No, the strength of the experimental method was that I could set up an isolated, artificial situation in which the theory would have the maximum chance of success. If it had succeeded, in subsequent studies I could have added variables to increase ecological validity. Would the theory work only under certain very limited circumstances, or would it turn out to be robust, operating well under a variety of conditions?

A program of research that begins by obtaining an important effect under highly artificial circumstances can gradually incorporate more ecologically valid features in later studies. In other words, a program of research can attain a level of generalizability impossible for an individual experiment or case study to obtain. Consider one such program.

Confirmation, Disconfirmation and the "Search for Truth"

Whether each subgroup obtained positive or negative results just prior to meeting made little difference in the outcome of the confrontation. This surprising result made me want to find out more about the effect of confirmation and disconfirmation on scientific reasoning.

At that point, a fellow graduate student handed me a paper by several researchers at Bowling Green State University (see Mynatt, Doherty, & Tweney, 1977, 1978). These researchers studied whether subjects could follow Popper's advice and falsify on tasks that model scientific reasoning. The Bowling Green researchers constructed an "artificial universe" in which subjects fired particles at objects to determine the laws of motion. Subjects exhibited a kind of confirmation bias which even instructions to disconfirm could not combat (for a thorough review of the confirmation bias literature, see Greenwald, Pratkanis, Leippe, & Baumgardner, 1986).

Popper himself might not have been interested in these results, preferring an analytical method for making normative generalizations about science (Barker, 1989). But I felt that philosophical norms ought to have testable consequences. What good is a norm if it does not work even under ideal laboratory conditions?

Mynatt et al. (1977, 1978) worked exclusively with individuals. I wondered whether groups could follow a Popperian strategy (Gorman, Gorman, Latta, & Cunningham, 1984). Gardner (1977) described a card game, New Eleusis, that models the scientific or mathematical search for truth. Each player attempts to guess a rule by playing cards to test his/her hypotheses. I realized that it would be an ideal simulation for falsification: One could look at whether students proposed cards designed to confirm or disconfirm their hypotheses.

Mynatt et al. (1977) suggested one independent variable: comparing confirmatory and disconfirmatory instructions. I was also interested in a second variable from the literature on group problem solving: whether interacting groups perform better than equal numbers of individuals working separately. Perhaps groups could make better use of instructions to disconfirm than individuals could.

Design

I compared groups of four people working together (interacting condition) to groups of four not permitted to communicate (coacting condition). In interacting groups, subjects took turns playing cards but talked openly about what they were doing and why. The experimenter gave them feedback on whether each card fit the rule. Interacting groups were told to reach consensus on the final rule on each of four Eleusis problems. Coacting groups took turns playing cards and saw the results obtained by others but could not talk. Each group member wrote his/her guess down separately.

The group conditions were crossed with two instructions: confirmatory or disconfirmatory. Confirmatory groups were told to propose "as many correct cards as possible." Disconfirmatory groups were told to "deliberately play cards you think will be wrong" and to study cards that did not fit the rule. Subjects were given examples of strategies, using "red and black cards must alternate" as a sample. Confirmatory groups were told that to verify this red-and-black pattern, they should play cards that alternated in color; disconfirmatory groups were told to play all possible red cards after a red card and all possible black cards after a black card.

Pilot studies showed that four rules were needed to give groups time to work together and master their strategies. An example of each rule is shown below (C = clubs, D = diamonds, H = hearts, S = spades, A = ace, J = jack, Q = queen, K = king). The horizontal line of cards follows a rule; the descending vertical lines are cards that were incorrect when played after the card above them.

Rule 1:	AD	2D	3D	4D	5S	6D	7S	6H	7H	8S	7D	
					JH	JS		8C		QC		
								2H		3S		
Rule 2:	AD	2D	3D	4D	6H	8D	10H	10H	QC	JS	9C	JH
					QC	KC		AH	8S	KD		
								2D		6S		

Rule 3:	AD	2D	3D	4D	KS	QH	7C	10S	3S	8D	AS	QH	
						10H	3C	2C	7C	QD			
							5D	6H	JD				
Rule 4:	AD	2D	3D	4D	4S	10H	2C	7C	QC	10H	3H	KC	AD
				4D			8S		8C		KD		
							QC		2C		7H		

The rules are:

1. Adjacent cards must be separated by a difference of 1.
2. Adjacent cards must be separated by a difference of 0, 1, or 2.
3. Odd and even cards must alternate.
4. Cards must alternate in terms of parity (odd vs. even), color (red vs. black), or both.

A simple sequence like ascending diamonds produces correct cards on every rule but to find the experimenter's rule, subjects must play some cards that should be wrong if their hypothesis is correct. The rules gradually increased in complexity. The first three rules emphasized differences between numbers. The fourth rule introduced color, but subjects were sensitized to this possibility by the examples that accompanied their strategy instructions.

All four rules correspond to a Eureka problem (Steiner, 1972), which demands a sudden insight. On a Eureka problem, group performance typically equals that of the best of an equal number of individuals. On a divisible problem, group members can be delegated subtasks and so perform better than individuals working separately. I wondered whether disconfirmatory instructions would transform Eleusis into a divisible problem: One or two group members could concentrate on trying to refute a hypothesis while others looked for evidence that supported it.

Subjects were introductory psychology students at the University of New Hampshire who were required to participate in experiments of their choice. Forty groups were run, 10 in each cell of the 2 × 2 design.

Replicating Tweney and Steiner

Basically, results replicated earlier studies: Interacting groups performed as well as coacting groups, and disconfirmatory instructions failed to improve performance. Disconfirmatory groups did follow their instructions, obtaining significantly more incorrect cards than did confirmatory groups: an average of 20.3 incorrect cards in disconfirmatory and 18.3 in confirmatory across all four tasks. But this difference is small, and shows that confirmatory subjects were obtaining a fair amount of disconfirmatory information. The four best groups, in terms of solving the most rules in the least time, were in the interacting disconfirmatory condition, suggesting that groups could potentially implement instructions to falsify. Once again I had a negative result, one that could be linked to other negative results and suggested the externally valid generalization that falsification is not effective on tasks that model scientific reasoning.

Again, one cannot infer from this that falsification would not be effective in at least some ecologically valid settings. But it seemed to me that a strategy that failed to work under ideal circumstances was unlikely to work in the noise and confusion of the 'real world.'

One thing still puzzled me, however. Why did confirmatory groups obtain so many incorrect cards? The answer was that the Eleusis rules allowed subjects to be given only 13 cards each, and to receive 2 more cards only when they played an incorrect card. A confirmatory group using an ascending diamonds hypothesis would quickly run out of confirmatory cards and would have to play a card out of sequence. This limited the generalization of my results to situations in which serendipitous disconfirmations were likely. I wanted to explore situations in which falsification required a deliberate effort.

A Second Eleusis Experiment

In a follow-up, the design of the first experiment was altered in three ways:

1. Each subject was given a full deck of cards with which to play, and each card was replaced. This would permit me to generalize to situations in which a subject or scientist could propose any experiment. Now a confirmatory group could play ascending diamonds ad nauseam.
2. In addition to confirmatory and disconfirmatory instructions, one third of the groups, randomly selected, were urged to confirm until they had a hypothesis and then to disconfirm. Mynatt et al. (1978) suggested that this might be the optimal strategy.
3. All subjects were run in interacting groups.

Otherwise, the second study was the same as the first.

Disconfirmation as a Successful Heuristic

In this study, disconfirmatory instructions produced the best performance, then instructions that combined confirmation and disconfirmation, and then confirmatory instructions. Table 4.1 shows the overall pattern of results. As groups gained experience, differences between conditions grew. These differences were statistically significant, indicating that instructions to disconfirm can be very effective (for more details, see Gorman et al., 1984).

Differences in solutions were mirrored by differences in proportion of incorrect cards obtained. Disconfirmatory groups obtained an average proportion .41 incorrect cards, combined .33, and confirmatory only .2. Apparently confirmatory and disconfirmatory groups followed their strategies.

Consider, for example, the cards played by a confirmatory group on Rule 3:

AD	2H	3D	4S	5C	6H	7D	8C	9C
7D								

The group concluded that the rule was "Cards must go in numerical order without regard to suit." The group obtained only one incorrect card—the 7D after the AD. The group had followed this pattern on their first two rules and were checking to see if it was violated. After this cursory attempt to falsify, group members settled on the ascending pattern and confirmed it with eight cards. Many confirmatory groups played more cards but settled on patterns like this that were sufficient to produce correct cards but were not correct rules.

In contrast, consider the cards played by a disconfirmatory group on Rule 3:

AD	2D	3D	4S	3H	2D	5D	6D	5H	6S	3H	QH	JC	10D	3D
KS		9H	4D	3C			10D					7S		JH
9H				3C								7C		5C
AH												3D		7H
												9H		JD
														9S

This group also sought a different pattern after the AD, but when three cards did not yield any positive results, members tried "diamonds ascending by one" for two cards. So even disconfirmatory groups needed some positive results to begin with. After the 4S, they tested for the two previous rules. After the 2D, they tried "something different": the 5D is

TABLE 4.1. Number of groups of eight that solved each rule

Rule	Disconfirmatory	Combined	Confirmatory
1	3	3	2
2	6	7	3
3	8	4	1
4	6	2	2

correct after the 2D. When one subject said, "It's been even–odd, even–odd all along" another subject said, "So I'll try another odd card. That should be wrong."

Group processes facilitated falsification. In one disconfirmatory group, a member complained, "I have a hard time guessing wrong." Another subject tried to explain how to disconfirm: "If you think the series goes like this (pointing to a sequence of cards ascending by ones), try to prove it wrong by putting down a card that doesn't go with the series." The first subject took the second's advice, and soon both were falsifying; the group solved all but the first rule.

Group processes could also interfere with falsification. In one combined group, a leader expressed Rule 2 in terms of "skips": "I think you might be able to skip cards by a certain number—like go from a four to a six but not from a four to a seven." Another subject ("L") tried to restate the rule later: "Could it be the same card, or within two?" But he was ignored. "L" also proposed the correct solution on Rule 3 but was overruled by the leader, who kept focusing on skips and decided that an Ace (1) could follow a King (13). By the end of Rule 3, "L" admitted that he did not understand the group's final guess.

Consider another combined group in which one subject took the leadership role and focused on suit on the last two rules. On Rule 4, the group's final guess read, "In clubs, hearts, spades after first two cards in descending order as many cards after were allowed. In diamonds four descending cards then any card was allowed." This guess resembles Lakatos's "degenerating research program," in which the "hard core" was that the rule must involve suit and auxiliary assumptions were created to maintain that "hard core" (See Gorman, 1992, for a lengthier discussion).

Combined groups fell between confirmatory and disconfirmatory instructions in terms of incorrect cards, which they obtained on one third of their plays. Subjects in these groups attended to the confirmation part of the combined instructions. One combined group began Rule 2 by proposing a long string of ascending cards, from Ace to King. They knew they should propose more incorrect cards, so one subject played a King.

When it was right, they knew that their "ascending by ones" hypothesis was wrong. They solved Rules 2 and 3 but ran out of energy on Rule 4, in part because proposing long strings of confirmations takes time. Most combined groups obtained positive evidence more two thirds of the time. Even disconfirmatory groups sought and obtained positive evidence more than half of the time; perhaps disconfirmatory instructions are necessary for groups to adopt a combined strategy.

Discussion

Why did disconfirmatory instructions improve performance in Gorman et al.'s (1984) Eleusis study but not in Mynatt et al.'s (1977) artificial universe? I considered two possibilities: (1) perhaps this positive effect of disconfirmatory instructions was limited to Eleusis; and/or (2) perhaps groups falsify more effectively than do individuals.

I then designed a study to eliminate both alternatives 1 and 2. I ran a similar design on individuals using Wason's (1960) 2-4-6 task. The result was that disconfirmatory instructions significantly improved the performance of individuals (Gorman & Gorman, 1984).

On a visit to Bowling Green, Tweney and I discovered the key difference between the effect of disconfirmatory instructions in Mynatt et al.'s studies and mine: They always told subjects whether each hypothesis was right or wrong, whereas I withheld feedback on hypotheses. The Bowling Green group, therefore, could generalize their results to situations in which subjects or scientists were told whether their hypotheses were right or wrong. In these cases, one could obtain disconfirmatory information from the experimenter. In contrast, I simulated a situation in which subjects or scientists had to design further experiments to test hypotheses, and demonstrated that falsification was an effective heuristic in these situations.

Klayman and Ha (1987) showed that a disconfirmatory heuristic is most likely to work when the subject's hypothesis is more narrow than the rule he/she is seeking. This was certainly true in my Eleusis studies; subjects would begin with a hypothesis such as "ascending diamonds" on a rule such as "odd and even cards must alternate." Given these studies, one could normatively hypothesize that scientists should adopt a disconfirmatory strategy if they think their current hypothesis may be a narrow subset of a more general law. Whether this norm works in ecologically valid situations is a matter for further research.

A sociologist of science might criticize this research by arguing that I am presuming that there are scientific laws. I would counter by arguing that these experiments were designed to generalize to situations in which there was an underlying law or rule. One could design experiments to test the effect of a disconfirmatory strategy in situations in which there is no target rule. Would groups construct or create laws in the way that Sherif's (1936) subjects created norms? To what extent do people behave differently when there are and are not patterns to be discovered?

The Possibility of Error and the Search for Truth

Scientists are always aware of error, but laboratory tasks to model scientific reasoning typically do not incorporate error (for an exception, see Kern, 1982). In the studies noted above, subjects always got reliable feedback. One way of increasing ecological validity of tasks like Eleusis is to add variables such as the possibility of error that simulate important features of science.

In a new experiment, I added instructions that told subjects that 0%–20% of their trials might yield erroneous feedback. In fact, no actual error was introduced: I wanted to explore whether the mere possibility of error would interfere with falsification. (In addition, a control condition was substituted for the combined condition and Rule 2 was eliminated). Would groups still disconfirm effectively?

The answer was no: The possibility of error eliminated the positive effect for disconfirmatory instructions. Indeed, only four groups over all conditions solved the odd–even rule, and none solved the final rule that combined color and parity (Gorman, 1986). Clearly, the mere possibility of error had a highly disruptive effect on problem solving.

Assigning Errors

On the odd–even rule, only one disconfirmatory group had any cards still turned over at the end of the session. (Groups turned over cards they thought were errors.) Half the no-strategy groups and all but two of the confirmatory groups left cards turned over. This difference is significant ($X^2 = 9$, Of $= 1$ $p < .01$) and indicates that, on the "odd and even" rule, disconfirmatory groups did a better job of discovering that there were no actual errors.

Overall, of the cards that were turned over at the end of each rule, 81% would have falsified the group's final guess if they had not made errors. This result varied little across strategy conditions and rules; groups used errors to immunize their hypotheses against disconfirmation. Twelve groups that failed to solve Rule 3 and sixteen that failed on Rule 4 did so because they declared one or more cards that would have disconfirmed their hypotheses to be errors.

Groups That Distorted Error to Fit Their Hypotheses

Five groups reconstructed the error instructions in ways that allowed them to preserve their hypotheses. Four of these replaced cards they had turned over with other cards that fit their current hypothesis even though they realized that this violated their instructions. Consider the cards played by a confirmatory group up to the point where they began an active discussion of error:

1D	2D	3D	4D	5D	10S	7D	8D	9D
13C					6S			
1C					6D			

The 10S was the only card that broke their confirmatory sequence of ascending diamonds. One subject thought the 10S could be treated as though it were a 6D. Another disagreed "because if there is error, it is as if the flipped card weren't there." A third subject agreed with both of them because "the six could also be error." Clearly one group member understood the error manipulation and realized that both the 10S and the 6D should be errors—but he did not try to convince the others to turn the 6D over.

In a group, members do not necessarily share the same representations of either the task or the rule. The majority can become convinced of one representation of error, and the single member with a different view does not push his/her challenge. The most radical misinterpretation of error occurred this way, when one control group studied the following sequence of cards:

1D	2C	1H	2S	7D	8S	7H	8C
1C			2D				

One subject said, pointing to cards one through four and cards seven through nine, "See—these are decreasing and increasing by one." Another subject pointed to the 7D and said, "But what explains this big jump here?" The first responded by turning over the 7D. He then persuaded the group that any card could follow an error, making it possible for the group to retain his hypothesis.

In conclusion, a few groups altered their representation of error to fit their hypothesis. In the present study, the source of potential error was defined to prohibit such representations; even so, a few groups could not resist. In science, where the potential sources of error are numerous and not always known, it may be even easier to immunize a hypothesis by invoking error. Replication is not sufficient to eliminate the possibility of error if what constitutes error is subject to negotiation within or between scientific groups.

Groups That Assigned No Errors

The possible-error manipulation disrupted even the performance of those groups that realized there were no errors. Five of these groups were disconfirmatory. Four of the five used simple hypotheses involving a difference of 1 between adjacent cards and did not try much to disconfirm their hypotheses, partly because they were so concerned with making sure there was no error.

This reinforces Klayman and Ha's (1987) observation that confirmation is particularly useful when results may be probabilistic. If a group can produce a long series of positive instances, the possibility of error can be virtually eliminated because errors would break the pattern. Similarly, scientists often persist in efforts to produce many confirmatory experiments. Tweney (1985) documented how Faraday struggled to obtain confirmations of electromagnetic induction, dismissing early disconfirmations as errors.

Once confirmation has been used to pinpoint errors, however, one must try to disconfirm. Faraday was not content to confirm that electricity could be induced from electromagnets; he also carefully eliminated alternate explanations for this phenomenon.

Disconfirmation plus Replication

The two disconfirmatory groups that solved the second rule illustrate the importance of combining disconfirmation with replication when the possibility of error contaminates results. The first of these successful disconfirmatory groups began with a simple ascending diamonds sequence, Ace through four. Then one subject said, "Try a diamond, but not a five—try a nine." It was correct. They tested for error by playing cards they thought would be right if the 9D were an error—a 3D, a 5C, a 5D. They realized that if 9D were an error, the next card in the sequence should follow 4D.

As of card 10, their sequence looked like this:

1D	2D	3D	4D	9D	10H	3H
				3D		
				5C		
				5D		
				11S		

One subject said, "It could just be any red card." They tried to disconfirm playing a 2C (correct) and a 4S (incorrect). They decided to guess that the rule was "any red card" and assigned errors to cards that disconfirmed that rule.

Many groups would have then stopped, but this group continued. They played a 9D next, to replicate the situation in which they first obtained incorrect red cards. It is difficult to replicate on Eleusis: It is not enough to play the same card twice, one must play a card like 9D in context with the same surrounding cards. So this replication was not exact—now 9D came after 4S, whereas earlier it came after 4D.

The 9D was correct. The group had recreated the situation in which red cards were wrong before. They combined replication with disconfirmation and got eight cards wrong in a row, seven of which were red. The average amount of error was 20%, so seven errors in a row was highly improbable.

The disconfirmatory instructions emphasized looking at incorrect cards, so one subject said, "Why don't we just look at these incorrect cards for a moment?" Shortly after another subject said, "I see something . . . odd–even." This rule fit all cards, and the group made their final guess after checking a few more cards.

This group carefully followed its suggested heuristic, seeking and studying incorrect cards. They replicated situations in which they uncovered disconfirmatory information. The best heuristic under possible-error conditions seems to be a combination of disconfirmation and replication.

A Normative Generalization

A tentative normative prescription from this research would be: When error is possible, replication should be combined with disconfirmation. This is a commonsense generalization, but it often is not followed. Replication accounts for only 5% of articles in social sciences journals, and confirmatory findings are more likely to be published (Mahoney, 1987). We do not know how many replications are submitted, but three experimental replication manuscripts I have submitted were rejected, and one editor refused to referee one because it was a replication (see Gorman, 1992, for a discussion). If so, social sciences journals might contain many positive results that are not replicable.

But have we really established the normative value of replication and disconfirmation? Were my findings peculiar to a particular task and situation? In further research, I used a variety of tasks and found out more about the exact circumstances under which replication and disconfirmation could be used to discover rules when there might be errors in the data (Gorman, 1989).

These studies made the normative generalization more situation specific. I found, for example, that the possibility of error has a large effect on Eleusis because results of the current card depend on previous cards. On rules where results of trials are independent, the possibility of error has a much smaller effect. Whether there are scientific tasks that have this feature of Eleusis is a matter that needs to be explored in ecologically valid settings.

Conclusion

I hope this chapter has demonstrated that one can derive normative generalizations from a program of experimental research directed at issues relevant to science studies. Furthermore, experiments incorporating error suggest how one can gradually add ecological validity to such a program without sacrificing control. Further research should add more such features, and ought to be more closely linked to history and sociology of science (see Gorman, 1992, for suggestions as to how this might be accomplished). Experiments cannot exist in isolation; they must complement other methods. The great strength of experiments is that they let us explore factors affecting science under ideal, artificial circumstances. We can manipulate and compare variables freely. But the bridge from experiment to reality is tenuous. At the end of the final chapter in this volume, I propose a program of research designed to create such a bridge.

References

Barker, P. (1989). The reflexivity problem in the psychology of science. In Gholson, B., Shadish, W. R., Neimeyer, R. A., & Houts, A. C. (Eds.), *Psychology of Science: Contributions to Metascience.* Cambridge: Cambridge University Press, 92–114.

Berkowitz, L., & Donnerstein, E. (1982). External validity is more than skin deep: Some answers to criticisms of laboratory experiments. *American Psychologist, 37,* 245–257.

Cook, T. D., & Campbell, D. T. (1979). *Quasi-experimentation: Design and analysis issues for field settings.* Chicago: Rand McNally.

Gardner, M. (1977). On playing New Eleusis, the game that simulates the search for truth. *Scientific American, 237* (4), 18–25.

Gorman, M. E. (1986). How the possibility of error affects falsification on a task that models scientific problem solving. *British Journal of Psychology, 77,* 85–96.

Gorman, M. E. (1989). Error, falsification and scientific inference: An experimental investigation. *Quarterly Journal of Experimental Psychology, 41A,* 385–412.

Gorman, M. E. (1992). Simulating science: Heuristics, mental models and technoscientific thinking. Bloomington: Indiana University Press.

Gorman, M. E., & Gorman, M. E. (1984). A comparison of discomfirmatory, confirmatory and combined strategies affect group problem-solving. *British Journal of Psychology, 75,* 65–79.

Gorman, M. E., Lind, E. A., & Williams, D. C. (1977). The effects of previous success or failure on a majority–minority confrontation. (ERIC Document Reproduction Service No. ED 177 257).

Greenwald, A. G., Pratkanis, A. R., Leippe, M. R., & Baumgardner, M. H. (1986). Under what conditions does theory obstruct research progress? *Psychological Review, 93,* 216–229.

Houts, A., & Gholson, B. (1989). Brownian notions: One historicist philosopher's resistance to psychology of science via three truisms and ecological validity. *Social Epistemology, 3,* 139–146.

Kern, L. (1982). *The effect of data error in inducing confirmatory inference strategies in scientific hypothesis testing.* Unpublished doctoral dissertation, Ohio State University.

Klayman, J., & Ha, Y–W. (1987). Confirmation, disconfirmation and information in hypothesis-testing. *Psychological Review, 94,* 211–228.

Mahoney, M. J. (1987). Scientific publication and knowledge politics. *Journal of Social Behavior and Personality, 2,* 165–176.

Moscovici, S. (1974). Minority influence. In C. Nemeth (Ed.), *Social psychology: Classic and contemporary integrations* (pp. 217–250). Chicago: Rand-McNally.

Mynatt, C. R., Doherty, M. E., & Tweney, R. D. (1977). Confirmation bias in a simulated research environment: An experimental study of scientific inference. *Quarterly Journal of Experimental Psychology, 29,* 85–95.

Mynatt, C. R., Doherty, M. E., & Tweney, R. D. (1978). Consequences of confirmation and disconfirmation in a simulated research environment. *Quarterly Journal of Experimental Psychology, 30,* 395–406.

Sherif, M. (1936). *The psychology of social norms.* New York: Harper & Bros.

Steiner, I. D. (1972). *Group process and productivity.* New York: Academic Press.

Tweney, R. D. (1985). Faraday's discovery of induction: A cognitive approach. In D. Gooding & F. James (Eds.), *Faraday rediscovered: Essays on the life and work of Michael Faraday, 1791–1867.* New York: Stockton Press.

Wason, P. C. (1960). On the failure to eliminate hypotheses in a conceptual task. *Quarterly Journal of Experimental Psychology, 12,* 129–140.

CHAPTER 5

The Social–Cognitive Bases of Scientific Knowledge

ARIE W. KRUGLANSKI

In late twentieth century (arguably dating from Kuhn's [1962] *Structure of Scientific Revolutions*), a significant trend in conceptualizing science has stressed its affinity to everyday modes of knowledge acquisition. Popper (1959, p. 22) pronounced science to be "common sense writ large," a view reiterated variously by subsequent authors (e.g., Quine & Ullian, 1970; Bloor, 1982). Such views imply that a good approximation to a model of science is a model of lay knowledge acquisition. Developing such a model seems highly relevant to work in which cognitive and social psychologists have been engaged all along.

Popper, Kuhn, and others (Feyerabend, 1976; Lakatos, 1970) admit the importance to science of such social psychological variables as goals, perceptions, beliefs, and attitudes. Neither they nor the psychologists who directly study such processes have done much thus far to furnish a comprehensive view of scientific activity based on extant understanding of how naive knowledge is formed. In an attempt to partially redress this state of affairs, I would like to explore the implications for science of a model of lay epistemics that I recently developed. The lay epistemic model (Kruglanski, 1989a) integrates much research in social cognition carried out in the last several decades and is based on cumulative social psychological knowledge about motivated–cognitive processes.

I first describe the lay epistemic model and briefly mention the kind of research support it has received. Then, I explore its implications for (1) the psychological interpretation of scientific activity and (2) theoretically guided research on psychology of science.

Process of Lay Epistemics

I define knowledge broadly as bodies of propositions that a person holds with a given degree of confidence. It should be obvious that my definition equates knowledge with belief; hence it differs from the prevalent epistemological position (cf. Chisholm, 1966) in which "knowledge" refers only to beliefs congruent with some external standard. But in my analysis (Kruglanski, 1989b), such standard is merely some "standard setter's" (e.g., a scientist's or the scientific community's consensual) *belief*. In presuming that the standard setter's beliefs are formed in essentially the same way as everybody else's, a theory of knowledge boils down to understanding how beliefs in general are formed or modified.

The view of knowledge as confidently held propositions (i.e., beliefs) imposes two functional requirements on a model of knowledge formation: (1) Propositional contents must be engendered, requiring a phase of *hypothesis generation*; and (2) a degree of confidence must be bestowed on the generated hypotheses, requiring a phase of hypothesis validation. Hypothesis-generation and hypothesis-validation models have been employed previously to depict epistemic activities in perception (Bruner, 1951), concept formation (Levine, 1975), problem solving (Newell & Simon, 1972) and scientific discovery (Popper, 1972). Beyond its general affinity to those earlier formulations, the present model makes some unique assumptions about how hypothesis validation and generation are performed.

Hypothesis Validation

Hypotheses are validated by relevant evidence. Relevance is determined by existing inference rules that, by the knower's assumption, link together different cognitive categories. Some linkages are logical in form as in the statement, "*if* an interviewee emerges smiling, *then* the interview must have been a success," which renders smiling relevant evidence for a success inference. Other linkages may be probabilistic, as with "20% of College Park residents are students" according to which residence in College Park is relevant evidence for inference of a probable student status.

Statistical and logical inferences are mediated via the same general process in which appropriate inference rules or heuristics are accessed and applied to a problem. Occasionally, only one such rule is considered. Often, however, competing rules may be cognized in which the same category of evidence (e.g., smiling) is tied to possible alternative inferences (e.g., pleased with the interview vs. trying to appear brave). To the extent that such alternatives are in competition (are mutually exclusive), they define a logical inconsistency that, if unresolved, prevents confident inference concerning either alternative. Choice between plausible competing alternatives (inconsistency resolution) is often accomplished via further inference rules incorporating *diagnostic* evidence, capable of differentiating among the rival hypotheses (e.g., if the interviewee continues to smile in private, the pleasing interview hypothesis is true and the brave appearance hypothesis, false).

Hypothesis Generation

As Campbell (e.g., Campbell, 1969/1988) and Popper (1959) have long asserted, the knower may generate further linkages in which the same category of evidence is tied to competing hypotheses. Because we often possess definite knowledge on various topics, such generation of alternatives must end somehow. Two kinds of conditions may bring this about, relating respectively to a person's cognitive capability and epistemic motivation.

Cognitive Capability: Availability and Accessibility

Recent social cognition literature identifies two notions relevant to the individual's mental *capability* to generate hypotheses on a given topic. Long-term capability relates to availability of constructs in memory (Higgins, King, & Mavin, 1982), and short-term capability relates to their momentary accessibility (Higgins & King, 1981). Consider a person whose car suddenly stalled on the highway. The possible explanations he/she may be able to generate for the mishap depend on the set of automotive concepts stored in this individual's memory (availability), and the subset of such notions recently activated by some event, for example, an article about defective carburetors in the morning paper (accessibility).

Epistemic Motivations: Needs for Nonspecific or Specific Closure and the Avoidance of Closure

The knower's tendency to generate hypotheses (and search for information concerning those hypotheses) also depends on his/her epistemic motivations, whose objects are particular states of knowledge. Epistemic motivations are classified on two orthogonal dimensions: closure *seeking* versus *avoidance* and *nonspecificity* versus *specificity*. This yields four epistemic needs, notably those for (1) nonspecific closure, (2) avoidance of nonspecific closure, (3) specific closure, and (4) avoidance of a specific closure.

Nonspecific closure refers to definite knowledge on a topic, irrespective of the particular content of such knowledge. Possession of definite knowledge may be deemed desirable or undesirable in various circumstances. Some perceived advantages of definite knowledge relate to predictability, affordance of a basis for action, and social status accorded the possessors of knowledge (i.e., experts). Other perceived advantages derive from the disadvanteges of lacking closure: Under time pressure or where the costs of information processing loom large, the individual may crave any closure because it curtails processing. In other circumstances, however, lack of closure may seem attractive, as when judgmental commitment to any opinion threatens to elicit damaging criticism, when one wishes to remain neutral with respect to conflicting views, or when a lack of closure preserves a valued sense of mystery and romance. Finally, people may desire knowledge with some special properties. This desire may stem from heterogeneous needs with which various knowledge contents may be congruent (e.g., esteem needs, needs for physical or material welfare, or needs for the welfare of significant others). An individual who desires to arrive at a particular knowledge (e.g., knowledge that he/she did well on an exam) needs specific closure, whereas an individual who wishes to avoid given knowledge (e.g., knowledge that he/she failed) needs to avoid specific closure.

How are epistemic motivations related to hypothesis generation? Hypothesis generation is motivated by a discrepancy or a *mismatch* between actual and desired epistemic states (e.g., by a lack of desired closure). Under these conditions, knowers generate numerous hypotheses and are highly sensitive to relevant stimulus information. One might say that their mind is open to new evidence and ideas. Conversely, lack of motivational discrepancy (a *match* between actual and desired epistemic states) puts the system to rest. In those conditions, the epistemic process is frozen (Lewin, 1943), hypothesis generation is arrested, and the individual becomes less sensitive to relevant stimulus information.

What research supports those motivational ideas? Ample support exists concerning the biasing effects of needs for specific closure or the avoidance of specific closure. Much motivational work in attribution (Zuckerman, 1979) demonstrates such effects, as does much work on cognitive dissonance (Cooper & Fazio, 1984). A comprehensive review of research on directional motivational effects on reasoning has been recently offered by Kunda (1990).

In contrast, nonspecific needs for closure and closure avoidance, were of historic interest to personality researchers (e.g., Rokeach, 1960) but have been largely ignored by social cognitive psychologists. Nonetheless, recent research on nonspecfic closure needs supported the lay epistemic model in contexts of (1) knowledge formation, (2) knowledge utilization, and (3) social interaction. Thus, in forming knowledge, subjects under high (vs. low) need for closure (e.g., induced via time pressure) processed less information before reaching a judgment, exhibited higher judgmental confidence, and tended to base judgments on information encountered early rather than late in the process (Kruglanski & Freund, 1983). Furthermore, under high (vs. low) need for closure, judgments were driven more by existing prejudices, stereotypes, and attitudes than by individuating stimulus information (Jamieson & Zanna, 1989; Kruglanski & Freund, 1983). Finally, subjects motivated to avoid (or postpone) closure tended more to socially compare with dissimilarly (vs. similarly) minded others, whereas those motivated to have closure tended more to compare with similarly minded others (Kruglanski & Mayseless, 1987) and to reject or devalue dissimilarly minded others (Kruglanski & Webster, 1991).

Consider the implications of the lay epistemic model for our understanding of how *scientific* knowledge is acquired.

Cognitive and Motivational Bases of Scientific Knowledge

Overview

Just like the layperson, the scientist generates hypotheses and evaluates them on the basis of apparently relevant evidence. Furthermore, scientific hypothesis generation is affected by cognitive capability and epistemic motivation, just as the layperson's.

Official and Unofficial Scientific Methodologies

Simple and inoccuous though such assumptions may seem, they have important implications for a psychologically based theory of science. To see why, it is first useful to distinguish between official and unofficial scientific methodologies. Official methodology is explicitly stated rules and techniques for conducting research. In psychology, official methodology consists of suggested approaches to research design as well as data collection and analysis. In contrast, unofficial methodology is "scientific folk wisdom" garnered by researchers through years of personal experience. It includes ways of being an influential scientist that transcend what one reads in the methodology texts. Both the official and the unofficial methodologies are aspects of the same process whereby scientists and laypersons alike form their knowledge.

Official Methodological Concepts: Issues of Research Design

A major question for official methodology is how to design research so that validity or interpretability of findings is maximized. Examination of these issues from the vantage of lay epistemics stresses the considerable flexibility research design should possess.

Consider the idea of design hierarchies whereby some methods (e.g., an experiment) are inherently superior to others (e.g., a case study). Such a viewpoint, typically adopted by methodology texts, seems to assume a finite list of "threats to validity" and the notion that some designs control for more threats than do others. But, according to the present model, no list of threats can be near-exhaustive. Threats represent (alternative) hypotheses generated by scientists; the number and range of such hypotheses are potentially limitless. Any list of threats is relative to a state of knowledge in a research domain. It is in reference to such knowledge that linkages between data and various hypotheses are considered plausible or implausible.

For instance, Campbell and Stanley (1963) listed the following as plausible threats to "internal validity": (1) history, (2) maturation, (3) testing, (4) instrumentation, (5) statistical regression, (6) selection biases, (7) experimental mortality, and (8) selection–maturation interaction. Subsequently, however, Cook and Campbell (1979) added to the list several further threats, namely, (9) diffusion or imitation of the treatment, (10) compensatory equalization of treatment, (11) compensatory rivalry, and (12) local history. Presumably, with further evolution of a science, further threats to validity could be added and others might be dropped from the list. Thus, at one time Campbell (with Stanley, 1963) expressed concern about the reactive effects of the pretest; his concerns were subsequently allayed (Campbell, 1969/1988) by Lana's (1969) reassuring review of research on this topic.

Yet if lists of threats are relative to states of knowledge, so must be design hierarchies founded on those lists. Thus, a case study could be quite interpretable (Campbell, 1975), no less so and occasionally more so than a randomized experiment. The same holds true for correlational designs often regarded as methodologically inferior to experimentation. It is true that random assignment of subjects to experimental conditions in principle controls over alternatives to a putative causal relation (i.e., that B is causing A rather than A causing B, or of both A and B being caused by a third variable). But the real issue in scientific inference is typically one of practice rather than of principle. And what matters in practice are the specific plausible alternatives advanced to account for a set of findings, rather than some general categories of possible alternatives that may or may not be applicable to those findings. For example, most developmental studies are correlational; this does not necessarily render them any less valid than many experimental studies. The reason is that general direction-of-causality problems are irrelevant to much developmental research. Thus, the issue whether the development of sufficient attentional capacity is the cause or the effect of the ability to perform complex conceptual learning hardly arises. We know that developmentally the former precedes the latter; hence it could not possibly constitute its effect.

Generally, then, the issue is not as much whether a given research design is of a particular type, but whether it controls for currently plausible interpretations of the data. A research design controlling for specific threats out of a fixed list could be controlling for the "wrong" (i.e., currently implausible) threats, while failing to control for threats of genuine concern. There can be an experiment that invites several alternative interpretations of the data, and a single-shot case study that invites none. For instance, a physician may be quite confident that a patient was poisoned, say, by antimony and be unable to engender any alternative explanations for the observed symptoms.

Not only is validity (or interpretability) of results relative to a state of knowledge within a scientific community, but it may also vary across communities. A research design that appears valid to the psychiatric community may not seem so to the community of clinical psychologists, and vice versa. A social psychologist may worry about demand characteristics (Orne, 1962) or reactivity of research procedures (Webb, Campbell, Schwartz, & Sechrest, 1966); such threats may be somewhat foreign to a cognitive psychologist. To achieve scientific acceptance of one's views, one may need to furnish different types of evidence to different scientific audiences. In this vein, Cronbach (1982) recently commented that the methodology of scientific inference may well be moved "from the realm of pure logic into the realm of rhetoric, or persuasion" (p. 109). In short, considerations of research design are highly flexible or contextually contingent, depending as they do on specific competing hypotheses that come to scientists' minds.

Unofficial Scientific "Methodologies"

Cronbach's references to rhetoric and persuasion naturally tie to the present notion of unofficial methodology, the social and psychological processes that determine acceptability of conclusions to a community of scientists. Three categories of such unofficial influence are (1) unofficial evidential bases, (2) cognitive bases, and (3) motivational factors affecting scientific persuasion.

Unofficial Evidential Bases: Heuristic Processing in Science

A theory of persuasive communication proposed recently by Chaiken, Liberman, and Eagly (1989) claims that attitudes are often deduced from simplifying heuristics unrelated to the message substance. Similar processes plausibly affect the formation of scientific conclusions. A case in point is "source effects" in scientific inference, that is, credibility boosts to arguments due to endorsement by prestigeous scientific sources. Examples of the efficacy of such endorsements include the endorsement of Darwin by Sir Charles Lyal, the famed geologist, of Einstein's theory of general relativity by Eddington, or of quantum optics by Henri Poincare.

Cognitive Bases of Scientific Persuasion: Accessibility Effects

To exert impact, researchers need to make their own ideas accessible to others, which in this age of information explosion might not be easy. One way of accomplishing this is through the work of prolific coworkers or graduate students that may amplify the accessibility of a scientist's notions. Some investigators may attempt to render their research noticeable by choosing faddish problems. This strategy may be excessively reactive to shifting trends, representing a scientific "knee-jerk relevance" at the expense of a more serious commitment to a topic.

Motivational Effects: Needs for Specific Closure

Finally, persuasion in science may be greatly affected by scientists' epistemic motivations, the needs for specific and nonspecific cognitive closure discussed earlier.

History of science is rife with instances in which scientists seemed biased toward conceptions compatible with their values or interests and against incompatible ones. Such effects may reflect the influence of needs for specific closure. Resistance of the psychiatric establishment to Freud's psychoanalysis is often explained by his emphasis on human sexuality, strikingly out of step with middle-class values in domains of religion, morals, and family. Mendel's genetic theorizing is compatible with his Augustinian perspective that stresses the heritability of the human nature (in particular, the original sin). And according to the "Forman thesis," the motivation to be in good standing within their larger intellectual milieu prompted German physicists in the Weimar republic to embrace the indeterminacy interpretation of quantum mechanics (Forman, 1971; Fuller, 1988).

A scientific work may exert greater impact if it appears to serve the prevailing needs of the audience. It may be better received when it communicates information that "the collegues deem . . . useful in the pursuit of their own studies" (Mulkay, 1979, pp. 69–70) than when it is perceived as irrelevant or contrary to their studies and underlying assumptions. Different groups of colleagues might find different information useful to their respective pursuits. Theoretical researchers may cherish the conceptual implications of a work; empirical researchers, its ramifications for future research; and practical colleagues, its applied implications. Persuasive presentations of work to different audiences stress different aspects in accord with the audiences' differing values and interests.

Also, the notion of specific closure highlights the importance of interpersonal relations in science and their impact on scientific persuasiveness. Congeniality and interpersonal trust may increase the likelihood that a colleague will appreciate one's point of view and find fault with the notions of critics. Thus, good interpersonal rapport among faculty in a research institution may be of more than just social significance. It may contribute significantly to members' ability to benefit from each other's ideas and to generate productive cross-stimulation. It also suggests that in the training of graduate students, it is important to provide ample opportunities for informal exchanges in a convivial atmosphere where evaluation concerns are at a minimum and ideas are openly explored.

Admittedly, the perfect motivational mix for optimal scientific productivity may be complex and depend to a large extent on the specific needs and values of particular scientists. Some researchers may be challenged to creativity by uncompromising criticism of their work; others may be threatened and discouraged by criticism. Some scientists may be motivated to engender supportive arguments for friends' theses; others may "bend over backwards" and be especially critical toward friends in an attempt to be and appear fair. Friendly overture may be occasionally interpreted as ingratiation, may evoke a negative reaction, and so on. All such possibilities are compatible with the notion that specific closure needs may bias scientific receptivity. However, the specific needs that are activated and their exact impact should be determined in reference to a particular research context in which the personal values, interests, and desires of particular scientists may play a role.

Needs for Nonspecific Closure

In their role as experts, scientists often feel the need to possess confident and definite closures in their specialty. This could be one reason why scientists are strongly committed to extant paradigms and are loathe to abandon them until definite alternatives existed (Kuhn, 1962). This motivational interpretation allows one to speculate when paradigmatic fixations are more or less likely to occur. For example, a scientist to whom "expert" image is very important (perhaps an applied scientist in frequent contact with the public) may evolve a stronger need for closure and hence cherish and defend the paradigm more than a scientist for whom the expert image is less critical.

Some scientists may evolve a strong "need to avoid closure," fearing an evaluation of their work by colleagues. They may shun commitment to potentially controversial paradigms, or avoid paradigmatic commitments altogether if no generally accepted paradigms exist.

Fear of invalidity may be an important psychological mechanism in promoting paradigmatic shifts (Kuhn, 1962), movements from one scientific framework toward another. Such fear may sensitize scientists to accumulated anomalies and hence pave the way to sympathetic consideration of an alternative paradigm free of threatening anomalies.

Kuhn (1962) stressed the accumulation of anomalies, but the present analysis implies that this may not suffice to produce change even if an alternative paradigm were available. When a group of scientists is motivated by a need for closure, or where the ruling paradigm represents those scientists' desired closure (e.g., because of career investments in the ruling paradigm), many anomalies might be disregarded or reinterpreted. Conversely, when the need to avoid closure predominates, scientists might abandon a framework threatened by few and/or relatively slight anomalies.

In a different analysis of conditions under which the accumulation of anomalies may or may not result in a paradigm shift, Bloor (1979) utilized an anthropological framework developed by Douglas (1973). According to Bloor, the readiness to reassess the paradigm in light of the anomalies is particularly likely when the scientific community has a low degree of hierarchical organization and low degree of group cohesion. On the other hand, continued adherence to a paradigm may occur where either the hierarchical organization is high, group cohesion is high, or both organization and cohesion are high. Although pitched at a different level of discourse, Bloor's anthropological notions are not incompatible with the present motivational–cognitive account. In fact, the effect on reactions to anomalies of different types of social structure may well be mediated by induction in individual group members of a high need for closure (in a social structure characterized by high cohesion and high organization) or a low need of closure (in a social structure with permeable boundaries and low organization). Those issues could be profitably pursued via further research.

Official and Unofficial Methodologies Compared and Contrasted

Whereas official methodology may seem logical and reasonable and unofficial methodology may seem unwarranted and psychologically biased, the two sources of influence on scientific judgments are inevitably present in each instance of inference. Specifically, unofficial variables as motivation and accessibility may generate the contents to which official principles are applied, thus, in a sense, providing "grist" for the logical "mill." It makes little sense, therefore, to equate official methodology with valid and unofficial with invalid scientific inference. For instance, a motivational influence stemming from the investigator's ego concerns (reflecting an unofficial influence on judgments) may sensitize him/her to anomalies within the ruling paradigm that the investigator may desire to overthrow, and promote the formulation of a new paradigm that advances scientific knowledge. A careful design attention to a potential confound (an official influence on judgments) may fail to result in a valid interpretation if another, critical confound was left unattended.

In the present account, both unofficial and official sources of influence on judgment play a part in the process whereby scientific (or lay) inferences are fashioned. This view contrasts with analyses (e.g., Collins, 1985) in which the demonstration of unofficial effects in scientific judgments is taken as tantamount to saying that official methodological principles are not applied. Furthermore, the effects of motivations on judgments are tempered by the scientist's available knowledge (i.e., his/her topic-relevant beliefs) (Kunda, 1990). Thus, not anything goes, and researchers do not simply embrace conclusions because these are in accord with their needs. Ultimately, they adopt judgments that make sense to them given what they believe to be true, even if the latter is somewhat (but not totally) affected by motivations.

The official and unofficial methodologies do differ in the following sense. The former refers to overt reasons for (scientific or lay) judgments and the latter relates to covert causes of judgment. Thus, official methodology concerns are likely to explicitly figure in arguments by which scientists justify their conclusions to themselves and others. Such concerns are typically represented in the "collective consciousness" of a scientific community reflected (e.g., in its published discourse).

In contrast, the various unofficial influences on judgment are typically tacit and do not form part of the scientist's conscious considerations. In fact, few scientists would probably admit that their judgments were partially affected by how desirable the judgments seemed, or by their compatibility with the scientist's interests or values. They might regard such interpretations not only as a challenge to their judgment's validity but as an affront to their scientific competence and perhaps their integrity.

Actually, scientists need not be defensive about the unofficial sociopsychological bases of their judgments. The bases of scientific judgments are in a sense irrelevant to the scientific dialogue as such. What matters are the contents of the judgments and the extent to which critics are capable of countering them with plausible contradictory arguments (i.e., inconsistent evidence or competing interpretations). Furthermore, as suggested earlier, awareness of the unofficial causes of scientific judgments may serve to improve one's delivery of scientific messages to different scientific audiences, to facilitate communication, and to lower the resistance to new ideas.

Research Implications

Thus far, the lay epistemic interpretation of scientific knowledge acquisition has been speculative and anecdotal. Whereas the model's propositions have received support in lay inference (Kruglanski, 1989a), much research remains to be done to substantiate its implications for scientific inference. How might such research be carried out? Ideally, it should study actual scientists addressing genuine scientific problems. This renders the experimental approach rather impractical. A methodological compromise is having lay subjects address scientific judgment tasks. Lord, Lepper, and Ross (1979) selected subjects who were for or against capital punishment and exposed them to two research studies using different methodologies, one supporting and one opposing the conclusion that capital punishment deterred crime. Subjects were more critical of methods in the study that disconfirmed their beliefs than they were of methods used in the study that confirmed their beliefs. Presumably these subjects' need for specific closure (affirmation of their existing view) biased their interpretation of evidence against need-incongruent conclusions. This and similar studies (Kunda, 1990) indicate the possible need for specific closure effects in the evaluation of scientific work. However, the lay epistemic model has numerous further implications for scientific knowledge acquisition. Those pertain to research *production* as well as to research *assessment*.

Research production

The psychological context in which research is planned, executed, and reported may have important consequences for its content. For instance, an investigator under high need for closure may give less thought to competing alternative hypotheses and tend less to incorporate them into the research design. In academia, need for closure may arise under promotion and tenure pressures, when extraneous demands are high (teaching, committee work), when the investigator is fatigued, and so forth. A need to avoid closure may be elevated when evaluation apprehension is salient, as when the investigator routinely deals with or discusses evaluative issues (e.g., in the role of a journal editor or a member of a grant review panel).

Production of research may depend on investigators' reactions to criticism (e.g., of one's journal articles submitted for publication). Despite the considerable value of criticism in science, many investigators fail to derive full benefits from it because of the epistemic motivations it may arouse. Criticism with an ad hominem flavor can arouse defensiveness, a need to avoid the specific closure propounded by the critic. Relative imperviousness to criticism may occur when the investigator has high need for closure (e.g., due to current absorption in alternative projects), or when he/she is in a highly ego-involved position. A need to avoid closure may foster open-mindedness to criticism. Such motivation may be introduced by editorial encouragement (lessening one's defensiveness), a sympathetic reaction to one's work, or the belief that incorporation of criticism may improve the paper's quality and likelihood of subsequent acceptance.

One might also explore the unofficial inference rules or heuristics to which members of different scientific communities, or different generations within the same community, subscribe. For instance, some say that social psychologists possess a rather brief collective memory (Berkowitz & Devine, 1990) regarding recently published evidence as more relevant to current work than similar early evidence. Subscription to such a "recency" heuristic might be particularly true of professionally younger (vs. older) investigators, because younger investigators' education is biased toward recent research whereas older investigators' education may be biased toward early research. Those issues are researchable and of considerable consequence to a social psychology of science. For instance, populating editorial boards with young investigators may amplify the impact of the recency heuristic, whereas including many older-generation researchers may reduce its impact.

Research Assessment

Parameters of the epistemic process also play an important a part in assessing scientific research. The following true story illustrates this. An editor of a major journal in social psychology reached many editorial decisions during daily jogs. One wonders if jogging systematically affected the decisions he reached. The lay epistemic model suggests that during any type of physical exercise, the costs of information processing are high since great effort and energy are expended on strenuous activity. This should increase one's need for closure because closure removes the need for further processing. The editor would tend to embrace the most accessible evaluation of the paper, perhaps the most recent one, the recommendation made by the largest number of reviewers, or the most negative recommendation, given the generally negative base rates of manuscript acceptance at major journals.

Low acceptance base rates may also arouse corresponding needs for specific closure, that is, reviewers' motivationally based tendencies to reach negative publication recommendations. Such motivation may stem from a desire to have one's suggestions accepted by the editor as well as to manage the impression one is making on the editor as a tough and demanding critic. Research by Amabile (1983) demonstrated that the negative evaluator is perceived as more intelligent and discerning ("brilliant though cruel," p. 146) than a positive evaluator. Thus, there may exist various needs for negative closure in the editorial review process. It is possible to systematically investigate moderating variables for such negativity. One may ask whether new investigators who have not yet established their reputation in a field, and so are more strongly motivated to create a respectable impression, may be more negative than are the more established investigators whose reputation is less heavily dependent on how their reviews are perceived by editors.

Implications for the Goals of Science

The present analysis suggests that just as with lay inference, the process of scientific inference is subject to various cognitive and motivational biases. What are the broad implications of this view? A major implication is that any attempt to rid scientific knowledge of bias is utopian and unrealistic. Biases are a fundamental part and parcel of the reasoning process. They are here to stay. But maybe we need to rethink our conception of science and redefine its goals.

In the traditional view, the ultimate goal of science has been the discovery of "truth." The scientific method was thought to provide an efficient tool for attaining that goal. However, it increasingly appears that truth or even meaningful progress toward the truth are chimerical ideals. Philosophers and historians of science (Feyerabend, 1976; Kuhn, 1962; Lakatos, 1970; Popper, 1972) have suggested that science has no meaningful way of assuring or even probabilifying the objective validity of its conclusions. Even if progress in a field of science seemed evident for a time, the possibility of toppling the latest, most "progressive" scientific view in the course of a scientific revolution is ever present. Accepted scientific truths reflect consensus in a community of scholars. In turn, consensus represents a tentative social psychological state, capable of changing under the influence of all those motivational and cognitive factors that affect belief formation in general.

This hardly implies that science is hopeless and unworthy of pursuit. What it implies is that truth should probably be relinquished as a legitimizing ideal for science. An equally powerful justification could be the *construction of new knowledge,* even though the fruits of our labor may be only temporary. Whether ultimately adjudged valid or invalid, knowledge is a necessity without which human action is unthinkable. Given the essential indispensability of knowledge, the issue of ultimate validity becomes almost immaterial, and the appeal of participation in the knowledge construction process remains as powerful as ever. Science is our institutional way of producing knowledge. Its unique historical characteristics as an institution, including the repudiation of dogmatism, openness to criticism, and the indefatigable quest for discovery, render it a particularly valuable vehicle for progress toward new ways of comprehending the world.

Conclusion

A model of lay knowledge acquisition may furnish a useful framework in which to understand the processes of scientific knowing. Such a perspective has many implications for the social psychological study of how scientific research is produced and assessed. It also has implications for making one's scientific endeavors more effective, more likely to produce new, communally shared knowledge. From this perspective, cognitive and motivational biases that influence scientific conclusions are fundamentally inevitable and are an integral part of how all knowledge is acquired. Rather than regarding them as impediments to truth, it may be more practical to take them into account to improve the quality and persuasiveness of one's research.

References

Amabile, T. M. (1983). Brilliant but cruel: Perceptions of negative evaluators. *Journal of Experimental Social Psychology, 19,* 146–156.

Berkowitz, L., & Devine, P. G. (1990). Research traditions, analysis, and synthesis in social psychological theories: The case of dissonance theory. *Personality and Social Psychology Bulletin, 15,* 493–507.

Bloor, D. (1979). Polyhedra and the abominations of Leviticus. *British Journal of the History of Science, 13,* 254–272.

Bloor, D. (1982). Durkheim and Mauss revisited: Classification and the sociology of knowledge. *Studies in the History and Philosophy of Science, 13,* 267–298.

Bruner, J. S. (1951). Personality dynamics and the process of perceiving. In R. Blake & G. V. Ramsey (Eds.), *Perception—an approach to personality* (pp. 94–98). New York: Ronald Press.

Campbell, D. T. (1975). Degrees of freedom and the case study. *Comparative Political Studies, 2*, 178–293.

Campbell, D. T. (1988). Prospective: Artifact and control In E. S. Overman (Ed.), *Methodology and epistemology for social science: Selected papers* (pp. 167–190). Chicago: University of Chicago Press. (Original work published 1969)

Campbell, D. T., & Stanley, J. (1963). Experimental and quasi-experimental designs for research on teaching. In N. L. Gage (Ed.), *Handbook of research on teaching* (pp. 171–246). Chicago: Rand McNally.

Chaiken, S., Liberman, A., & Eagly, A. H. (1989). Heuristic and systematic processing within and beyond the persuasion context. In J. S. Uleman & J. A. Bargh (Eds.), *Unintended thought* (pp. 212–252). New York: Guilford Press.

Chisholm R. (1966). *Theory of knowledge*. Englewood Cliffs, NJ: Prentice-Hall.

Collins, H. M. (1985). *Changing order: Replication and induction in scientific practice*. Newbury Park, CA: Sage.

Cook, D. T., & Campbell, D. T. (1979). *Quasi-experimentation: Design and analysis issues for field settings*. Chicago: Rand McNally.

Cooper, J., & Fazio, R. H. (1984). A new look at dissonance theory. In L. Berkowitz (Ed.), *Advances in experimental social psychology* (Vol. 17, pp. 229–266). New York: Academic Press.

Cronbach, L. J. (1982). *Designing evaluations of educational and social programs*. San Francisco: Jossey-Bass.

Douglas, M. (1973). *Natural symbols: Explorations in cosmology*. Beaminster, England: Harmondsworth.

Feyerabend, P. (1976). *Against method*. New York: Humanities Press.

Forman, P. (1971). Weimar culture, causality and quantum theory, 1918–1927 adaptation by German physicists and mathematicians to a hostile intellectual environment. In R. McCormach (Ed.), *Historical studies in the physical sciences*. Philadelphia: University of Pennsylvania Press.

Fuller, S. (1988). *Social epistemology*. Bloomington: University of Indiana Press.

Higgins, E. T., & King, G. A. (1981). Accessibility of social constructs: Information processing consequences of individual and contextual variability. In N. Cantor & J. F. Kihlstrom (Eds.), *Personality, cognition, and social interaction*. Hillsdale, NJ: Erlbaum.

Higgins, E. T., King, G. A., & Mavin, G. H. (1982). Individual construct accessibility and subjective impressions and recall. *Journal of Personality and Social Psychology, 43*, 35–47.

Janneson, D. W., & Zanna, M. P. (1989). Need for structure in attitude formation and expression. In A. R. Pratkanis, S. J. Breckler, & A. G. Greenwald (Eds.), *Attitude structure and function* (pp. 73–89). Hillsdale, NJ: Erlbaum.

Kruglanski, A. W. (1989a). *Lay epistemics and human knowledge: Cognitive and motivational bases*. New York: Plenum Press.

Kruglanski, A. W. (1989b). The psychology of being "right": The problem of accuracy in social cognition and perception. *Psychological Bulletin, 106*, 395–409.

Kruglanski, A. W., & Freund, T. (1983). The freezing and unfreezing of lay inferences: Effects on impressional primacy, ethnic stereotyping, and numerical anchoring. *Journal of Experimental Social Psychology, 19,* 448–468.

Kruglanski, A. W., & Mayseless, O. (1987). Motivational effects in the social comparison of opinions. *Journal of Personality and Social Psychology, 53,* 834–842.

Kruglanski, A. W., & Webster, D. M. (1991). Group members' reactions to opinion deviates and conformists under varying degrees of proximity to decision deadline and environmental noise. *Journal of Personality and Social Psychology, 61,* 212–225.

Kuhn, T. S. (1962). *The structure of scientific revolutions.* Chicago: Chicago University Press.

Kunda, Z. (1990). The case for motivated reasoning. *Psychological Bulletin, 108,* 480–498.

Lakatos, I. (1970). Falsification and the methodology of scientific research programs. In I. Lakatos & A. Musgrave (Eds.), *Criticism and the growth of knowledge* (pp. 91–196). Cambridge, England: Cambridge University Press.

Lana, R. E. (1969). Pretest sensitization. In R. Rosenthal & R. L. Rosnow (Eds.), *Artifact in behavioral research* (pp. 121–140). New York: Academic Press.

Levine, M. A. (1975). *A cognitive theory of learning.* Hillsdale, NJ: Erlbaum.

Lewin, K. (1943). Forces behind food habits and methods of change. *Bulletin of the National Research Council, 108,* 35–67.

Lord, C. G., Ross, L., & Lepper, M. R. (1979). Biased assimilation and attitude polarization: The effects of prior theories on subsequently considered evidence. *Journal of Personality and Social Psychology, 37,* 2098–2109.

Mulkay, M. J. (1979). *Science and the sociology of knowledge.* London: Allen & Unwin.

Newell, A., & Simon, H. A. (1972). *Human problem solving.* Englewood Cliffs, NJ: Prentice-Hall.

Orne, M. (1962). On the social psychology of the psychological experiment: With particular reference to demand characteristics and their implication. *American Psychologist, 17,* 776–783.

Popper, K. R. (1959). *The logic of scientific discovery.* New York: Harper.

Popper, K. R. (1972). *Objective knowledge.* Oxford, England: Oxford University Press.

Quine, W. V. O, & Ullian, J. S. (1970). *The web of belief.* New York: Random House.

Rokeach, M. (1960). *The open and closed mind: Investigations into the nature of belief systems and personality systems.* New York: Basic Books.

Webb, E. J., Campbell, D. T., Schwartz, R. D., & Sechrest, L. B. (1966). *Unobtrusive measures: Nonreactive research in the social sciences.* Chicago: Rand McNally.

Zuckerman, M. (1979). Attribution of success and failure revisited, or: The motivational bias is alive and well in attribution theory. *Journal of Personality, 47,* 245–287.

CHAPTER 6

On Being One's Own Case Study: Experimenter Effects in Behavioral Research—30 Years Later

ROBERT ROSENTHAL

Over 30 years ago I began to study the unintended social psychological effects of psychological experimenters on the results of their research. My special interest was in the effects of psychological experimenters' research hypotheses on the responses they obtained from their human and animal research subjects. It is not my intent to describe, or even to summarize, that body of literature today. Instead, I want to illustrate some mechanisms of the psychology of science by drawing on this program of research on the social psychology of science (SPS). The technical details of the many experiments on this topic are available elsewhere and have been summarized periodically (Rosenthal, 1966, 1976, 1985, 1991b; Rosenthal & Jacobson, 1992; Rosenthal & Rubin, 1978). Accordingly, I have been authorized to wax more anecdotal and more speculative than is my wont.

Experimenter Expectancy Effects and an Unnecessary Statistical Analysis

As a graduate student at UCLA in the mid-1950s I was much taken with the work of two giants of personality theory, Freud and Murray. I was taken with Freud, as were so many others, for the richness and depth of his theory. I was taken with Murray, as were not enough others, for similar reasons but also because of Murray's brilliant way of inventing whatever tool was needed to get him along in his inquiry. Thus, the

Thematic Apperception Test (TAT) was invented simply as a tool to further his research, although it has become recognized as a major contribution in its own right. My dissertation was to depend on the work of both these great theorists.

Sigmund Freud's Projection

As a graduate student in clinical psychology, I was (and still am) interested in projective techniques. Murray's TAT, Shneidman's Make a Picture Story Test, and, of course, the Rorschach, were exciting methods for understanding people better. Shneidman, a brilliant researcher and clinician, was my first clinical supervisor during my Veteran's Administration clinical internship. Bruno Klopfer, one of the all-time Rorschach greats, was the chair of my doctoral committee. It was natural, therefore, for me to be concerned about the defense mechanism of projection for the part it might play in the production of responses to projective stimuli.

Harry Murray's Party Game

Freud's defense mechanism of projection, the ascription to others of one's own states or traits (Freud, 1953; Rosenthal, 1956) is only one of the mechanisms that has been isolated as contributing to the process of producing responses to projective stimuli. Another mechanism is complementary apperceptive projection, that is, finding in another the reasons for one's own states or traits. It was this mechanism that Harry Murray (1933) investigated in his classic paper. At his 11-year-old daughter's house party, Murray arranged a game called "Murder," which frightened delightfully the five party-going subjects. After the game, Murray found that the children perceived photographs as more malicious than they did before the game of Murder. Murray's wonderfully direct and deceptively simple procedure of assessing projective processes by means of assessing changes in perceptions of photographs was the basic measuring device I adopted for my dissertation:

"An Attempt at the Experimental Induction of the Defense Mechanism of Projection" (Rosenthal, 1956)

With the foregoing as its almost unbearable title, my dissertation employed a total of 108 subjects: 36 college men, 36 college women, and 36 patients hospitalized with paranoid symptomatology. Each of these three groups was further divided into three subgroups receiving success, failure, or neutral experience on a task structured as and simulating a standardized test of intelligence. Before the subjects' experimental conditions were imposed, they were asked to rate the degree of success or failure of persons pictured in photographs. Immediately after the experimental manipulation, subjects were asked to rate an equivalent set of photos on their degree of success or failure. The dependent variable was the magnitude of the difference scores from pre- to postratings of the photographs. It was hypothesized that the Success condition would lead to the subsequent perception of other people as more successful while the Failure condition would lead to the subsequent perception of other people as having failed more as measured by the pre- post difference scores.

In an attack of student compulsivity, an attack that greatly influenced my scholarly future, I did a statistical analysis that was quite extraneous to the main purpose of the dissertation. In this analysis I compared the mean *pre*treatment ratings of the three experimental conditions. These means were: Success = −1.52, Neutral = −0.86, and Failure = −1.02. The pretreatment rating mean of the Success condition was significantly lower than the mean of either of the other two conditions. It must be emphasized that these three treatment groups had not yet undergone their treatment; they were only destined to become the subjects of the three conditions. If the Success group started out lower than the other conditions, then, even if there were no differences among the three conditions in their posttreatment photo ratings, the Success group would show the greatest gain, a result favoring one of my hypotheses, namely, that projection of the good could occur just as well as projection of the bad. Without my awareness, the cards had been stacked in favor of obtaining results supporting one of my hypotheses. It should be noted that the Success and Failure groups' instructions had been identical during the pretreatment rating phase of the experiment. (Instructions to the Neutral group differed only in that no mention was made of the experimental task, because none was administered to this group.)

The problem, apparently, was that I knew for each subject which experimental treatment he or she would subsequently be administered. As I noted in 1956 with some dismay:

> The implication is that in some subtle manner, perhaps by tone, or manner, or gestures, or general atmosphere, the experimenter, although formally treating the success and failure groups in an identical way, influenced the Success subjects to make lower initial ratings and thus increase the experimenter's probability of verifying his hypothesis. (Rosenthal, 1956, p. 44)

As a further check on the suspicion that Success subjects had been differentially treated, the conservatism–extremeness of pretreatment ratings of photos was analyzed. (The mean extremeness-of-rating scores were as follows: Success = 3.92, Neutral = 4.41, and Failure = 4.42.) The Success group rated photos significantly less extremely than did the other treatment groups. Whatever I did differently with those subjects whom I knew were destined for the Success condition, it seemed to affect not only their mean level of rating but their style of rating as well.

The Search for Company

When I discussed these strange goings-on with some faculty members they seemed not overly surprised. A not very reassuring response was "Oh yes, we lose a few Ph.D. dissertations now and then because of problems like that." There followed a frantic search of the literature for references to this phenomenon, which I then called unconscious experimenter bias. As far back as Ebbinghaus (1885/1913) psychologists had been referring to something like this phenomenon, including such notables as Oskar Pfungst (1911/1965), Ivan Pavlov (cited in Gruenberg, 1929), and Saul Rosenzweig (1933). Unfortunately, none of these investigators (or even later ones) had explicitly designed and conducted an experiment to test the hypothesis of unconscious experimenter bias; that remained to be done.

There is something I want to add about the paper by Rosenzweig (1933), which appeared the same year that Harry Murray's (1933) paper appeared, and, incidentally, which appeared the same year that I appeared. In my own reviews of the literature (e.g., Rosenthal, 1956, 1966) I had completely missed the Rosenzweig paper. I believe it was my good friend, my longtime collaborator, and my scholarly tutor, Ralph Rosnow, who called my attention to Rosenzweig's extraordinarily insightful and prophetic paper. Not only did Rosenzweig anticipate the problem of unconscious experimenter bias, but he also anticipated virtually the entire area now referred to as the social psychology of the psychological experiment. The Rosenzweig paper makes good reading even today, a half century later. There is a superb appreciation of the Rosenzweig paper in Rosnow's (1981) brilliant book about the methodology of social inquiry.

The Production of Company

If it was my "unconscious experimenter bias" that had led to the puzzling and disconcerting results of my dissertation, presumably we could produce the phenomenon in our own laboratory and with several experimenters rather than just one. Producing the phenomenon in this way would yield not only the scientific benefit of demonstrating an interesting and important concept, but also the very considerable personal benefit of showing that I was not alone in having unintentionally affected the results of my research by virtue of my "bias" or expectancy.

There followed a series of studies employing human subjects in which we found that when experimenters were led to expect certain research findings, they were more likely to obtain those findings from their subjects. These studies were met with incredulity by many investigators who worked with human subjects. However, investigators who worked with animal subjects often nodded knowingly and told me that it was this kind of phenomenon that encouraged them to work with animal subjects. In due course, then, my students and I began to work with animal subjects and found that when experimenters were led to believe that they were working with maze-bright rats, the rats learned faster than did the rats randomly assigned to experimenters who had been led to believe their rats were dull. That result surprised many psychologists who worked with animal subjects, but it would not have surprised Pavlov, Pfungst, or Bertrand Russell (1927), who said:

> Animals studied by Americans rush about frantically, with an incredible display of hustle and pep, and at last achieve the desired result by chance. Animals observed by Germans sit still and think, and at last evolve the solution out of their inner consciousness. (pp. 29–30)

Our experiments on the effects of investigators' expectancies on the behavior of their research subjects should be distinguished from the much older tradition of examining the effects of investigators' expectations, theories, or predilections on their observations or interpretations of nature. Examples of such effects have been summarized elsewhere (Rosenthal, 1966, see especially chaps. 1 and 2 on observer effects and interpreter effects) and there is continuing lively interest in these topics (Gorman, 1986; Mahoney, 1989; Mitroff, 1974; Rudwick, 1986; Tweney, 1989).

Teacher Expectation Effects and an Essential Principal

If rats became brighter when expected to, it should not be farfetched to think that children could become brighter when expected to by their teachers. Indeed, for years Kenneth Clark (1963) said that teachers' expectations could be important determinants of intellectual performance. Clark's ideas and our research should have sent us right into the schools to study teacher expectations, but that is not what happened.

What did happen was that after we had completed about a dozen studies of experimenter expectancy effects (we no longer used the term *unconscious experimenter bias*), I summarized our results in a paper for the *American Scientist* (Rosenthal, 1963). (As an aside, I should note that although this research began in 1958, and although there had been more than a dozen papers, none of them found its way into an American Psychological Association [APA] publication. I recall an especially "good news–bad news" type of day when a particular piece of work was simultaneously rejected by an APA journal and awarded the American Association for the Advancement of Science [AAAS] Socio-Psychological Prize for 1960. During these years of nonpublication, there were three "psychological sponsors" who provided enormous intellectual stimulation and personal encouragement: Donald T. Campbell, Harold B. Pepinsky, and Henry W. Riecken. I owe them all a great deal.)

I concluded the 1963 paper by wondering whether the same interpersonal expectancy effects found in psychological experimenters might not also be found in physicians, psychotherapists, employers, and teachers (subsequent research showed that indeed it could be found in all these practitioners). "When the master teacher tells his apprentice that a pupil appears to be a slow learner, is this prophecy then self-fulfilled?" (Rosenthal, 1963, p. 280) was the closing line of this paper.

Among the reprint requests for this paper there was one from Lenore F. Jacobson, the principal of an elementary school in South San Francisco. I sent her a stack of unpublished papers and thought no more about it. On November 18, 1963, Lenore Jacobson wrote me a letter telling of her interest in the problem of teacher expectations. She ended her letter with the following line: "If you ever 'graduate' to classroom children, please let me know whether I can be of assistance" (L. F. Jacobson, personal communication, November 18, 1963).

On November 27, 1963, I accepted Jacobson's offer of assistance and asked whether she would consider collaborating on a project to investigate teacher expectancy effects. A tentative experimental design was suggested in this letter as well.

On December 3, 1963, Jacobson replied mainly to discuss concerns over the ethical and organizational implications of creating false expectations for superior performance in teachers. If this problem could be solved her school would be ideal, she felt, with children from primarily lower-class backgrounds. Jacobson also suggested gently that I was "a bit naive" to think one could just *tell* teachers to expect some of their pupils to be "diamonds in the rough." We would have to administer some new test to the children, a test the teachers would not know.

Telephone calls and letters followed, and in January 1964, I made a trip to South San Francisco to settle on a final design and to meet with the school district's administrators to obtain their approval. This approval was forthcoming because of the leadership of the school superintendent, Dr. Paul Nielsen. Approval for this research had already been obtained from Robert L. Hall, Program Director for Sociology and Social Psychology for the National Science Foundation which supported much of the early work on experimenter expectancy effects.

The Pygmalion Experiment

All the children in Jacobson's school were administered a nonverbal test of intelligence, which was disguised as a test that would predict intellectual "blooming." The test was labeled as the Harvard Test of Inflected Acquisition. There were 18 classrooms in the school, 3 at each of the six grade levels. Within each grade level, the 3 classrooms were composed of children with above-average ability, average ability, and below-average ability, respectively. Within each of the 18 classrooms, approximately 20% of the children were chosen at random to form the experimental group. Each teacher was given the names of the children from his or her class who were in the experimental condition. The teacher was told that these children had scored on the Test of Inflected Acquisition such that they would show surprising gains in intellectual competence during the next 8 months of school. The only difference between the experimental group and the control group children, then, was in the mind of the teacher (Rosenthal & Jacobson, 1966, 1968).

At the end of the school year, 8 months later, all the children were retested with the same test of intelligence. Considering the school as a whole, the children from whom the teachers had been led to expect greater intellectual gain showed a significantly greater gain than did the children of the control group. The magnitude of this experimental effect was .30 standard deviation units, equivalent to a point biserial r of .15 (Cohen, 1988).

Some Substantive Consequences: Processes of Social Influence

Among the most interesting and important implications of the research on interpersonal expectancy effects have been those for the study of subtle processes of unintended social influence. The early work in this area has been summarized in detail elsewhere (e.g., Rosenthal, 1966, 1969). When we look more particularly at the mediation of teacher expectancy effects, we find early summaries by Brophy and Good (1974), whose contributions to this area have been enormous, and by Rosenthal (1974). More recent summaries of this domain are by Brophy (1985) and Harris and Rosenthal (1985). There is space here only to illustrate the type of research results that have been accumulating. A preliminary four-factor "theory" of the communication of expectancy effects suggests that teachers (and perhaps clinicians, supervisors, and employers) who have been led to expect superior performance from some of their pupils (clients, trainees, or employees) tend to treat these "special" persons differently from the way they treat the remaining "less special" persons in the following four ways (Rosenthal, 1971, 1973, 1974):

1. *Climate.* Teachers appear to create a warmer socioemotional climate for their special students. This warmth appears to be at least partially communicated by nonverbal cues.
2. *Feedback.* Teachers appear to give their special students more differentiated feedback, both verbal and nonverbal, as to how these students have been performing.
3. *Input.* Teachers appear to teach more material and more difficult material to their special students.
4. *Output.* Teachers appear to give their special students greater opportunities for responding. These opportunities are offered both verbally and nonverbally (e.g., giving a student more time in which to answer a teacher's question).

A recent simplification of the four-factor theory of the mediation of teacher expectation effects has been proposed (Rosenthal, 1989). This simplification, called the affect/effort theory, states that a change in the level of expectations held by a teacher for the intellectual performance of a student is translated into (1) a change in the affect shown by the teacher toward that student and, relatively independently, (2) a change in the degree of effort exerted by the teacher in the teaching of that student. Specifically, the more favorable the change in the level of expectation held by the teacher for a particular student, the more positive the affect shown toward that student and the greater the effort expended on behalf of that student. The increase in positive affect is presumed to be a reflection of increased liking for the student for any of several plausible reasons (Jussim, 1986). The increase in teaching effort is presumed to be a reflection of an increased belief on the part of the teacher that the student is capable of learning so the effort is worth it (Rosenthal & Jacobson, 1968; Swann & Snyder, 1980).

Some of the aspects of affect/effort theory currently under investigation with Nalini Ambady have exciting implications. For example, we have been able to predict student ratings of a college instructor's effectiveness over the course of an entire semester from an examination of a 30-second slice of teaching behavior in which we have access only to the silent videotape or to the tone of voice (not the content) in which the instructors are communicating with their students. These predictive correlations are often in the range of .6 to .7 (Ambady & Rosenthal, 1993).

Similarly, work with Sarah Hechtman has shown the potential for affect/effort theory to help explain the traditional sex differences in cognitive functioning. We found that teachers teaching verbal material to males and quantitative material to females (the so-called sex-inappropriate materials) showed greater hostility to their students in the nonverbal channels (video only) than did teachers teaching the so-called sex-appropriate materials to these same students. These bias effects were smaller for female than for male teachers and they were smaller for more androgynous than for more sex-typed teachers (Hechtman & Rosenthal, 1991).

Some Methodological Consequences: Replication and the Benefits of Meta-analysis

Unfriendly reactions to the research on interpersonal expectancy effects and claims of failures to replicate the effects led me to examine closely and, no doubt, defensively, the concept of replication in behavioral research (Rosenthal, 1966, 1990). These early reactions also led me to an early meta-analytic *Weltanschauung* (Rosenthal, 1963, 1969) in which all of the research studies bearing on a particular research question would be assembled and summarized quantitatively as to both the typical size of the effect investigated and its overall statistical significance (Rosenthal, 1991a).

More Obvious Benefits

Some beneficial consequences of a meta-analytic view of data and of data analysis are fairly obvious. Our summaries are likely to be more complete, more explicit, more quantitative, and more powerful in the sense of decreasing Type II errors. There are also some less obvious benefits.

Less Obvious Benefits

Moderator Variables

Moderator variables are more easily spotted and evaluated in a context of a quantitative research summary. This aids theory development and increases empirical richness.

Cumulation Problems

Meta-analytic procedures address, in part, the chronic complaint that social sciences cumulate so poorly compared to the physical sciences. It should be noted that recent historical and sociological investigations have suggested that the physical sciences may not be all that much better off when it comes to successful replication (Collins, 1985; Hedges, 1987; Pool, 1988). For example, Collins (1985) has described the failures to replicate the construction of TEA lasers despite the availability of detailed instructions for replication. Apparently TEA lasers could be replicated dependably only when the replication instructions were accompanied by a scientist who had actually built a laser.

Enshrinement of Single Studies

One not so obvious benefit that will accrue to the social sciences is the gradual decrease in the overemphasis on the results of a single study. There are good sociological grounds for our monomaniacal preoccupation with the results of a single study. Those grounds have to do with the reward system of science where recognition, promotion, reputation, and the like depend on the results of the single study, also known as the smallest unit of academic currency. The study is "good," "valuable," and above all "publishable" when $p < .05$. Social science disciplines would be further ahead if we adopted a more cumulative view of science in which the impact of a study were evaluated less on the basis of p levels, and more on the basis of its own effect size and on the revised effect size and combined probability that resulted from the addition of the new study to any earlier studies investigating the same or a similar relationship.

Differentiation Drive, Renomination, and Concept Capture

Related to the problem of overemphasis on single studies is the problem of "differentiation drive," a motivational state (and trait) frequently found among scientists. This is the drive to be different, to be more first, to be more right, to be more unique than others. Priority strife is one reflection of the differentiation drive. Another reflection is the occurrence of "renomination," the mechanism by which a well-known process is given a new name in hopes of effecting "concept capture." Concept capture is the mechanism by which ownership of a concept is claimed by virtue of the renaming of the concept. Differentiation drive keeps us from viewing the world meta-analytically, i.e., in a more Bayesian way, by keeping us from seeing the similarity of our work to the work of others.

B. F. Skinner (1983) has spoken eloquently, if indirectly, on this matter:

> In my own thinking, I try to avoid the kind of fraudulent significance which comes with grandiose terms or profound "principles." But some psychologists seem to need to feel that every experiment they do demands a sweeping reorganization of psychology as a whole. It's not worth publishing unless it has some such significance. But research has its own values, and you don't need to cook up spurious reasons why it's important. (p. 39)

The New Intimacy

This new intimacy is between the reviewer and the data. Reviewers cannot do a meta-analysis by reading abstracts and discussion sections. They are forced to look at the numbers and, very often, to compute the correct ones themselves. Meta-analysis requires us to cumulate *data*, not *conclusions*. "Reading" a paper is quite a different matter when we need to compute an effect size and a fairly precise significance level—often from a results section that never heard of effect sizes or precise significance levels (or the APA publication manual).

Dichotomous Significance Testing and Other Statistical Issues

Far more than is good for us, social and behavioral scientists operate under a dichotomous null hypothesis decision procedure in which the evidence is interpreted as anti-null if $p < .05$ and pro-null if $p > .05$. If one's dissertation p is $< .05$, it means joy, a Ph.D., and a tenure-track position at a major university. If one's p is $> .05$ it means ruin, despair, and one's advisor's suddenly thinking of a new control condition that should be run. That attitude really must go. God loves the .06 nearly as much as the .05. Indeed, I have it on good authority that God views the strength of evidence for or against the null as a fairly continuous function of the magnitude of p. As a matter of fact, two .06 results are much stronger evidence against the null than one .05 result; and 10 p's of .10 are stronger evidence against the null than 5 p's of .05. Other more statistical benefits include the overthrow of the diffuse or omnibus test of significance, the increased recognition of contrast analyses (Rosenthal & Rosnow, 1985), and an increased understanding of the meaning of interaction effects (Rosnow & Rosenthal, 1989; Rosenthal & Rosnow, 1991).

Meta-analytic Procedures Are Applicable beyond Meta-Analyses

Many of the techniques of contrast analyses among effect sizes, for example, can be used within a single study (Rosenthal & Rosnow, 1985).

Computing a single effect size from correlated dependent variables and comparing treatment effects on two or more dependent variables serve as illustrations (Rosenthal & Rubin, 1986).

The Decrease in the Splendid Detachment of the Full Professor

Meta-analytic work requires careful reading of research and moderate data analytic skills. We cannot send an undergraduate research assistant to the library with a stack of 5 × 8 cards to bring back "the results." With narrative reviews, that often seems to have been done. With meta-analysis, the reviewer must get involved with the actual data, and that is all to the good.

Conclusion

When first conducted, experiments on the effects of experimenters' expectations on the results of their human subjects' performance were met with incredulity. That inhospitable reception might well have closed off that area of research had it not been for the support of a handful of highly visible scholars, including Campbell, Pepinsky, and Riecken. Subsequent work with animal subjects was similarly received. The "pygmalion" experiments on teachers' expectations as determinants of pupils' intellectual development might, in fact, have remained unpublished had it not been for Campbell.

Let me close with a research episode that, in the immortal words of Yogi Berra, seemed to be "déjà vue all over again." This time, however, the episode did not involve my own research but my evaluation of the research of others. I was asked by a committee of the National Research Council (NRC) to evaluate, on methodological grounds, the research on the effectiveness of several procedures employed to enhance human performance (Swets & Bjork, 1990). One of the research areas investigated was the body of research on the ganzfeld experiments in which subjects typically are asked to guess which of four stimuli had been "transmitted" by an agent or sender, with these guesses made under conditions of sensory restriction.

Of the five areas of human performance technology that Monica Harris and I evaluated in the background paper we were commissioned to prepare for the NRC committee, we found the typical methodological quality of the ganzfeld experiments to be superior to the typical quality of the other four areas we considered (i.e., the Suggestive-Accelerative Learning and Teaching (SALT) method, neurolinguistic programming, mental practice, and biofeedback) and with an average effect size equivalent to the typical effects found in biofeedback research (Harris & Rosenthal, 1988).

Some members of the NRC committee objected to our relatively favorable treatment of the ganzfeld research and we were asked to remove all reference to the ganzfeld research from our background paper. We refused, of course, and, despite that, our background paper is available from the National Academy Press. Once again, perhaps, all's well that ends well. My favorable evaluation of the ganzfeld literature (Rosenthal, 1986) was based on a meta-analytic debate between the late Charles Honorton, a leading investigator in this research area, and Ray Hyman, a leading longtime critic of this area and a member of the NRC committee (Honorton, 1985; Hyman, 1985; Hyman & Honorton, 1986). The scholarly debate between Honorton and Hyman makes good reading for any social psychologist of science.

Based on this controversy, in part, let me express my optimism for the continued promise of a meta-analytic perspective; that it will continue to comfort the afflicted by providing better procedures for doing social science and that it will continue to afflict the comfortable by reminding us that we could be doing social science a good deal better.

Acknowledgments

This chapter is based in part on the Donald T. Campbell Address presented at the annual meeting of the American Psychological Association, New Orleans, LA, August 14, 1989, and on my chapter titled "From Unconscious Experimenter Bias to Teacher Expectancy Effects" in *Teacher Expectancies* (pp. 37–65) edited by J. B. Dusek, 1985, Hillsdale, NJ: Erlbaum. Much of the work reported has been supported over the years by the National Science Foundation and more recently by the Spencer Foundation. Views expressed are solely the responsibility of the author.

References

Ambady, N., & Rosenthal, R. (1993). Half a minute: Predicting teacher evaluations from thin slices of nonverbal behavior and physical attractiveness. *Journal of Personality and Social Psychology, 64*, 431–441.

Brophy, J. E. (1985). Teacher–student interaction. In J. B. Dusek (Ed.), *Teacher expectancies* (pp. 303–328). Hillsdale, NJ: Erlbaum.

Brophy, J. E., & Good, T. L. (1974). *Teacher–student relationships*. New York: Holt, Rinehart, & Winston.

Clark, K. B. (1963). Educational stimulation of racially disadvantaged children. In A. H. Passow (Ed.), *Education in depressed areas* (pp. 142–162). New York: Columbia University Press.

Cohen, J. (1988). *Statistical power analysis for the behavioral sciences* (2nd ed.). Hillsdale, NJ: Erlbaum.

Collins, H. M. (1985). *Changing order: Replication and induction in scientific practice*. Newbury Park, CA: Sage.

Ebbinghaus, H. (1913). *Memory*. (H. A. Ruger & C. E. Bussenius, Trans.). New York: Columbia University Press. (Original work published in 1885)

Freud, S. (1953). *Collected papers* (Vol. 4.). London: Hogarth.

Gorman, M.E. (1986). How the possibility of error affects falsification on a task that models scientific problem solving. *British Journal of Psychology, 77*, 85–96.

Gruenberg, B. C. (1929). *The story of evolution*. Princeton, NJ: Van Nostrand.

Harris, M. J., & Rosenthal, R. (1985). The mediation of interpersonal expectancy effects: 31 meta-analyses. *Psychological Bulletin, 97*, 363–386.

Harris, M. J., & Rosenthal, R. (1988). *Human performance research: An overview*. Background paper commissioned by the National Research Council. Washington, DC: National Academy Press.

Hechtman, S., & Rosenthal, R. (1991). Teacher sex and nonverbal behavior in the teaching of sexually stereotyped materials. *Journal of Applied Social Psychology, 21*, 446–459.

Hedges, L. V. (1987). How hard is hard science, how soft is soft science? *American Psychologist, 42*, 443–455.

Honorton, C. (1985). Meta-analysis of psi ganzfeld research: A response to Hyman. *Journal of Parapsychology, 49*, 51–91.

Hyman, R. (1985). The ganzfeld psi experiment: A critical appraisal. *Journal of Parapsychology, 49*, 3–49.

Hyman, R., & Honorton, C. (1986). A joint communiqué: The psi ganzfeld controversy. *Journal of Parapsychology, 50*, 351–364.

Jussim, L. (1986). Self-fulfilling prophecies: A theoretical and integrative review. *Psychological Review, 93*, 429–445.

Mahoney, M. J. (1989). Participatory epistemology and psychology of science. In B. Gholson, W. R. Shadish, Jr., R. A. Neimeyer, & A. C. Houts (Eds.), *Psychology of science: Contributions to metascience* (pp. 138–164). Cambridge, England: Cambridge University Press.

Mitroff, I. I. (1974). *The subjective side of science*. Amsterdam: Elsevier.

Murray, H. A. (1933). The effect of fear upon estimates of the maliciousness of other personalities. *Journal of Social Psychology, 4*, 310–329.

Pfungst, O. (1965). *Clever Hans*. (C. L. Rahn, Trans.). New York: Holt, Rinehart, & Winston. (Original work published 1911)

Pool, R. (1988). Similar experiments, dissimilar results. *Science, 242*, 192–193.

Rosenthal, R. (1956). An attempt at the experimental induction of the defense mechanism of projection. Unpublished doctoral dissertation, UCLA.

Rosenthal, R. (1963). On the social psychology of the psychological experiment: The experimenter's hypothesis as unintended determinant of experimental results. *American Scientist, 51*, 268–283.

Rosenthal, R. (1966). *Experimenter effects in behavioral research*. New York: Appleton-Century-Crofts.

Rosenthal, R. (1969). Interpersonal expectations. In R. Rosenthal & R. L. Rosnow

(Eds.), *Artifact in behavioral research* (pp. 181–277). New York: Academic Press.

Rosenthal, R. (1971). The silent language of classrooms and laboratories. *Proceedings of the Parapsychological Association, 8,* 95–116.

Rosenthal, R. (1973). The mediation of Pygmalion effects: A four-factor "theory." *Papua New Guinea Journal of Education, 9,* 1–12.

Rosenthal, R. (1974). *On the social psychology of the self-fulfilling prophecy: Further evidence for Pygmalion effects and their mediating mechanisms* (pp. 1–28) New York: MSS Modular Publications, Module 53.

Rosenthal, R. (1976). *Experimenter effects in behavioral research* (enl. ed.). New York: Irvington.

Rosenthal, R. (1985). Nonverbal cues in the mediation of interpersonal expectancy effects. In A. W. Siegman & S. Feldstein (Eds.), *Multi-channel integrations of nonverbal behavior* (pp. 105–128). Hillsdale, NJ: Erlbaum.

Rosenthal, R. (1986). Meta-analytic procedures and the nature of replication: The ganzfeld debate. *Journal of Parapsychology, 50,* 315–336.

Rosenthal, R. (1989, August 14). Experimenter expectancy, covert communication, and meta-analytic methods. Donald T. Campbell Award presentation at the meeting of the American Psychological Association, New Orleans, LA. (ERIC Document Reproduction Service No. TM014556)

Rosenthal, R. (1990). Replication in behavioral research. *Journal of Social Behavior and Personality, 5,* 1–30.

Rosenthal, R. (1991a). *Meta-analytic procedures for social research* (rev. ed.). Newbury Park, CA: Sage.

Rosenthal, R. (1991b). Teacher expectancy effects: A brief update 25 years after the Pygmalion experiment. *Journal of Research in Education, 1,* 3–12.

Rosenthal, R., & Jacobson, L. (1966). Teachers' expectancies: Determinants of pupils' IQ gains. *Psychological Reports, 19,* 115–118.

Rosenthal, R., & Jacobson, L. (1968). *Pygmalion in the classroom.* New York: Holt, Rinehart, & Winston.

Rosenthal, R., & Jacobson, L. (1992). *Pygmalion in the classroom* (enl. ed.). New York: Irvington.

Rosenthal, R., & Rosnow, R. L. (1985). *Contrast analysis: Focused comparisons in the analysis of variance.* New York: Cambridge University Press.

Rosenthal, R., & Rosnow, R. L. (1991). *Essentials of behavioral research: Methods and data analysis* (2nd ed.). New York: McGraw-Hill.

Rosenthal, R., & Rubin, D. B. (1978). Interpersonal expectancy effects: The first 345 studies. *Behavioral and Brain Sciences, 3,* 377–386.

Rosenthal, R., & Rubin, D. B. (1986). Meta–analytic procedures for combining studies with multiple effect sizes. *Psychological Bulletin, 99,* 400–406.

Rosenzweig, S. (1933). The experimental situation as a psychological problem. *Psychological Review, 40,* 337–354.

Rosnow, R. L. (1981). *Paradigms in transition.* New York: Oxford University Press.

Rosnow, R. L., & Rosenthal, R. (1989). Definition and interpretation of interaction effects. *Psychological Bulletin, 105,* 143–146.

Rudwick, M. J. S. (1986). The group construction of scientific knowledge: Gentlemen-specialists and the Devonian controversy. In E. Ullmann-Margalit (Ed.), *The kaleidoscope of science* (pp. 193–217). Dordrecht, The Netherlands: Reidel.

Russell, B. (1927). *Philosophy*. New York: W. W. Norton.

Skinner, B. F. (1983, August). On the value of research. *APA Monitor*, 39.

Swann, W. B., Jr., & Snyder, M. (1980). On translating beliefs into action: Theories of ability and their application in an instructional setting. *Journal of Personality and Social Psychology, 38*, 879–888.

Swets, J. A., & Bjork, R. A. (1990). Enhancing human performance: An evaluation of "new age" techniques considered by the U.S. Army. *Psychological Science, 1*, 85–96.

Tweney, R. D. (1989). A framework for the cognitive psychology of science. In B. Gholson, W. R. Shadish, Jr., R. A. Neimeyer, & A. C. Houts (Eds.), *Psychology of science: Contributions to metascience* (pp. 342–366). Cambridge, England: Cambridge University Press.

CHAPTER 7

Meta-Analysis and Some Science-Compromising Problems of Social Psychology

NORMAN MILLER
VICKI E. POLLOCK

Meta-analysis is a collection of procedures for quantitatively integrating the outcomes of an array of individual studies. In this chapter, we do two things. First, we identify and discuss values, practices, and biases that characterize current social psychology. No hard empirical evidence confirms these as problems, but we believe that many contemporary social psychologists concur that they are issues of concern. Our second and major purpose is to consider how the use of a particular technological advance, methodological "fix," or innovation, namely, *meta-analysis*, might have the ancillary benefit of providing remediation for each of the issues. The possibility that a specific technological improvement might also yield side benefits and bear on issues separate from those it primarily was designed to redress is not unique to meta-analysis (e.g., Salomon, Perkins, & Globerson, 1991).

Shadish (1989) discusses the issue of improving science. He notes three general strategies for improvement. An obvious starting point is to clarify what is troublesome, to specify problems. These problems might include questionable features of the peer-review process for evaluating manuscripts and grant proposals (e.g., Cicchetti, 1991), instances of fraud in science (e.g., Joynson, 1989), or the uneven allocation of resources to scientists (e.g., National Research Council, 1988). Within and across fields scientists will, of course, differ among themselves about the relative importance of entries to such a list of concerns. And those who work in *science studies* may note concerns not commonly mentioned by practicing scientists. Whatever the eventual set of problems, however, a higher-order task is to identify issues and rank them in terms of their importance. Once the issues are identified and ranked, a second step is to generate solutions.

Separate from this approach to remediation, however, is the notion of developing novel options for changing the way in which science is conducted—suggestions for change, per se, rather than change in response to an identified problem. Here, as examples, Shadish cites Mitroff's (1974) advocacy model, which urges proponents of competing views to take the strongest opposing stands possible, and Shadish, Cook, and Houts' (1986) *critical multiplist* model (as well as Campbell's [1986] somewhat similar orientation), which espouses seeking as much critical feedback from as many different sources as possible.

A third approach is the introduction of innovations that change scientific practice. Obviously, some of these will overlap with both of the preceding categories. Nevertheless, this category differs in that its entries are more specific, rather than consisting of general principles applicable in all of science. Shadish (1989) suggests that successful innovations *namely, those that are accepted,* tend to be technological or methodological advances that "address and solve a salient problem with an easy, cheap, and robust algorithm or technology" (p. 401).

This brings us back to the major point of our chapter. The first concern, however, is whether psychology will accept meta-analysis as a research tool. A field often resists technological advance. The reluctance of assessment psychologists to accept the clear advantage of actuarial statistical prediction over clinical prediction (Dawes, 1979) provides a striking illustration. Greenwald, Pratkanis, Leippe, and Baumgardner (1986) list several methodological procedures designed to improve the quality of scientific reporting in psychology. These procedures include using a more stringent alpha criterion than the customary .05 (Selvin & Stuart, 1966; Sterling, 1959), standard reporting of effect sizes (Hays, 1981), avoiding bias against publishing null results by making editorial decisions solely on basis of method sections (Walster & Cleary, 1970), and, finally, using statistical procedures other than the Fisherian model of *rejection of the null hypotheses*—e.g., Bayesian techniques) (Bakan, 1966; Edwards, Lindman, & Savage, 1963; Grant, 1962; Greenwald, 1975). None of these procedures seems to have been widely accepted in psychology.

Such rejection by the field, however, does not seem likely with respect to meta-analysis. Cooper and Lemke (1991) report that the use of meta-analysis in the social sciences is already very substantial, and has increased at a rate of 20 additional meta-analytic reviews per year in each of the past 3 years. Myers (1991) notes there were fewer than 10 entries of the root word *meta-anal* (as in *meta-analysis*) in *Psychological Abstracts* in 1980, but more than 170 entries in 1989, with no evidence of any decrease in the rate of growth.

Problems in Social Psychology

Most psychologists in the research community are unquestioning positivists (Gergen, 1982). They routinely assume that the one feature that allegedly sets science apart from other scholarly work is its cumulative quality. They argue that other forms of knowledge creation, more so than science, will shift in their conclusion from one point in time to the next. Perhaps more often than is true in science, they may revert back to an earlier discarded view, or combine a new approach with elements of a previously rejected one. Even among those who are more sophisticated and who acknowledge the socially constructed aspects of science (Collins, 1983), however, some seem to believe that its changes over time tend to be more unidirectional and in that sense, more "progressive" than other forms of knowledge (Phillips, 1987; Campbell, 1986; Collins, 1989).

Holding this latter view, we list five problems that we believe impede the progressive, cumulative feature of social psychology as a science. Although some of them are relevant to other fields of scientific psychology as well, we see them as having particular salience in contemporary social psychology. These problems are (1) inventing new names for old concepts, (2) rejecting theoretically obvious findings; (3) selectively choosing citations, (4) a bias toward hypothesis and theory confirmation, and (5) journal publication policies. Most of these tendencies act to personalize science and to increase the visibility and distinctiveness of its contributors. We do not recommend that an equally extreme opposing bias be substituted for current practice. Rather, we believe that social psychology would be healthier if current trends and policies in these directions were less extreme.

In this section we discuss each of these issues. In the subsequent section we consider the potential impact that the use of meta-analytic procedures might have on them.

New Names for Old Concepts

It sometimes appears that psychologists are prone to "rediscovering the wheel." Often, a researcher reintroduces a process well studied in the past, but attaches an idiosyncratic label to it so as to give it a more distinctive (and self-referring) quality than it might otherwise have had.

Put more generally, in its practice, scientific psychology is more concerned with differentiation among concepts or principles than with their integration.

To take a specific example, one of the most pervasive social psychological findings is assumed similarity—the tendency to perceive or exaggerate similarity between oneself and others. It applies to one's attitudes, personality traits, interests, and values. The phenomenon is sufficiently reliable that one can count on it as a classroom demonstration in introductory social psychology courses. Francis Bacon (1620/1853) described this prominent bias in human social perception in 1620—projecting one's own world view onto others. One form of it, from the domain of personality trait attribution, is Freud's (1915) defense mechanism, projection—the unconscious attribution of one's own negative traits onto others. Indeed, research on assumed similarity was so common that there were detailed discussions of methodological issues with respect to its quantitative analysis over 3 decades ago (e.g., Cronbach, 1955).

In 1977, Ross, Greene, and House published research on the *false consensus effect*. Although the paradigm they used to study it differed slightly from some of those used previously to study assumed similarity effects, there is little reason to think that the core of the phenomenon they reported differed from that which underlay assumed similarity findings and other reported social projection effects. Yet, if unfamiliar with the field, readers of the chapter in the prestigious Berkowitz series that summarizes this initial false consensus research (Ross, 1977) would be unlikely to become aware of the extensiveness of prior related work and its relation to false consensus. Indeed, some of that earlier research employed an experimental paradigm exactly identical to that of Ross et al. (1977; Travers, 1941; Wallen, 1943). In introducing the presentation of the false consensus research, the entirety of previous literature is mentioned once as follows:

> References to "egocentric attribution" (Heider, 1958; Jones & Nisbett, 1970) to "attributive projection" (Holmes, 1968) and to specific findings and phenomena related to false consensus biases *have appeared sporadically* [italics added] in the social perception and attribution literatures (cf. Katz & Allport, 1931; Kelley & Stahelski, 1970). Perhaps the most compelling evidence, however, is provided in a series of studies by Ross, Greene, and House (1977) . . . . (Ross, 1977, p. 188)

Table 7.1 provides a list of synonyms for the term *false consensus* with references for each. Clearly, the assumption of similarity between self and others has been well researched for over 5 decades.

TABLE 7.1. References for Similarity Projection, Assumed Similarity, False Consensus, and Other Synonyms

Assimilative projection (Benjafield & Adams–Webber, 1975; Lay & Thompson, 1968; Lundy, 1956a; Mintz, 1956)

Assumed similarity[a] (Alfert, 1958; Billig & Tajfel, 1973; Campbell, Miller, Lubetsky, & O'Connell, 1964; Codol, 1975; Cronbach, 1955; Cronbach & Gleser, 1953; Fiedler, 1953; Fields & Schuman, 1976; Gage & Cronbach, 1955; Granberg & Seidel, 1976; Kipnis, 1961; Kitt & Gleicher, 1950; Korte, 1972; Lundy, 1956b, 1958; Lundy, Katkovsky, Cromwell, & Shoemaker, 1955; McArthur, 1972; Miller & Ross, 1975; Rokeach, 1945; Snyder, Stephan, & Rosenfield, 1976; Tajfel, Billig, Bundy, & Flament, 1971; Tajfel, Sheikh, & Gardner, 1964; Tversky & Kahneman, 1973; Weisskopf-Joelson, & Wexner, 1970; Weisskopf-Joelson, Zimmerman, & McDaniel, 1970; Wingate & Hamre, 1967)

Attributive projection (Bellak, 1956; Bender & Hastorf, 1950, 1953; Bennett & Holmes, 1975; Bieri, 1953, 1955; Bieri, Blacharsky, & Reid, 1955; Bramel, 1962; Brehm & Cohen, 1962; Cameron, 1947, 1951; Cameron & Magaret, 1951; Cattell, 1944; Cowden, 1955; Dymond, 1950; Edlow & Kiesler, 1966; Epstein & Baron, 1969; Fabian, 1954; Feshbach & Feshbach, 1963; Feshbach & Singer, 1957; Festinger & Bramel, 1962; Fiedler, 1951, 1953; Ford & Singer, 1968; Freud, 1913/1955; Friedman, 1955; Goldings, 1954; Griffitt, 1973; Halpern, 1955; Halpern & Goldschmitt, 1976; Hastorf & Bender, 1952; Hastorf, Bender, & Weintraub, 1955; Holmes & Houston, 1971; Horney, 1939; Jones & Davis, 1965; Katz & Allport, 1931; Lemann, 1952; Lindzey & Kalnins, 1958; Lundy, 1956a; Markowitz & Ford, 1967; Mintz, 1956; Munn, 1946; Murstein, 1957; Rokeach, 1945; Sears, 1937; Secord, Backman, & Eachus, 1964; Sherwood, 1979; Singer, 1963; Singer & Feshbach, 1962; Smith, 1960; Suchman, 1956; Thomsen, 1941; Wallen, 1941; Weingarten, 1949; Wells & Goldstein, 1964; Wittich, 1956; Wright, 1942; Zemore & Greenough, 1973)

Autistic projection (Brozek, Guetzkow, & Baldwin, 1951; Johnson, 1937a, 1937b; Levine, Chein, & Murphy, 1942; McClelland & Atkinson, 1948; Murphy, 1947; Sanford, 1936, 1937; Sears, 1943)

Classical projection (Ackerman & Jahoda, 1950; Adorno, Frenkel-Brunswik, Levinson, & Sanford, 1950; Bettleheim & Janowitz, 1950; Epstein & Baron, 1969; Freud, 1913/1955; Healy, Bronner, Augusta, & Bowers, 1930; Hoffman, 1935; Holmes, 1974; Jelgersma, 1926; Kaufman, 1934; Knight, 1940; Lecky, 1951; Lorber, 1973; Lundy & Berkowitz, 1957; Markowitz & Ford, 1967; Murstein, 1956; Norman & Ainsworth, 1954; Norman & Leiding, 1956; Novick & Hurry, 1969; Schafer, 1954; Symonds, 1949; Warren, 1934; Zimmer, 1955; Zucker, 1952)

Comparison projection (Bowerman, 1975)

Defensive projection (Edlow & Kiesler, 1966; Heilbrun, 1972; Hochreich, 1975; Holmes & Houston, 1971; Kreines & Bogart, 1974; Mondy, 1967)

False consensus effect (Jones & Nisbett, 1971)

Projection (Bellak, 1950, 1954, 1956; Goss & Brownell, 1957; Holmes, 1968; Murray, 1938, 1951; Rapaport, 1942, 1952; Rapaport, Schafer, & Gill, 1945; Schachtel, 1950; Smith, 1974; Van Lennep, 1957)

(*cont.*)

TABLE 7.1 (*cont.*)

Rationalized projection (Allport, 1939; Baldwin, 1955; Bellack, 1944; Fenichel, 1945; Freud, 1913/1955; Murstein, 1956, 1957; Piaget, 1926; Posner, 1940; *The Psychiatric Dictionary*, 1940)

Reference projection (Bowerman, 1975)

Similarity projection (Feshbach & Feshbach, 1963; Halpern & Goldschmitt, 1976; Hornberger, 1960; Sears, 1936)

Social comparison (Allport, 1924; Festinger, 1950; Ford & Singer, 1968; Weisskopf–Joelson, Zimmerman, & McDaniel, 1970)

Social judgment (Travers, 1941; Wallen, 1943)

Social projection (Allport, 1924; Ford & Singer, 1968)

Supplementary projection (Peabody, 1970)

Thematic projection (Holmes, 1974)

Note. We only list publications that temporally preceded Ross, et al. (1977).
[a]From a logical standpoint, studies of assumed dissimilarity might also have been included in this table. Also omitted from this table are content similarity and elevation similarity (Cronbach, 1955) and studies that use direct measures of similarity (cf. Marks & Miller, 1987).

Erdelyi (1990) provides an example from clinical and cognitive psychology. He suggests that repression is analogous or equivalent to thought inhibition. He cites Harry Stack Sullivan (1956), who spoke of "selective inattention" when discussing thought avoidance; Dollard and Miller (1950), who used the term *thought stopping*; and Eriksen and Pierce (1968), who discussed "cognitive avoidance." Erdelyi (1990) emphasizes that theorists who have examined aspects of this phenomenon emerged with different verbal labels for it, although all were probably discussing the same basic concept.

In an example from research on psychopathology and creativity, Prentky (1980) discusses the dichotomization of cognitive styles into two opposing types—the focused and narrow minded versus the general and broad minded. He starts with Pascal's (1670) depiction in *Pensees* and also includes Pavlov (1955) and Eysenck (1957) in his discussion. Among other authors described as using a similar dichotomy are Löwenfeld (1939), Holton (1978), Chapman (1961), Payne (1961), and Claridge (1967). Bogen (1975) lists 39 scholars who have postulated this same dichotomy (see Table 7.2).

Contemporary social psychology is rife with other examples. Since Festinger's (1957) original statement of cognitive dissonance theory, researchers became concerned with delineating the conditions necessary

TABLE 7.2. Authors and their Synonymous Dichotomous Terms for Two Postulated "Ways of Knowing," "Types of Intelligence," or "Cognitive Styles"

Akhilinanda	buddhi	manas
Assagioli	intellect	intuition
Austin	convergent	divergent
Bateson & Jackson	digital	analogic
Blackburn	intellectual	sensuous
Bronowski	deductive	imaginative
Bruner	rational	metaphoric
Cohen	analytic	relational
De Bono	vertical	horizontal
Deikman	active	receptive
Dieudonne	discrete	continuous
Freud	secondary	primary
Goldstein	abstract	concrete
Hilgard	realistic	impulsive
Hobbes (per Murphy)	directed	free
Humphrey & Zangwill	propositional	imaginative
W. James	differential	existential
A. Jensen	transformational	associative
Kagan & Moss	analytic	relational
D. Lee	lineal	nonlineal
Levi-Strauss	positive	mythic
Levy & Sperry	analytic	gestalt
Lomax & Berkowitz	differentiation	integration
Maslow	rational	intuitive
McFie, Piercy (from Spearman)	relations	correlates
McKellar	realistic	autistic
Neisser	sequential	multiple
Oppenheimer	historical	timeless
Ornstein	analytic	holistic
Pavlov	second signaling	first signaling
C. S. Pierce	explicative	ampliative
Polanyi	explicit	tacit
Price	reductionist	compositionist
Radhakrishnan (per H. Smith)	rational	integral
Reusch	discursive	eidetic
Schopenhauer	objective	subjective
Sechenov (per Luria)	successive	simultaneous
C. S. Smith	atomistic	gross
Wells	hierarchical	heterarchical

Note. From Bogen (1975). Copyright 1975 by UCLA Educator. Reprinted by permission. For the sources of most of the entries, Bogen refers the reader to Bogen (1969), Bogen, DeZure, TenHouten, & Marsh (1972), and Cameron (1970).

to produce dissonance. Over the next several decades, qualifications were elaborated, leading to the conclusion that a necessary ingredient is choice, which in turn has implications for self-esteem (Aronson, 1969; Greenwald & Ronis, 1978; Goethals & Cooper, 1972; Cooper & Fazio, 1984).

In 1988, Steele published his theory on the psychology of self-affirmation, which argues that self-affirmation processes underlie dissonance-produced attitude change. There appears to be substantial overlap between the conceptual content of *self-affirmation* and that which has evolved for *dissonance.* It is now believed that dissonance is aroused after inconsistent behavior because of the meaning of the inconsistency—not because inconsistency between or among cognitions is intrinsically aversive. Thus, dissonance is aroused when one has brought about an aversive event for which one feels responsible (Scher & Cooper, 1989). Scher and Cooper describe the internal state produced by these circumstances as an impaired sense of self-efficacy. This view of the conditions necessary for producing dissonance, however, corresponds directly to those views that Steele describes as inducing self-affirmation processes. Although theorists search for nuances that distinguish one model from the next, the commonality between them often remains more striking than the distinctions. Whether discriminative construct validity for *dissonance, self-affirmation, responsibility for a negative event, guilt, impaired self-efficacy,* and other related terms will eventually emerge to justify distinct theories for each is still unclear.

Sometimes, instead of rediscovering the wheel, psychologists temporally codiscover it, and each simultaneously gives it its own name. This seems to be true for two contemporary theories of attitude change: the elaboration likelihood model (Petty & Cacioppo, 1986) and the heuristic model (Chaiken, 1987). Although the names of the models emphasize opposite sides of the dichotomy around which each theory frames its postulates, both models rest on the observation that sometimes persuasive messages are processed carefully whereas other times people rely on peripheral cues, unrelated to message content, in deciding whether to accept a persuasive appeal. The proponents of both models note in common the research trends that were steering the field of persuasion in the directions they jointly advocate, citing similar literatures in support of this dichotomy. Again, of course, despite much in common, nuances of difference between the two models can be noted (Chaiken & Stangor, 1987).

Rosenthal (Chapter 6, this volume) labels this phenomenon *concept capture*—a mechanism by which ownership of a concept or phenomenon is claimed by virtue of renaming it. Instances of concept capture are so numerous that it is clear that the examples cited above exemplify a general phenomenon that is not limited to social psychology. Even within established and accepted domains of research, greater rewards of research support, career advancement, and public recognition tend to fall to the researcher who investigates and confirms a new phenomenon than to those who pursue a more familiar, well-researched idea (see discussions by Armstrong, 1982; Fishman & Neigher, 1982; Greenwald, 1975). Nevertheless, the prevalence of concept capture undermines the cumulative approach to knowledge integration and constitutes one way in which theory impedes scientific progress.

Nonobvious Findings

Within social psychology, reviewers and editors alike frequently reject articles because the results are "obvious." This problem, the obviousness of social science results, dates back well into its history. In discussing its bearing on educational research, Gage (1991) compared two reviews of the landmark volume *The American Soldier*, each published about 4 decades ago. One reviewer, the well-known historian Arthur Schlessinger Jr. (1949), commented: "One can find little in the 1200 pages of text and the innumerable surveys which is not described more vividly and compactly and with far greater psychological insight in a small book entitled *Up Front* by Bill Maulden." Driving the nail home, he added, "What Maulden may have missed will turn up in the pages of Ernie Pyle" (p. 854). His point was that any astute observer, such as these well-recognized and highly popular news journalists, can discern obvious truths about human behavior. According to this perspective, there is no special need for the ponderous tools of social science.

Gage (1991) also describes the contemporaneous review by sociologist Paul Lazarsfeld (1949), who, like Schlessinger, was concerned with the issue of obviousness. Lazarsfeld wrote:

> It is hard to find a form of human behavior that has not already been observed somewhere. Consequently, if a study reports a prevailing regularity, many readers respond to it by thinking, "Of course, that's the way things are." Thus . . . the argument is advanced that surveys only put into complicated form observations which are already obvious to everyone. (pp. 379–380)

Then, as prefatory to a short list of research outcomes, Lazarsfeld stated:

> The reader may be helped in recognizing this attitude if he looks over a few statements which are typical of many survey findings and carefully observes his own reaction. A short list of these, *with brief interpretive comments* [italics

> added], will be given here in order to bring into sharper focus probable reactions of many readers.

Lazarsfeld's list of some survey results concerning World War II soldiers follows:

> 1. Better educated men showed more psycho-neurotic symptoms than those with less education. (The mental instability of the intellectual as compared to the more impassive psychology of the man-in-the-street has often been commented on.)
> 2. Men from rural backgrounds were usually in better spirits during their Army life than soldiers from city backgrounds. (After all, they are more accustomed to hardships.)
> 3. Southern soldiers were better able to stand the climate in the hot South Sea Islands than Northern soldiers. (Of course. Southerners are more accustomed to hot weather.)
> 4. White privates were more eager to become non-coms than Negroes. (Because of their having been deprived of opportunity for so many years, the lack of ambition among Negroes was quite understandable.)
> 5. Southern Negroes preferred Southern to Northern white officers (because Southerners were much more experienced in having interpersonal interactions with Negroes than Northern officers were).
> 6. As long as the fighting continued, men were more eager to be returned to the States than they were after the Germans surrendered (because during the fighting, soldiers were in danger of getting killed, but after the surrender there was no such danger). (pp. 379–380)

Then, after facetiously asking why so much money is needed to establish findings if they are indeed obvious, Lazarsfeld revealed that each of those he had earlier reported in his review of the *American Soldier* was directly opposite to what actually had been found and reported in it. Persuasive as Lazarsfeld's rhetorical ploy might have been, however, its success depends on a reader's willingness to have accepted his false version of the *American Soldier* research results as valid. Perhaps, even in the 1950s, many readers did not believe most or any of the false statements.

Gage reviewed subsequent research that explored these issues more systematically. Baratz (1983) presented statements that were either a true research finding or an opposite false finding, with or without an explanation of it. Each of 85 subjects evaluated 16 statements: Thus, 4 statements reported true findings with an explanation; 4 contained an opposite, false finding plus explanation; 4 were true statements without explanation; and 4 contained an opposite, false finding without an explanation. Subjects indicated how predictable (or obvious) they found each result. No matter which version of an item they received (A, true or B, false, either with or without an explanation), the majority of subjects thought they would have predicted most of the statements (approximately 80% in form A, and 66% in form B). Thus, contradictory pairs of findings, when rated independently, were each judged as obvious, irrespective of their veracity.

In a similar study by Wong (1987), over 1,200 respondents evaluated true and false versions of 12 findings previously cited in the third edition of the *Handbook of Research on Teaching* (Wittrock, 1986). On one form, subjects judged which of two reported results were correct. For example, given the stem, "When first-grade teachers work on reading with a small group of children, some call on the children in a fixed order whereas others call on children in a random order," respondents could indicate that reading achievement is higher "when children are called on in a fixed order" or "when children are called on in a random order." After choosing, subjects rated their choice on a scale anchored with end points labeled "extremely obvious" and "extremely unobvious." Other subjects received a form that contained one of the two versions of the items on the first form and only rated their obviousness. There were two counterbalanced versions of this form, each with half of the items phrased in the true and half in the false direction.

On 4 of the 12 items, subjects chose the research-supported finding more often, but on the other 8 they chose the false finding more frequently. Among four groups of respondents: undergraduates in engineering, undergraduates in psychology, teacher trainees, and experienced teachers, the latter were no more accurate than were other groups. Moreover, the more persons choosing the finding, the higher was its mean rating of obviousness ($r = .66$).

Wong (1987) concluded:

> Judging by the smaller proportions of students choosing the actual finding as the real findings, and the mean rating of obviousness on the presented (both actual and opposite) finding statements, we can say reasonably that people cannot distinguish true findings from their opposites. (p. 86)

He added that "[the results] clearly confirm the idea that knowledge of outcome increases the feeling of obviousness. Thus, when people claim to have known it all along when an event is reported to them, their claim is often not warranted" (p. 88). This effect, termed *hindsight bias,* is now confirmed by an extensive literature (see Hawkins & Hastie, 1990, for a review).

The criticism that psychology or social science is nothing more than common sense continues to receive contemporary support from social scientists themselves. Specifically, Lamal (1991) summarized 15 conclusions from Lock and Latham's theory of work motivation and job satisfaction and emphasized that all of them are nothing more than common sense. In a companion article, Lock and Latham (1991) argued that psychologists are subject to a form of bias in which a principle or finding that has been developed from arduous scientific research will, after the fact, routinely be judged obvious—a conclusion empirically confirmed by the substantial literature on *hindsight bias* (Hawkins & Hastings, 1990).

Billig (1990), a social psychologist, argues that the goal of social psychology is to show that ordinary common sense is unsatisfactory, and that social psychologists must labor hard to correct its imperfections. Better still, they must replace it with a new form of knowledge.[1] With this self-imposed requirement that its content exceed or replace common sense, it is understandable that submitted social psychological experiments and research proposals are approached by reviewers and editors alike with high vigilance, so as to guard against the rediscovery of mere common sense. As argued above, however, when viewed after the fact, any finding appears obvious.

Selective Citation

To obtain information on what features of review articles they deemed most important, Becker (1991) interviewed editors of the *Psychological Bulletin* and the *Review of Educational Research* who had served since 1980. Each editor was asked to indicate the three most important manuscript characteristics and the single most influential feature of a review. Most prominent among the categories named was *a scholarly or systematic approach to the review*. The editors also spontaneously mentioned breadth of coverage of the literature. With respect to primary-level research, Becker (1991) cites data (Frantz, 1968; Wolff, 1970) that show that *review of the literature* received ranks of 7.2 and 8.1 among respective 14- and 15-item sets that listed criteria important for judging its quality. Thus, although important in primary-level research too, this dimension apparently receives greater weight for review articles. Clearly, however, experts think that the nature and thoroughness of the cited literature bear on the quality of an article.

It should not be surprising that the inventor of a "new" concept does not cite substantial literature confirming its prior study under another label, or that an author of theoretically oriented research selectively cites

[1] Billig goes on to argue, however, that as social psychology progresses, it develops sets of mutually contradictory relationships that in form are no different from the contradictory aphorisms that form the core of commonsense psychology (e.g., shift to risk, shift to conservatism and false consensus, false uniqueness). Structurally, these phenomena parallel commonsense maxims such as: "*absence makes the heart grow fonder*"; "*out of sight, out of mind*"; "*many hands make light work*"; and "*too many cooks spoil the broth.*"

prior work that bears positively on the theorizing. Perhaps more surprising, however, is that chapters meant to be nonbiased reviews of the research in a given area may exhibit markedly different citation patterns.

In the most recent edition of the *Handbook of Social Psychology*, two chapters exemplify this point. In his chapter on attitudes, McGuire (1985) states that its main purpose is to provide understanding of the effects of the mass media on opinions. A parallel chapter by Roberts and Maccoby (1985) is titled "Mass Communication." Thus, these lengthy chapters by distinguished experts ostensibly review research on the same substantive issue. Yet, they exhibit almost no overlap in their cited references. Inspection shows that 9% of the references cited by Roberts and Maccoby are common to both reviews, whereas only 2.8% of those cited by McGuire are common to both. Experts from different disciplines who work on the same problem do not cite each other's literatures and, apparently, do not know them.

Another bias in citation is the tendency to ignore older references. Marketing concerns account for its occurrence in introductory textbooks. It makes financial sense to view existing scientific principles as out of date every 2 or 3 years and to include among cited work the recent publications of younger members of the field. They are the ones who tend to be saddled with teaching the introductory courses. But this cannot account for the similar practice seen in our empirical journals. Perhaps it is desirable to cite recent work in practical or applied areas where the latest up-to-date data are, indeed, more correct for current purposes. For basic theoretical work, however, this latter argument does not apply. And although more recent work may conform better to the latest statistical or methodological prescripts, as Stinchcombe (1984) emphasized in a section titled "Why are generalizations forgotten?" (pp. 55–56), despite methodological shifts in fashion—from correlation to path analysis to structural equation modeling to loglinear analysis—all techniques rely on the same basic underlying data and procedures, namely, the reporting and recording of correlational relationships.

An analysis of issues of *Journal of Personality and Social Psychology* (JPSP) and *Psychological Review* (PR) confirms this tendency to cite most recent work. Within each of three randomly selected issues of each journal for the years 1965 and 1990, we examined the reference sections of three randomly selected articles, with the constraint that each of the three sections of JPSP be sampled in the 1990 issues. The mean percentage of citations for which the publication date fell within five years of the article's submission date was 61% (JPSP) and 53% (PR) in 1965 and 40% for each journal in 1990. In contrast, the percentage of citations that predated the submission date by 20 or more years was 8% for both journals in 1965 and in 1990 was 12% for JPSP and 17% for PR. The main effect for recent versus old citations is highly significant: $F_{(1,56)} = 215.03, p < .001$.

Although these figures suggest a decline in the strength of this bias in more recent years (as confirmed by the interaction of year of publication with recent vs. old citations, $F_{(1,56)} = 19.7, p < .001$) their general implication is that research from the past tends to slip out of awareness, having had limited impact for a relatively short span of time. Perhaps a more important concern, however, is the degree to which all relevant references are cited, and if they are not, whether omissions are evenly balanced with respect to their theoretical positions and empirical outcomes.

In sum, the comprehensiveness and appropriateness of citations contribute to the judged quality of primary empirical research, as well as to literature reviews. The explosion of social science research in this century may, to some degree, account for citation differences among those pursuing the same question, yet when the content of the topic areas addressed should show substantial overlap, as in the previously described comprehensive mass communication reviews, it is difficult to understand the lack of similarity in their reference citations.

Theory Confirmation Bias

Social psychologists Greenwald and colleagues (Greenwald et al., 1986; Greenwald & Pratkanis, 1988) and Nisbett and Ross (1980) are among the many who are concerned with confirmation bias in science. The premise is that the scientist is not immune to basic biases that characterize all humans. Confirmation bias is merely one of these. It includes an array of well-documented psychological processes: primacy effects in impression formation and persuasion, positive and negative transfer effects, expectancy effects, belief perseverance in the face of discrediting information, selective retrieval of information confirming various beliefs, and other, related effects such as self-fulfilling prophecies. Thus, the argument is that scientists who hold theoretical predilections must naturally exhibit these biases which, in turn, interfere with optimal progress in science.

"Researchers' dispositions to confirm hypotheses support their use of methods that are demonstrably prone to misinterpretation and, because of that, obstruct scientific progress" (Greenwald et al., 1986, p. 222). To take one example, despite the well-established evidence for one source of confirmation bias, experimenter bias effects (Rosenthal, Chapter 6, this volume), there is little systematic use of procedures designed to ensure that experimenters lack knowledge of the specific hypotheses under investigation, or that precautions are taken to remove expectancy effects by the use of technologically sophisticated instrumentation that precludes their intrusion. Inspection shows that less than 1 out of 10 articles in JPSP explicitly mentions use of such precautionary procedures.

Greenwald and his colleagues (Greenwald et al., 1986; Greenwald & Pratkanis, 1988) propose an alternative, which they label result-centered research methods. In this approach, the researcher takes a neutral attitude toward the phenomenon being studied—seeking the circumstances under which it is and is not found, rather than attempting to confirm a particular theory. McGuire's (1989) arguments for *perspectivism* call for a similar research strategy. Although good sense argues in favor of this approach, there has been little evidence to suggest its acceptance in social psychology.

Journal Publication Policies

A dysfunctional feature of social psychology is the tendency to publish only that research that confirms theory. Editors of our best journals in social psychology, *Journal of Personality and Social Psychology, Journal of Experimental Social Psychology, Personality and Social Psychology Bulletin, Social Psychology Quarterly,* and probably also the better European journals, *European Journal of Social Psychology* and *British Journal of Social Psychology,* believe that journal space should not be spent on disconfirmations of prior work. Thus, publication policy works hand in hand with the confirmation biases that operate at the level of the individual researcher.

When coupled with the advice given in the most visible formal set of prescriptions for how to write a social psychological experiment for publication (Bem, 1987), this bias becomes more complicated. Bem's chapter undoubtedly represents pervasive thinking within the field. It advocates that if an initial theoretical premise receives no empirical support, one should concoct a plausible theoretical story and present it in the introduction as theory that guided the research. Thus, the beginner is advised against reporting the initial formulation, along with the disconfirming facts that emerged. Nor is a new data-compatible theoretical account to be offered only in the discussion section, where it can be correctly identified as a post hoc explanation. (Editors and reviewers routinely reject experiments with post hoc explanations.) In sum, when a researcher cannot produce theory confirmation, post hoc theory invention is substituted and paraded in the introduction of the article as its driving force. Unfortunately, this prevents science from knowing that the formulation is post hoc and one that may well be paradigm specific, rather than being a more general formulation that received support within the particular experimental paradigm.

The examination of publication policy bias in greater detail will show how it operates in combination with these prescriptions for writing social psychology articles to mislead a field. For this purpose, we first distinguish among several different types of replication of an experimental finding. An *exact replication* repeats a published experiment in as faithful a manner as possible. Its isomorphism with the original study is compromised only by the degree to which the method section of the original is incomplete or ambiguous. Thus, exact replications of what is really a post hoc theoretical account of an obtained relationship between independent variable p and dependent variable q will always contain the paradigm-specific features of the original experiment, and thus, assuming a reliable effect, appropriate implementation, and statistical power, will confirm it, implementation, along with its theoretical account. Fortunately, journal policy strongly precludes publication of exact replications thereby preventing these paradigm-specific outcomes from contributing to the apparent support of a more general relationship. Although they do not raise the specific argument we make above, editors feel that the contribution to knowledge made by exact replications is minimal.

In a *conceptual replication,* the researcher uses new operationalizations of the theoretical variables thought to underlie the independent or dependent variable of the original study. Thus, the experimental procedures and/or measuring instruments clearly deviate from those of the original study. Additionally, the population sampled may differ from that of the original sample. Conceptual replications are somewhat more likely to be published than are exact replications, but in keeping with the general publication principle, only if they are confirmatory. Consequently, if the finding of the original experiment is paradigm specific, only those conceptual replications that contain the necessary paradigm-specific features will be publishable.

A *mediational replication* examines a variable that accounts for or explains why a previous relation was found. Thus, it postulates a causal chain in which a third variable is causally linked to both the independent and the dependent variables of a previously published study. The mediating variable simultaneously occupies the status of a dependent and an independent variable. As a dependent variable, it must be shown to be a consequence of the independent variable of the original study. As an independent variable, it must be shown to causally elicit the effect initially observed on the dependent measure of the original study. As is true of other research, a mediational study is only likely to be published if its outcome is positive.

With respect to the problem of perpetuating the false belief that a paradigm-specific finding is a general effect, this approach can be instructive if it attempts to test the mediating role of a key paradigm-specific feature that is linked to the nature of the operationalization of the independent variable in the original study and is also responsible for the obtained effect in the original study. In such cases, the key paradigm-specific factor will have been identified and confirmed. Most studies aimed at testing mediators, however, are far more likely to examine the role of some other variable, and consequently, they are unlikely to confirm a mediational role for this other variable (or to confirm the original finding) unless the key paradigm-specific feature that was necessary for the original result is also present in the design of the mediational study. Thus, most published mediational studies will be constrained unknowingly to the paradigm-specific features necessary to the original outcome.

Like a mediator study, a moderator study replicates a previously published experiment, either exactly or conceptually, but attempts to show that it is replicable only under specific conditions, or that the strength or direction of the original relationship is altered by the level of a third variable. This third variable is not causally linked to both the independent and the dependent variables, as is the case with a mediational study. Instead, it is linked only to the dependent variable. Types of subject (e.g., male or female), task (e.g., simple or complex), or culture in which the study is conducted are examples of variables that may function as moderators of any particular experimental finding. As was true for mediational studies, a researcher may experimentally examine the effect a key moderator variable that was responsible for the effect found in the original study. Again, in such a case the key paradigm-specific variable will be revealed. As is true for mediator studies, however, researchers are more likely to examine moderators other than a specific one necessary for the effect that was obtained in the original study. And given the bias toward publishing only positive results, again, as was true for mediator studies, only those studies that contain the paradigm-specific features necessary for the outcome of the original (as well as a relevant moderator variable) will be publishable.

The point of the preceding discussion is that most publishable replications of the original finding are likely to possess the identical paradigm-specific characteristics that were necessary to elicit the outcome of the original experiment. Under what conditions, then, other than the relatively rare exceptions mentioned above, might research that disconfirms the original results be publishable? The answer is, only when an alternative theoretical account of what affects dependent variable *q* happens to include *p* among its independent variables. When manipulated in a different theoretical context that does not contain the paradigm-specific features of the original study, the failure of variable *p* to elicit or covary with q may not constitute nonreplication or failure to obtain an effect. Instead it may show a confirmation of the new Theory K under some conditions (either high or low levels of p) but not under others. Of course, Theory K's findings and explanations may be as post hoc and paradigm specific as those of original Theory J, but its presence allows publication of results that are disconfirming of the results for Theory J. The presence of post hoc theories L, M, N, etc., each with their kernels of truth, eventually provides numerous tests of the relation between p and q, some confirming and others disconfirming. Unfortunately, the discovery of the limited range of conditions within which the p–q relation is constrained is slowed by the combination of (1) journal policy that thwarts publication of disconfirming conceptual replications, and (2) the standard misleading reporting style urged for social psychology articles, which discourages presentation of post hoc explanation as such and, instead, urges that it be presented as the theory that guided the conceptualization of the research.

Application of Meta-analysis as an Innovation

In this section we discuss the potential impact of meta-analysis on each of the problems identified.

New Names for Old Concepts

One of the dysfunctional by-products of concept capture is that it impedes the development of broad, integrative principles, such as, for example, that concerned with the conglomerate of related research confirming the interaction frequency/group identity/conformity/normative influence relationships:

> The longer, more intensely, and more exclusively persons interact with each other, the more they will identify with one another as a group, and the more pressure they will exert and feel for conforming to local patterns of behavior and belief, provided that they are not unequals in power or competitors for scarce resources. (Collins, 1989, p. 125)

The research supporting this principle includes Homans' studies in the 1950s of informal groups in industrial settings, Durkheim's (1912/1954) analysis of religious rituals and their role in producing conformity, Goffman's (1981) extension of this work to social conversations, the Asch (1951) experiments on conformity, the work on expectation states theory (Berger and his colleagues 1983), and Whyte's (1943) study of street gangs.

We cannot claim that meta-analysis will change the human spirit and eradicate needs for self-aggrandizement, perpetuation of one's contribution to the science by being known for a phenomenon, or whatever aspects of egocentrism are represented by concept capture. Meta-analysis can, however, perform a useful service for science by distinguishing instances of apparent concept capture from instances in which the new name for a concept reflects something empirically distinctive. To a degree better than is possible in a single experiment, meta-analysis can provide an assessment of the discriminant construct validity of the relevant concepts.

The meta-analytic approach to this issue is to assess the relation of other theoretically relevant variables to the concept in question and then to compare each of these relations to that obtained between each of its alleged synonyms and these other relevant variables. If, for example, each of 10 variables exhibits the same functional relationship with concept 1a and 1b, there is little reason to view 1a and 1b as distinct. On the other hand, if some of the 10 disordinally interact with (show different directions of effect on, or as a function of) 1a and 1b, that is evidence for viewing them as distinct constructs (e.g., as 1 and 2, not as 1a and 1b).

This strategy was used in conjunction with meta-analytic procedures to assess the functional relation between indices of verbal and written aggression with each of four relevant independent variables (Carlson, Marcus-Newhall, & Miller, 1989). This was of value because it had been suggested that these two measures of aggression reflect distinct concepts rather than an underlying unitary one (e.g., Tedeschi, Smith, & Brown, 1974; Kane, Joseph, & Tedeschi, 1976; Tedeschi, 1983). The results showed that, apart from scaling differences, measures of verbal and written aggression were affected similarly by anger, frustration, personal attack, and directness of aggression, implying that each of the two types of measures of aggression assesses the same underlying construct.

A limitation of this approach arises when there are insufficient numbers of studies available to achieve adequate power for tests of discriminant construct validity. Nonetheless, judgment procedures (as described in the subsequent section titled Journal Publication Policy) can be applied to meta-analysis to generate information on new variables whose functional relation to concepts 1a and 1b could then be examined. The net result is that application of meta-analytic procedures could serve as Occam's razor, slashing out unneeded terms from the field.

Nonobvious Findings

We argued previously that when viewed after the fact, many findings are likely to be judged by experts, as well as by ordinary persons, as obvious.

On the other hand, it is more difficult to assert that each of two contradictory findings is obvious. From this standpoint, meta-analysis provides an advantage that may not exist in ordinary experimental research. By considering simultaneously the entire literature on a topic, meta-analysis can examine the effects of many moderating and mediating variables. Thus, it provides a vehicle for determining the circumstances under which a hypothesis that is generally true is false.

This was shown, for instance, in Carlson and Miller's (1987) meta-analysis of the literature on negative mood and helping behavior. Helpfulness is usually increased by a negative as compared to a neutral mood. Under circumstances that Carlson and Miller labeled victimization, however, the helpfulness of those in a negative mood is significantly *reduced* compared to those in a neutral mood. Of course, even in examples such as this, which isolate opposing relationships, a cynic can view both findings, although opposite, as obvious.

Another countering force to the attribution of obviousness emphasizes the extent to which one can study that which has not been studied before. Just as this is possible in individual experiments, meta-analysis, too, provides an opportunity to test new relationships—ones that were not considered by those who conducted the primary-level research. In a meta-analysis of alcohol effects, for example, Steele and Southwick (1985) defined a new theoretical variable, inhibition conflict, and explained alcohol effects on an array of 12 different social behaviors. Surely, it is difficult to classify this conceptualization as lacking novelty. Similarly, if a meta-analysis produces a theoretical integration of several prior models, rejecting and retaining parts of each, the result cannot easily be labeled obvious. Of course, meta-analysis per se does not produce these outcomes. They depend on the ingenuity of the individual researcher.

In sum, meta-analysis cannot eliminate the preferential bias reviewers and editors possess for novel findings. However, because it allows examination of very large data sets in which there is variation on many dimensions, the capacity of meta-analysis to provide novel findings and to allow tests of broad theoretical integrations should enable its users to challenge attributions of obviousness.

Selective Citation

Will the use of meta-analysis reduce incompleteness and systematic selectivity in the citation of references? Beaman (1991) examined this issue by comparing narrative and meta-analytic reviews in *Psychological Bulletin* for two, periods: 1987–1988, and 1981–1983. The published reviews were evaluated in terms of 28 items taken from Jackson's (1978, 1980) list of criteria important to review articles.

Earlier, Jackson had reported that only one of 37 traditional narrative reviews specified the citation retrieval procedures that had been used. In Beaman's data set, the situation was basically the same: No article in his sample of traditional reviews specified a retrieval system, and only one gave a beginning and ending date for articles covered by the review. In contrast, 70% of the meta-analyses specified retrieval procedures and 60% reported beginning and ending dates. Of the traditional reviews, 95% did not discuss the full set of studies and gave no information about criteria for excluding studies. Over half of the traditional reviews cited only those studies that *supported* some generalization of concern. In contrast, of the 20 meta-analyses, 18 discussed all the cited studies and gave exclusion criteria. Meta-analytic reviews are also likely to include more studies than are traditional reviews, although that specific information—number of references cited—was not reported by Beaman.

These differences that distinguish the two modes of reviewing seem to reflect features inherent in their respective methods. The more objective methods that characterize the meta-analytic approach normatively require specification of retrieval procedures and dates that define the material covered. This results in less selective citation of sources in meta-analytic reviews than in traditional reviews.

Theory-Confirmation Bias

There is no reason to believe that the users of meta-analysis are immune to the substantial evidence for confirmation and other bias. Often, however, the possibility of determining whether confirmation bias has occurred in a given instance is greater for a meta-analysis than for other kinds of studies. The reason for this is that the raw data for a meta-analysis sometimes are published within the article, whereas this is rarely the case for primary-level research. In such cases, the outcomes of comparison tests within the meta-analysis can be verified by those who wish to do so. Although there is debate about the desirability of publishing the effect sizes for the individual studies or comparisons that comprise a meta-analytic review, their inclusion serves as a deterrent for confirmation bias.

To the degree that a meta-analysis can include tests in which an equal number of studies from opposing theoretical camps are allowed to contribute data, confirmation bias can be balanced. This, of course, assumes that confirmation bias intrudes equally into the published work of each camp. Meta-analysis also can assess whether unique effects are obtained within a given laboratory, as was shown for an effect in a meta-analysis of the persuasion literature (Johnson & Eagly, 1989).

Nevertheless, as is the case with primary-level studies where researchers might pick and choose among various data analyses to report those that provide outcomes supportive of one's theory, such biased reporting of outcomes is possible in meta-analysis too.

The National Institute of Education (NIE) commissioned a study that provides an interesting example of the effect of confirmation bias in meta-analysis (NIE, 1984). Six experts in desegregation research were invited to conduct a meta-analysis to examine the effects of desegregation on black achievement. The study was designed so that the effect of ideology could be assessed qualitatively. Two of the researchers were selected because it was believed they favored school desegregation as a social policy, two were selected because their attitudes were believed to be neutral, and two were selected because it was believed that they opposed desegregation. Before undertaking the meta-analysis, the researchers developed a single set of inclusion criteria for selecting a common set of studies and all agreed to perform a meta-analysis on the 19 core studies that met the agreed-upon criteria. Eventually, however, each researcher rejected a few studies within this commonly defined set.

Although they did not vary substantially, differences in the meta-analytic outcomes were in accord with the researchers' preexisting ideologies. Further, these differences emerged even though all the researchers knew beforehand that a prominent methodologist who was part of the team would write an overview to summarize the conclusions of the six substantive experts.

One might wonder why there would be any variation among their meta-analytic outcomes other than that due to calculation errors or to differences among researchers in the one or two studies that they excluded from their analyses. There was a window, however, for additional variation. One source was the dependent measures chosen for analysis. Some researchers included test scores from topic content areas besides reading and mathematics, whereas others did not. Another source of variation involved the control groups selected for comparison. In many reported studies, there was more than one control group. Sound but different arguments would augur for the selection of one or another—with each choice representing a different type of compromise. A third source of variation arose from how panelists calculated effect sizes.

In summarizing this project, Cook (1984) thought that the findings were remarkable in their convergence. On the basis of his reanalysis of Wortman's results, along with the reported effects of the three other panelists who did perform meta-analyses, Cook concluded that desegregation did not cause a decrease in black achievement. Further, the gain in black students' performance ranged from 0 to .08 standard deviation units for math performance and from .06 to .16 for reading achieving achievement. (These are typically judged to be extremely small effects. Cohen [1988] describes effects with a magnitude of .2 as small, whereas those that exceed .8 are described as large.) Nevertheless, the two reviewers who were most positively predisposed ideologically, Crain and Wortman, reported as their conclusion the largest effect sizes (e.g., Wortman, .25, on the basis of a subset of studies that he selected as most critical; Crain preferred to base his conclusion on an earlier meta-analysis that he performed on 93 studies). And Armor, who was most negative ideologically, reported the smallest effect size (.03, combined reading and math).

In sum, meta-analysis as a procedure contains some normatively based practices that to some degree act against the intrusion of the theory-confirmation bias that exists to an unknown degree in primary research. On the other hand, as suggested by the NIE study, meta-analysis potentially allows assessment of such bias, either by examining the meta-analytic results produced by those holding different perspectives or by comparing within a single meta-analysis, subsets of studies produced by different laboratories or theoretical camps.

Journal Publication Policy

In our discussion of journal publication policy we noted the dysfunctional aspects of the bias against publishing conceptual replications that fail to confirm a previously published finding. In addition, we pointed out with dismay that current guidelines for writing social psychological articles unduly encourage the parading of post hoc explanation as the theory that generated an unpredicted result. Moreover, we noted that most studies of factors that moderate or mediate such a finding, if published, are likely to unknowingly contain the same paradigm-specific features that limit or define the circumstances to which the finding is constrained. This state of affairs means that the science is led down more blind paths than is necessary, or put differently, its statements about what it knows are exaggerated.

Because meta-analysis can simultaneously examine a large collection of studies about dependent variable q, it can examine all cases in which the effect of p on q has been examined. By coding study qualities and/or analyzing subsets of them, the meta-analyst may discover certain circumstances under which q is related to p, and others under which it is not. In our earlier discussion of this issue, however, we raised reasons why such discovery is impeded for the meta-analyst as well as for the author of a typical narrative review. We argued that most of the *published* studies explicitly concerned with the p–q relationship will inadvertently contain nonidentified, paradigm-specific features that are necessary for its confirmation. Thus, there is likely to be a positive constant bias in the published literature. On the other hand, for social psychological research, judges can be used to generate information about the levels of variable p in experiments that are concerned with the dependent variable q, but in which p was never manipulated. To generate this information, judges rate the level of p for experimental (and control) subjects in each study concerned with q. This bootstrapped information is especially valuable because it is obtained with respect to a large array of studies with varied contextual features, rather than the constrained context that we argued is likely to characterize published reports of the p–q relationship. Thus, meta-analysis can counter the constant bias created by publication policies.

Recently, we used judges' ratings to expand the meta-analytic data base and examine competing theoretical models of mood, helping behavior, and aggression (Carlson & Miller 1987; Carlson, Charlin, & Miller, 1988; Carlson, Marcus-Newhall, & Miller, 1990). In Carlson and Miller (1987), for example, we compared three models that offer different explanations of the relation between negative mood and helping behavior: the negative state relief model of Cialdini and colleagues (Cialdini, Baumann, & Kenrick, 1981; Cialdini, Darby, & Vincent, 1973; Cialdini & Kenrick, 1976; Cialdini et al., 1987), the attentional focus model of Rosenhan, Salovey, and Hargis (1981), and the responsibility–objective self-awareness model (Rogers, Miller, Mayer, & Duval, 1982). Basically, we specified the theoretical variables relevant to each of the models, as well as theoretically relevant control variables. Then judges rated each of the 90 methods sections of the published studies to assess the contextual level of each of the theoretical variables for subjects in the experimental condition of each study. Partial correlations were used to control the effects of all other variables when examining the specific relation of any single variable to helpfulness effect sizes. These methods enabled examination of the degree to which each of the variables within each of the competing theories received empirical support.

Among the three theories, the negative state relief model is the most differentiated, having more propositions than the others. Nevertheless, in more than 20 different tests of this model, using various subsets of the data and framing the research questions with respect to it in alternative ways, none of its variables received any support (Miller & Carlson, 1990). This was especially interesting because it appears to be the more widely accepted theory among the three. There was, however, strong support for the attentional focus and the responsibility–objective self-awareness models.

As emphasized, the most important feature of this work is its bootstrapping nature. Judges' ratings were used to create information that was not available within specific studies and which was then applied to make systematic theoretical comparisons. Thus, when such judgments are combined with meta-analysis, it is possible to study relationships across a much broader array of contexts than those that will exist in primary-level research on that phenomenon. Such procedures have been used successfully in a number of other meta-analyses of social psychological research (for reviews, see Miller & Carlson, 1990; Miller & Pollock, in press). We argue that their use has special advantages for countering the constant positive bias for confirming an initial *p–q* relation that occurs as a consequence of publication policy.

Conclusion

As an innovation, meta-analysis has much to offer that can be applied to the five problems we cited as confronting contemporary social psychology. Although it is a powerful tool, meta-analysis is not a panacea. Lest we fall into the trap of being overly enthusiastic about meta-analysis, we close by mentioning some of its limitations.

Meta-analysis alone is insufficient to overcome the undesirable features of concept capture. Indeed, the failure to recognize redundancy, historical shifts in terminology, or substantial overlap in terms is a conceptual issue, not a methodological one. The individual researcher's recognition of the relevance of prior work is a qualitative skill that depends on knowledge of prior work. It is, however, a critical determinant of meta-analytic results. The increasing use of computerized search methods for retrieval of research literatures is likely to add to the problem. For instance, if a specific label, such as false consensus, is used to delimit the universe of potentially relevant citations, older relevant work bearing on the topic—such as projection—will not be cited (e.g., Mullen et al., 1985).

Nonobvious findings can be detected by the application of meta-analysis—as, for example, when it is applied to literatures in which few, if any single-study outcomes appear to yield reliable differences. Sometimes, in contradiction to prior individual conclusions, they collectively confirm a small but significant effect size. Yet, how is the practical significance of such findings to be judged? As emphasized by Rosenthal, the qualitative aspects of evaluating research significance require scrutiny of the variables under evaluation (Rosenthal & Rubin, 1982).

As seen in the suggestive evidence of the NIE study of desegregation effects on black academic achievement, the propensity toward theory confirmation or ideological bias may affect meta-analytic results as well as those of individual studies. Thus, meta-analyses too require replication. For instance, Dillard (1991) notes that two recent meta-analyses of research on sequential request compliance strategies (foot-in-the-door and door-in-the-face) yield different answers about whether the effectiveness of both strategies is constrained to cases that use a prosocial appeal.

The errors of inference that can occur in meta-analysis are exactly the same as those that can arise in primary-level research studies. And, just as is true in primary-level studies with small sample sizes, if the number of studies available for meta-analysis is small, the results may fail to yield precise or reliable estimates of the magnitude of effects. Additionally, in a meta-analysis based on few cases it might be difficult to demonstrate evidence of discriminant construct validity. Thus, replication is critical for meta-analysis, for without it, most meta-analytic reviews would not be possible. As we noted, however, as ordinarily implemented, meta-analysis is vulnerable to whatever constant biases may exist in published literature.

Finally, for any proposed improvement of science, a form of evaluation should go hand in hand with it. The criteria for specifying its improvement, however, as well as the procedures for assessing improvement, pose very serious difficulties that are not easily solved (Shadish, 1989). Nevertheless, such evaluative efforts should be made for meta-analysis in that it appears to be an innovation that continues to gain broader acceptance.

Acknowledgments

In the section on nonobvious findings, we unabashedly paraphrase N. L. Gage (1991). We thank Casey Kramer for compiling the data concerning the citations of McGuire and of Roberts and Maccoby. We thank Valerie Benveniste-Kirkus for compilation and Amy Marcus-Newhall for the analyses of the data concerning the citations in *Journal of Personality and Social Psychology* and *Psychological Review*. We thank Sharon Gross for preparing Table 7.1. The full citations for the listed references, as well as other related references extending through 1988, can be obtained by writing to the first author of this article or to Dr. Gross at the same address.

References

Armstrong, J. S. (1982). Research on scientific journals: Implications for editors and authors. *Journal of Forecasting, 1*, 83–104.

Aronson, E. (1969). The theory of cognitive dissonance: A current perspective. In L. Berkowitz (Ed.), *Advances in experimental social psychology* (Vol. 4, pp. 1–34). New York: Academic Press.

Asch, S. E. (1951). Effects of group pressure upon the modification and distortion of judgments. In H. Guetzkow (Ed.), *Groups, leadership, and men.* Pittsburgh: Carnegie Press.

Bacon, F. (1853). *The physical and metaphysical works of Lord Bacon* (J. Dewey, Trans.). London: Bohn. (Original work published 1620)

Bakan, D. (1966). The test of significance in psychology research. *Psychological Bulletin, 66,* 432–437.

Baratz, D. (1983). How justified is the "obvious" reaction. *Dissertation Abstracts International, 44/02B,* 644B. (University Microfilms No. DA 8314435)

Beaman, A. L. (1991). An empirical comparison of meta-analytic and traditional reviews. *Personality and Sociology Bulletin, 17,* 252–257.

Becker, B. J. (1991). The quality and credibility of research reviews: What the editors say. *Personality and Social Psychology Bulletin, 17,* 267–272.

Bem, D. J. (1987). Writing the empirical journal article. In M. P. Zanna & J. M. Darley (Eds.), *The complete academic: A practical guide for the beginning social scientist.* New York: Random House.

Berger, J., Wagner, D. G., & Zeiditch, M., Jr. (1983). *Expectation states theory: The status of a research program.* Stanford University: Technical Report No. 90.

Billig, M. (1990). Rhetoric of social psychology. In I. Parker & J. Shodder (Eds.), *Deconstructing social psychology* (pp. 47–60). London: Routledge.

Bogen, J. E. (1969). The other side of the brain II: An appositional mind. *Bulletin of Los Angeles Neurological Society, 34,* 135–162.

Bogen, J. E. (1975). Some educational aspects of hemispheric specialization. *UCLA Educator, 17,* 24–32.

Bogen, J. E., DeZure, R., TenHouten, W. D., & Marsh, J. F. (1972). The other side of the brain IV: The A/P ratio. *Bulletin of Los Angeles Neurological Society, 37,* 49–61.

Campbell, D. T. (1986). Science's social system of validity-enhancing collective belief change and the problems of the social sciences. In D. W. Fiske & R. A. Shweder (Eds.), *Metatheory in social science: Pluralisms and subjectivities in the social sciences* (pp. 108–135). Chicago: University of Chicago Press.

Carlson, M., Charlin, V., & Miller, N. (1988). Positive mood and helping behavior: A test of six hypotheses. *Journal of Personality and Social Psychology, 55,* 211–229.

Carlson, M., Marcus-Newhall, A., & Miller, N. (1989). Evidence for a general construct of aggression. *Personality and Social Psychology Bulletin, 15,* 377–389.

Carlson, M., Marcus-Newhall, A., & Miller, N. (1990). Effects of situational aggression cues: A quantitative review. *Journal of Personality and Social Psychology, 58,* 622–633.

Carlson, M., & Miller, N. (1987). Explanation of the relation between negative mood and helping. *Psychological Bulletin, 102,* 91–108.

Carmeron, A. (1970). Impaired utilization of kinesthetic feedback in right hemispheric lesions. *Neurology, 20,* 1033–1038.

Chaiken, S. (1987). The heuristic model of persuasion. In M. P. Zanna, J. M. Olsen, & C. P. Herman (Eds.), *Social influence: The Ontario symposium* (Vol. 5, pp. 3–39). Hillsdale, NJ: Erlbaum.

Chaiken, S., & Stangor, C. (1987). Attitudes and attitude change. *Annual Review of Psychology, 38,* 575–630.

Chapman, L. J. (1961). A reinterpretation of some pathological disturbances in conceptual breadth. *Journal of Abnormal and Social Psychology, 62*, 514–519.

Cialdini, R. B., Baumann, D. J., & Kenrick, D. T. (1981). Insights from sadness: A three-step model of the development of altruism as hedonism. *Developmental Review, 1*, 207–223.

Cialdini, R. B., Darby, B. L., & Vincent, J. E. (1973). Transgression and altruism: A case for hedonism. *Journal of Experimental Social Psychology, 9*, 502–516.

Cialdini, R. B., & Kenrick, D. T. (1976). Altruism as hedonism: A social development perspective on the relationship of negative mood state and helping. *Journal of Personality and Social Psychology, 34*, 907–914.

Cialdini, R. B., Schaller, M., Houlihan, D., Arps, K., Fultz, J., & Beaman, A. L. (1987). Empathy-based helping: Is it selflessly or selfishly motivated? *Journal of Personality and Social Psychology, 52*, 749–758.

Cicchetti, D. V. (1991). The reliability of peer review for manuscript and grant submissions: A cross disciplinary investigation. *Behavioral and Brain Sciences, 14*, 119–186.

Claridge, G. (1967). *Personality and arousal.* Oxford: Pergamon Press.

Cohen, J. (1988). *Statistical power analysis for the behavioral sciences* (2nd ed.). Hillsdale, NJ: Erlbaum.

Collins, H. M. (1983). The sociology of scientific knowledge: Studies of contemporary science. *Annual Review of Sociology, 9*, 265–285.

Collins, R. (1989). Sociology: Proscience or antiscience? *American Sociological Review, 54*, 124–139.

The Columbia Encyclopedia (1963). W. Bridgwater & S. Kurtz (Eds.). New York: Columbia University Press.

Cook, T. D. (1984). What have black children gained academically from school integration? Examination of the meta-analytic evidence. In T. Cook, D. Armor, R. Crain, N. Miller, W. Stephan, H. Walberg, & P. Wortman (Eds.), *School desegregation and black achievement*. Unpublished report. Washington, DC: National Institute of Education. (ERIC Document Reproduction Service No. ED 241 671)

Cooper, J., & Fazio, R. H. (1984). A new look at dissonance theory. In L. Berkowitz (Ed.), *Advances in experimental social psychology* (Vol. 17, pp. 229–266). New York: Academic Press.

Cooper, H. M., & Lemke, K. M. (1991). On the role of meta-analysis in personality and social psychology. *Personality and Social Psychology Bulletin, 17*, 245–251.

Cronbach, L. J. (1955). Processes affecting scores on "understanding of others" and "assumed similarity." *Psychological Bulletin, 52*, 177–193.

Dawes, R. M. (1979). The robust beauty of improper linear models in decision making. *American Psychologist, 34*, 571–582.

Dillard, J. P. (1991). The current status of research on sequential-request compliance techniques. *Personality and Social Psychology Bulletin, 17*, 283–288.

Dollard, J., & Miller, N. (1950). *Personality and psychotherapy.* New York: McGraw-Hill.

Durkheim, E. (1912/1954). *The elementary forms of religious life.* New York: Free Press.

Edwards, W., Lindman, H., & Savage, L. J. (1963). Bayesian statistical inference for psychological research. *Psychological Review, 70,* 193–242.

Erdelyi, M. H. (1990). Repression, reconstruction and defense: History and integration of the psychoanalytic and experimental frameworks. In J. L. Singer (Ed.), *Repression and dissociation: Implications for personality theory, psychopathology, and health* (pp. 5–28). Chicago: University of Chicago Press.

Eriksen, C., & Pierce, J. (1968). Defense mechanisms. In E. Borgatta & W. Lambert (Eds.), *Handbook of personality theory and research* (pp. 1007–1040). Chicago: Rand-McNally.

Eysenck, H. J. (1957). *The dynamics of anxiety and hysteria.* New York: Praeger.

Festinger, L. (1957). *A theory of cognitive dissonance.* Stanford, CA: Stanford University Press.

Fishman, D. B., & Neigher, W. D. (1982). American psychology in the eighties: Who will buy? *American Psychologist, 37,* 533–546.

Frantz, T. T. (1968). Criteria for publishable manuscripts. *Personnel and Guidance Journal, 47,* 384–386.

Freud, S. (1915). The unconscious. In J. Strachey (Ed. and Trans.). *The standard edition of the complete psychological works of Sigmund Freud* (Vol. 14). London: Hogarth Press and Institute of Psycho-analysis.

Gage, N. L. (1991). The obviousness of social and educational research results. *Educational Researcher, 20,* 10–16.

Gergen, K. (1982). *Toward transformation in social knowledge.* New York: Springer-Verlag.

Goethals, G. R., & Cooper, J. (1972). The role of intention and postbehavioral consequences in the arousal of cognitive dissonance. *Journal of Personality and Social Psychology, 23,* 293–301.

Goffman, E. (1981). *Forms of talk.* Philadelphia: University of Pennsylvania Press.

Grant, D. A. (1962). Testing the null hypothesis and the strategy and tactics of investigating theoretical models. *Psychological Review, 69,* 54–61.

Greenwald, A. G. (1975). Consequences of prejudice against the null hypothesis. *Psychological Bulletin, 82,* 1–20.

Greenwald, A. G., & Pratkanis, A. R. (1988). On the use of "theory" on the usefulness of theory. *Psychological Review, 95,* 575–579.

Greenwald, A. G., Pratkanis, A. R., Leippe, M. R., & Baumgardner, M. H. (1986). Under what conditions does theory obstruct research progress? *Psychological Review, 93,* 216–229.

Greenwald, A. G., & Ronis, D. L. (1978). Twenty years of cognitive dissonance: Case study of the evolution of a theory. *Psychological Review, 85,* 53–57.

Hawkins, S. A., & Hastie, R. (1990). Hindsight: Biased judgments of past events after the outcomes are known. *Psychological Bulletin, 107,* 311–327.

Hayes, W. L. (1981). *Statistics for psychologists* (3rd ed.). New York: Holt, Rinehart, & Winston.

Heider, F. (1958). *The psychology of interpersonal relations*. New York: Wiley.
Holmes, D. S. (1968). Dimensions of projection. *Psychological Bulletin, 69*, 248–268.
Holton, G. (1978). *The scientific imagination: Case studies*. Cambridge: Cambridge University Press.
Homans, G. C. (1950). *The human group*. New York: Harcourt Brace.
Jackson, G. B. (1978). *Methods for reviewing and integrating research in the social sciences* (Grant Report No. DIS 76–20398). Washington, DC: National Science Foundation.
Jackson, G. B. (1980). Methods for integrative reviews. *Review of Educational Research, 50*, 438–460.
Johnson, B. T., & Eagly, A. H. (1989). The effects of involvement on persuasion: A meta-analysis. *Psychological Bulletin, 106*, 290–314.
Jones, E. E., & Nisbett, R. E. (1970). The actor and the observer: Divergent perceptions of the causes of behavior. In E. E. Jones, D. E. Kanouse, H. H. Kelley, R. E. Nisbett, S. Valins, & B. Weiner (Eds.), *Attribution: Perceiving the causes of behavior*. Morristown, NJ: General Learning Press.
Joynson, R. B. (1989). *The Burt affair*. London: Routledge.
Kane, T. R., Joseph, J. M., & Tedeschi, J. T. (1976). Person perception and the Berkowitz paradigm for the study of aggression. *Journal of Personality and Social Psychology, 33*, 663–673.
Katz, D., & Allport, F. (1931). *Students' attitudes*. Syracuse, NY: Craftsman Press.
Kelley, H. H., & Stahelski, A. (1970). The social interaction basis of cooperators' and competitors' beliefs about others. *Journal of Personality and Social Psychology, 16*, 66–91.
Lamal, P. A. (1991). Psychology as common sense: The case of findings concerning work motivation and satisfaction. *Psychological Science, 2*, 129–130.
Lazarsfeld, P. F. (1949). The American soldier—an expository review. *Public Opinion Quarterly, 13*, 377–404.
Lock, E. A., & Latham, G. P. (1991). The fallacies of commonsense "truths": A reply to Lamal. *Psychological Science, 2*, 131–132.
Löwenfeld, V. (1939). *The nature of creative activity*. London: Kegan Paul, Trench, Trubner.
McGuire, W. J. (1985). Attitudes and attitude change. In G. Lindzey & E. Aronson (Eds.), *Handbook of social psychology* (3rd ed.) (Vol. 2, pp. 233–346). Hillsdale, NJ: Erlbaum.
McGuire, W. J. (1989). A perspectivist approach to the strategic planning of programmatic scientific research. In B. Gholson, W. R. Shadish, Jr., R. A. Neimeyer, & A. C. Houts (Eds.), *Psychology of science: Contributions to metascience* (pp. 214–245). Cambridge, England: Cambridge University Press.
Miller, N., & Carlson, M. (1990). Valid theory-testing meta-analyses further question the Negative State Relief model of helping. *Psychological Bulletin, 107*, 215–225.
Miller, N., & Pollock, V. E. (in press). Meta-analytic syntheses for theory development. In H. Cooper & L. Hedges (Eds.), *Handbook of research synthesis*.

New York: Russell Sage.

Mitroff, I. (1974). *The subjective side of science*. Amsterdam: Elsevier.

Mullen, B., Atkins, J. L., Champion, D. S., Edwards, C., Hardy, D., Storey, J. E., & Vanderklok, M. (1985). The false consensus effect: A meta-analysis of ISS hypothesis tests. *Journal of Experimental Social Psychology, 21*, 262–283.

Myers, D. (1991). Union is strength: A consumers view of meta-analysis. *Personality and Social Psychology Bulletin, 17*, 265–266.

National Institute of Education. (1984). *School desegregation and black achievement*. Unpublished report. Washington, DC: Author. (ERIC Document Reproduction Service No. ED 241 671)

National Research Council. (1988). *The behavioral and social sciences: Achievements and opportunities*. Washington, DC: National Academy Press.

Nisbett, R., & Ross, L. (1980). *Human inference: Strategies and shortcomings of social judgment*. Englewood Cliffs, NJ: Prentice-Hall.

Petty, R. E., & Cacioppo, J. T. (1986). The elaboration likelihood model of persuasion. In. L. Berkowitz (Ed.), *Advances in experimental social psychology* (Vol. 19, pp. 123–205). New York: Academic Press.

Phillips, D. C. (1987). *Philosophy, science, and social inquiry*. Oxford, England: Pergamon Press.

Prentky, R. A. (1980). *Creativity and psychopathology: A neurocognitive perspective*. New York: Praeger.

Roberts, D. F., & Maccoby, N. (1985). Effects of mass communication. In G. Lindzey & E. Aronson (Eds.), *Handbook of social psychology* (3rd ed.) (Vol. 2, pp. 539–598). Hillsdale, NJ: Erlbaum.

Rogers, M., Miller, N., Mayer, F. S., & Duval, S. (1982). Personal responsibility and salience of the request for help: Determinance of the relation between negative affect and helping behavior. *Journal of Personality and Social Psychology, 43*, 956–970.

Rosenhan, D. L., Salovey, P., & Hargis, K. (1981). The joys of helping: Focus of attention mediates the impact of positive affect on altruism. *Journal of Personality and Social Psychology, 40*, 899–905.

Rosenthal, R., & Rubin, D. B. (1982). A simple, general purpose display of magnitude of experimental effect. *Journal of Educational Psychology, 74*, 166–169.

Ross, L. (1977). The intuitive psychologist and his shortcomings: Distortions in the attribution process. In L. Berkowitz (Ed.), *Advances in experimental social psychology* (Vol. 10, pp. 173–220). New York: Academic Press.

Ross, L., Greene, E., & House, P. (1977). The "false consensus effect": An egocentric bias in social perception and attributional processes. *Journal of Experimental Social Psychology, 13*, 279–301.

Salomon, G., Perkins, D. N., & Globerson, T. (1991). Partners in cognition: Extending human intelligence with intelligent technologies. *Educational Researcher, 20*, 2–9.

Scher, S. J., & Cooper, J. (1989). Motivational basis of dissonance: The singular role of behavioral consequences. *Journal of Personality and Social Psychology, 56*, 899–906.

Schlessinger, A., Jr. (1949). The statistical soldier. *Partisan Reviews, 16,* 852–856.
Selvin, H. C., & Stuart, A. (1966). Data-dredging procedures in survey analysis. *American Statistician, 20,* 20–22.
Shadish, W. R. (1989). The perception and evaluation of quality in science. In B. Gholson, W. R. Shadish, Jr., R. A. Neimeyer, & A. C. Houts (Eds.), *Psychology of science: Contributions to metascience* (pp. 383–426). Cambridge, England: Cambridge University Press.
Shadish, W. R., Cook, T. D., & Houts, A. C. (1986). Quasi-experimentation in a critical multiplist mode. In W. M. K. Trochim (Ed.), *Advances in quasi-experimental design and analysis* (pp. 29–46). San Francisco: Jossey-Bass.
Steele, C. M. (1988). The psychology of self-affirmation: Sustaining the integrity of the self. In L. Berkowitz (Ed.), *Advances in experimental social psychology* (Vol. 21, pp. 261–302). New York: Academic Press.
Steele, C. M., & Southwick, L. (1985). Alcohol and social behavior. The psychology of drunken excess. *Journal of Personality and Social Psychology, 48,* 18–35.
Sterling, T. D. (1959). Publication decisions and their possible effects on inferences drawn from tests of significance—or vice versa. *Journal of the American Statistical Association, 54,* 30–34.
Stinchcombe, A. L. (1984). The origins of sociology as a discipline. *Acta Sociological, 27,* 51–61.
Sullivan, H. S. (1956). *Clinical studies in psychiatry.* New York: W. W. Norton.
Tedeschi, J. T. (1983). Social influence theory and aggression. In R. T. Geen & E. I. Donnerstein (Eds.), *Aggression: Theoretical and empirical reviews* (pp. 135–162). Orlando, FL: Academic Press.
Tedeschi, J. T., Smith, R. B., & Brown, R. C. (1974). A reinterpretation of research on aggression. *Psychological Bulletin, 81,* 540–562.
Travers, R. M. W. (1941). A study in judging the opinions of groups. *Archives of Psychology,* No. 266.
Wallen, R. (1943). Individuals' estimates of group opinion. *Journal of Social Psychology, 17,* 269–274.
Walster, G. W., & Cleary, T. A. (1970). A proposal for a new editorial policy in the social sciences. *American Statistician, 24,* 16–19.
Whyte, W. F. (1943). *Street corner society.* Chicago: University of Chicago Press.
Wittrock, M. C. (Ed.). (1986). *Handbook of research on teaching* (3rd ed.). New York: MacMillan.
Wolff, W. M. (1970). A study of criteria for journal manuscripts. *American Psychologist, 25,* 636–639.
Wong, L. (1987). Reaction to research findings: Is the feeling of obviousness warranted? *Dissertation Abstracts International, 48/12,* 370B. (University Microfilms No. DA 8801059)

CHAPTER 8

Social Influence in Science: Agreement and Dissent in Achieving Scientific Consensus

ROBERT E. ROSENWEIN

As Campbell (1988a) points out, science may be thought of as a social system vehicle for producing knowledge. In this sense, like any other social system, science is characterized by "group processes," for example, recruitment of new members, socialization to group norms, facilitation of group cohesiveness, maintenance of group loyalty, and leadership for continuity and coordination. Further, in this formulation, there is a conflict in science, as in any system, between the "conservative" tendency to preserve and protect those modes of thought and behavior that have been successful and the tendency to tolerate and absorb new ideas, to provide an atmosphere where creativity and criticism can flourish, in short, to provide the possibility of change and growth. Campbell argues that a virtue in science is its commitment to minimizing those psychological and social psychological processes that interfere with science's goal of valid knowledge production; others are less sure (see Campbell, Chapter 2, this volume).

In this chapter, I examine one approach to understanding group processes in science and its relation to the production of knowledge, namely, asking how a group influences its members and how the group, in turn, is influenced by its members. I review the literature on social influence in social psychology guided by (1985) Moscovici's depiction of groups as characterized by the "essential tension" between consensus-seeking tendencies and dissensus-seeking tendencies. In particular, I pay special attention to the role of the dissenter from a prevailing scientific consensus in the achievement of new scientific belief structures. The rationale for focusing on dissent is that new ideas in science begin as minority positions. If, however, we are serious about the essential tension, one needs to understand how majorities and minorities interact to produce scientific belief change.

Two Case Studies

The following two examples of controversy in science are illustrative of the kinds of issues this chapter addresses.

Astroidal Impact and Prehistoric Extinction

Some 65 million years ago, approximately 85% of all animal species then existing became extinct. Over the years, there has been much speculation about the reasons for this die-off. A popular idea is that dramatic climatic changes were responsible. Others conjectured that the extinction was a relatively prolonged affair, not the result of some terrible catastrophe.

In 1980, a Nobel prize-winning physicist, Walter Alverez, along with his colleagues Asaro and Michel published a paper in *Science* describing their finding that soil samples found in Italy from the time in question showed an abnormally high incidence of iridium. Since this element is rare in soil and since it occurs commonly in "large bodies" that impact the planet, they hypothesized that a large asteroid hit the earth 65 million years ago, throwing into the atmosphere massive amounts of pulverized rock and asteroid fragments. This dust blocked sunlight so much that photosynthesis was severely inhibited. Food chains broke down, which led to the extinction of animals.

Opposition from the majority of scientists interested in this problem was expressed in a series of articles. The arguments were of four types:

1. Questioning the empirical data (e.g., not enough known about the chemistry of iridium, small sample size, extinctions took place over longer period than the Alverez group seemed to admit).
2. Suggesting plausible rival hypotheses (e.g., iridium as a lag concentrate of something in the soil at that time, biological origin of iridium).
3. Questioning the logic of the argument (e.g., argument is ad hoc, why not massive extinction of plant life, and where is the hole left by the asteroid).
4. Questioning the credentials, style of presentation, and motives of the Alverez group (e.g., the group had no paleontological ex-

pertise, who were they to resurrect old debates in geology, and the group was too quick to call a press conference).

In short, reaction was strong, angry, and scornful. Still, within 2 years, as the Alverez group continued to refine their thesis and to present at conferences, work about their hypothesis was being generated. Geologists were looking closely at rocks of various origins and periods for evidence of iridium anomalies, a periodicity in mass extinctions over a 600-million-year time frame had been found, and explanations of how an asteroid could have impacted the earth were being formulated (the most notorious was the "nemesis" hypothesis, that a planet periodically transverses a path across earth's plane and into the asteroid belt, causing a deflected asteroid to impact the planet).

Paleontologists who thought the idea of a heavenly visitation sounded too much like a *deus ex machina* explanation, weighed in with a counterhypothesis focusing on a period of unprecedented volcanic activity in India which generated the climactic effects leading to the extinctions (Raup, 1986).

The controversy continues. The October 1990 issue of *Scientific American* featured an exchange of articles between Alverez and Asaro for astroidal impact, focusing on the iridium evidence, and Courtillot dismissing the iridium evidence and advancing the volcanic activity hypothesis.

Mechanism of Natural Selection

In 1900, the question of the mechanism underlying natural selection had not been settled. Two theories were extant. The first, by Darwin, was that evolution proceeded through continuous evolution, small individual differences accumulated over time. The opposing theory, called saltational, suggested that evolution proceeded by large, discontinuous variation. Mendel's results, with their emphasis on the role of mutation as a mechanism of change, provided an empirical basis for this theory. Galton, in the 1890s, persuasively argued in support of the theory as well.

The controversy crystallized in the writing of three British authors: Karl Pearson and Raphael Weldon on the side of continuous evolution and William Bateson on the other. These men had considerable stature and prestige in their field. Pearson and Weldon pioneered the use of statistical methods for correlations, a field they called biometry. This was untraditional biology and was objected to by Bateson, a zoologist. Both sides were highly invested in their positions and resorted to impugning the character of the other to assert their points. When Pearson submitted a paper to the Royal Society, Bateson forced the secretary of the society to circulate his criticism of Pearson's paper along with the paper itself. Annoyed by Bateson's machinations, Weldon, Pearson, and Galton left the society and founded the journal *Biometrika*. Over the next 10 years, the journal published polemics against Bateson and Bateson returned the favor in books and articles.

Neither Weldon nor Pearson fully converted to the Mendelian position even when that had become the consensus view. Two students of Weldon, Arthur Darbishire and Edgar Schuster, quietly conducted research that they thought would show the validity of the biometric position. Bateson (and Castle, an American biologist) corresponded extensively with Darbishire and Schuster, reanalyzing their data and pointing out how these data were more compellingly interpreted from a Mendelian standpoint. Unlike Weldon and Pearson, these two men were not as emotionally invested in the biometric position, nor did they harbor antagonism to Bateson. Indeed, Darbishire's favorite maxim was that the scientist's attitude should be "one of continual, unceasing, and active distrust of oneself" (quoted in Kim, in press).

The shift toward the Mendelian position did not come from science alone or from the conversion of middle-level biologists. Rather, it came from two groups that were relatively disinterested in the debate but supplied a wealth of evidence that was more clearly interpretable from a Mendelian perspective. These two groups were animal and plant breeders with much experience in hybridization and artificial selection of plants and animals and medical doctors interested in the familial origins of transmitted disease. Both groups found that Mendelian genetics supplied a more scientific basis for their data. Thus, midlevel scientists allied with interested onlookers, to create a consensus around the Mendelian position (Kim, in press).

These two case studies highlight the issues around which I structure the discussion of social influence in science. Since they were selected to highlight those issues, they distort the range of possible scientific controversy. Scientific controversy is not inevitably an either–or proposition. The average scientific controversy ends not with a winner and loser but with a compromise. Rigidity in minority behavior may encourage negotiations leading to greater compromise (Papastamou & Mugny, 1985). Campbell (1961) suggested that dependent minorities work toward "resolved composites," compromise between the evidence of one's own eyes and the need for social comparison (Festinger, 1954). We need to look at this more closely. Specifically, I address the following questions:

1. What behavioral or social devices do groups or individuals have to command the attention of the majority? (This question concerns the importance of consistent minority behavior.)

2. What resources are available to the dissenter in maintaining the attention of the majority (This question concerns the issue of norms and beliefs as resources, as well as the importance of source credibility.) Conversely, what resources are available to the majority in resisting the claims of the dissenter? (This question concerns the issue of pressure to uniformity and majority amplification.)
3. How do those espousing majority views come to terms with those who dissent from those views? Is it important to understand differences in personal and social investment in the majority opinion among those nominally sharing the consensus position? Is it important to understand similar differences among the minority? (This question concerns public compliance–private conversion as well as the internal dynamics of majority and minority groups.)
4. When is a minority view most likely to be influential; are there critical periods during which the minority is most effective in shifting scientific belief? (This question concerns Kuhn's (1970) discussion of conditions favoring a paradigm shift.)

Nature of Scientific Activity

I begin this discussion of the social psychology of science (SPS) by distinguishing between two traditions of the sociology of science, new and old. I have already evoked Campbell's structural–functional concept of science as a social system vehicle for the production of knowledge. In this volume, Campbell (Chapter 2) discusses at some length the new sociology of science and its contraepistemic and proepistemic dialectics. Therefore, I limit myself to the following comments about the new sociology of science. As Campbell points out, underlying these visions of science as a human activity are two assumptions that provide a rationale for the analysis of social influence processes in this chapter. The first is the assumption that, to some as-yet-unspecified degree, scientific beliefs are underjustified by the empirical data. The second, and allied, assumption is that the data themselves are "theory laden"; that is, they are given both structure and meaning through the theoretical suppositions scientists bring to the creation and analysis of data. Thus, a conceptual space is opened that allows us to examine the determination of scientific belief by psychological and social factors.

In addition, although I am enough of a realist to believe that the empirical world constrains explanation, the idea that explanation is also socially constructed is key to SPS. The anthropological stance of relativism allows us to consider scientists as any other group that favors certain epistemological criteria over others (in this case, controlled collection of data and subsequent statistical analysis) and to see this not as correct but as a shared community assumption about what constitutes "truth getting" (Collins, 1985).

An older sociology of science is identified with the sociologist Robert Merton (1973) and concerns the social structure of the scientific enterprise, as opposed to its "internal" operation. His emphasis is on the social stratification of scientific communities and the reward structure associated with acquiring status. Merton posits an ideal meritocratic system that places high priority on creativity and originality of thought. Individuals exhibiting these characteristics rise through the status structure and gain university professorships, large grants, journal editorships, or positions on review panels. These individuals then influence the direction of scientific work by allocating rewards within the community.

Further, Merton proposed rules or norms that presumably govern scientists working at their optimum. Science is universalistic, antiauthoritarian (in that persuasion is limited to egalitarian means), and communal (all members remain in the community and attend to each other's arguments). Merton characterized the scientific community by "organized skepticism" (although "organization" and "skepticism" seem in contradiction) (see Campbell, 1988b). These moral norms are required if science is to achieve its institutional goal, the acquisition of scientific knowledge.

Merton would be the first to agree that, in practice, science departs in some measure from this ideal (see Chubin & Hackett, 1990). One might further argue that violation of norms of universalism, emotional neutrality, and so on is socially functional for science as a social system. Mitroff (1974) argued that the "counternorms" of particularism and emotional commitment, the emotional investment in a view and a willingness to push these views obnoxiously, allow for debate in which all the relevant issues may be aired.

Merton, however, explicitly exempts his analysis from examining the role of these moral norms in the production and validation of scientific knowledge and the emergence of belief consensus in a scientific community. In this tradition, sociological studies identify the relationship between normative imperatives and the reward structure of science. However, for my purposes, I treat the norms of science as important elements in determining the outcome of the interaction between majority and minority positions in the production of scientific knowledge.

In summary, I begin with Campbell's agenda aimed at discovering and improving forms of social organization that would facilitate the achievement of new scientific belief consensuses. I reformulate this agenda to see where science is better, or could be made better, at dealing with the essential tension (Moscovici, 1985) between consensus seeking and dissensus seeking, both of which are necessary for science to survive as a social system vehicle for producing knowledge.

Literature on Minority and Majority Influence Characterized

We can distinguish an American tradition and a European tradition, which Levine and Russo (1987) label, respectively, the dependence perspective and the conflict perspective on social influence in groups. In the dependence perspective (e.g., Asch. 1951, 1955, 1956; Festinger, 1954; Schachter, 1951; Deutsch & Gerard, 1955), the issue is the effect of oppressive majorities on the the opinions and beliefs of those who differ (Moscovici, 1974). The focus is on the dissenter and his/her fate. Will he/she go along with (conform to) the view expressed by the majority? What factors, personal or structural, predict resistance? What can we do to strengthen a dissenter in the face of the overwhelming power of the majority to silence dissent? Factors of interest have included size of majority (Wolf & Latane, 1985; Tanford & Penrod, 1984), size of minority (Asch, 1955), motivational determinants of conforming responses (Deutsch & Gerard, 1955; Kelman, 1958), and causal attributions by majority and minority group members (Ross, Bierbauer, & Hoffman, 1976; Allen & Wilder, 1977).

The distinction between normative and informational influence (Deutsch & Gerard, 1955) is important to SPS. In normative influence, people are influenced because of a desire to be right. In informational influence, they are influenced because of a desire to be correct. While most situations have elements of both, one tends to think of scientists as being motivated primarily by the desire for knowledge and therefore more susceptible to informational influence. An important research task is to discover the circumstances in which normative influence, which favors consensus per se over knowledge, significantly affects the activities of scientists.

Several important assumptions exist within this tradition for SPS. First, majorities are typically conceptualized as if they were equally committed to the consensus position for similar reasons. Thus, the majority speaks, as it were, with one voice. But studies of scientific controversy usually reveal that those who espouse the majority are usually not committed at the same level of intensity or for the same reasons. Second, "consensus" must be carefully defined. For example, agreement with the majority varies depending on the subject matter. The analogue in science would be to ask: Consensus about what? Is it about the existence or nonexistence of a phenomenon (cold fusion, cognitive dissonance)? Is it about the facts? Is it about theoretical explanations? Is it about paradigmatic assumptions? Finally, the minority is conceptualized, either implicitly or explicitly, as a *threat* to the consensus of the majority. Majorities *must* absorb or eliminate the minority to survive.

In contrast to the American tradition, the European tradition for studying influence processes in groups focuses on the problem of social change. If consensus-seeking tendencies in groups silence dissent, what are mechanisms by which change in consensus can occur? Researchers in the dependence perspective study conflict resolution or consolidation processes, but those in the conflict perspective study conflict *creation* (Moscovici & Nemeth, 1974). Theoretical consensus in science does change. How can such change occur in the face of powerful consensus-seeking tendencies? Moscovici, with whose name much of this literature is associated, suggested that minority influence is the vehicle for innovation and change. New ideas originate with individuals or subgroups, and these minorities, although a threat to the consensus of the majority, exert influence through a variety of means (Nemeth & Kwan, 1987). How, and to what extent, does the majority change when confronted with minority views? What personal and situational conditions lead to movement toward minority positions? Factors of interest have included consistency of the minority position (for a review, see Levine, 1989; Kruglanski & Mackie, 1990), attributions to the minority (Maass, West, & Cialdini, 1987), and motivations for change (Chaiken & Stangor, 1987). Greater emphasis in this tradition is placed on relational dynamics between the majority and minority and on the essential tension between conformity and innovation (Moscovici, 1985; Levine & Russo, 1987), between social control and innovation (Nemeth & Staw, 1989), or between stability and change.

As with the dependence perspective, important assumptions exist in the conflict perspective for SPS. First, It is assumed that consensus operates to repress dissent. Thus, the struggle of the minority is to overcome this tendency. We can point to many examples in science where minority positions accepted at some later time were initially rejected and ignored. However, we can also point to examples of tolerance for minority views even though these are not shared by the majority. Indeed, the norms of science support expression and consideration of alternative beliefs.

Second, although Moscovici and others do understand that the tension between consensus and dissensus is essential, there is a tendency to evaluate pressures for uniformity as negative and dysfunctional, to identify with the plight of the lone dissenter. Yet, as Raup (1986) points out, science is an "unfair" game and this is critically important. A new idea *should* be considered guilty until proven innocent. As Raup (1986):

> Acceptance of any new theoretical framework depends on credibility and improvement on existing alternatives. In cases where the choice is not immediately obvious, the burden of proof lies with the new idea . . . . Given a choice, the scientific community invariably sticks with the conventional wisdom. Further, the older ideas have usually been around long enough to have accumulated supporting evidence, whereas the new idea rarely has much going for it, at least at first. It is not a fair game. (p. 195)

Raup then goes on to make this telling point:

> Despite the lack of "fairness" to new ideas, the traditional practice in science may serve us better than a more democratic mode. History shows that most new ideas fall. Science would be very confused much of the time if all new ideas were given precisely equal treatment. (p. 195)

Thus, while the norms of science may be antiauthoritarian, they are not necessarily democratic (see also Campbell, 1979/1988).

Two research paradigms do not fall neatly into either dependence or conflict perspectives but have implications for the study of influence in science. Sherif's (1936) work on the autokinetic effect creates a situation of extreme ambiguity in which two or more individuals making individual judgments over time have those judgments converge. This recreates an analogue to normal science; people with different interpretations come to a consensus on the meaning of those data (see discussion of this research and its extensions in Campbell, Chapter 2, this volume).

A second paradigm, which seems to fall somewhere between the dependence and conflict perspectives, is Schachter's (1951) work on communication, deviance, and rejection. In this experiment, subjects first argue to consensus concerning the desirability of different treatment for a troubled high school student. The minority confederate then joins the group and argues for a significantly different treatment. The number of communicative acts to this dissenter rises sharply and then drops off sharply. In other words, the majority talks to the dissenter, trying to change the dissenter's mind, but when the deviant refuses to change, abruptly ignores him/her (these results only obtain where the group is attractive to its members and the issue is relevant to the group's purposes). In this paradigm, communication between the participants is unconstrained; it therefore may have greater ecological validity as an analogue to communication and decision making in science.

Brief Critique of Majority and Minority Influence Literature

A number of authors have recently surveyed the influence literature (see Moscovici, 1985; Levine & Russo, 1987; Levine, 1989; Maass, West, & Cialdini, 1987; Nemeth & Staw, 1989; Kruglanski & Mackie, 1990). My analysis of majority and minority influence in achieving scientific consensus is based on certain points made in these excellent surveys. First, the findings are almost entirely based on research involving one or a brief series of laboratory experiments, which is problematic for understanding dissent in science. The doing of science is embedded in more molar social contexts, which are not easily duplicated in the laboratory. Laboratory studies are mostly done with strangers who have no further commitments to each other after the experiment. Yet the scientific community is composed of people with deep commitments, who know each other with differing degrees of intimacy. Scientific groups persist over long periods. Thus, it is difficult to know whether inferences from the experiments to the real world of science are valid.

Most theoretical and empirical attention has been directed to the cognitive mediators and their role in social influence, rather than relational issues between majority and minority over time (Maass & Clark, 1984; Chaiken & Stangor, 1987; Nemeth & Staw, 1989). Various speculative accounts embed minority influence in larger social contexts and their developmental history (Allen, 1975; Gerard, 1985; Levine & Russo, 1987; Moreland, 1987; Moscovici, 1985; Deconchy, 1985; Papastamou & Mugny, 1985; Nemeth & Staw, 1989). These authors conceptualize minority influence as interpersonal, intergroup, and intragroup, that is, as a multilevel phenemenon with different processes operating at different levels. The form influence processes take may reflect the specific context in which they are embedded.

This review of the literature on social influence is constructed in a way that might lead the reader to believe that there were two sets of qualitatively different processes, for majority and minority groups, that needed to be explicated. In a recent extensive critique of the literature, Kruglanski and Mackie (1990) address the question of the process and outcome distinctiveness of majority and minority influence. They assert that their review does not support a "strong" distinction between majority and minority influence effects, but rather a "weak" view that says that minority and majority influence *typically* covaries with certain psychological and social dimensions. For example, it is *typically* the case that the minority view is perceived as more extreme than the majority view, that minority positions are more cognitively salient, and that minorities are perceived as more consistent than majorities in their views. Moreover, the behavior of minority versus majority group members may not be as distinct as has previously been suggested.

I am not claiming a *necessary* distinction between majority and minority influence, or that majorities and minorities are inherently different by virtue of underlying psychological or social processes. Rather, I develop a developmental perspective on social influence in science that reflects my earlier comments on science and its essential tension. This perspective integrates some of the results of the experimental literature on both majority and minority influence, keeping in mind the caveats suggested by Kruglanski and Mackie (for other developmental models, see Tuckman, 1965; Moscovici, 1985; Levine & Moreland, 1985).

Attention, Minority Style, and Initial Majority Response

Suppose that there is a well-defined consensus view within a scientific community, and a challenge is put forward to some element of that consensus. Is the mere statement of the position enough to begin a process of influence? The answer is no. Any science is awash in dissenting hypotheses and theories and facts (Gleick, 1987). Gleick has charmingly discussed the mail he receives from people who are sure they have *the answer* to the secrets of the universe and are just as certain that they are being rejected for various nefarious reasons. It is unlikely that these letters are going to receive any hearing. What is it within a scientific community that *warrants* attention to a deviant idea?

First, there is the normative structure of science. In order for the essential tension to exist, there must be a normative resource within science on which new voices can draw to gain attention. As skeptical as scientists are, a hypothesis is a hypothesis. Virtually every science declares certain ideas to be unacceptable, crazy, or unscientific, only to find yesterday's craziness becoming today's truth. Raup (1986) may be correct in arguing that the deck is stacked against new hypotheses, but at least a norm of tolerance theoretically exists that can be a resource for those proposing the new idea.

Second, there is source credibility, defined as judgments made by an observer of the believability of a communicator. O'Keefe (1990) notes that two dimensions associated with credibility are trustworthiness and competence. Eagly, Wood, and Chaiken (1978) have suggested that what observers look for (or what scientists might attend to in the communication of a new idea) are the degree of knowledge bias and reporting bias on the part of the communicator. "Knowledge bias refers to a recipient's belief that a communicator's knowledge about external reality is nonveridical, and reporting bias refers to the belief that a communicator's willingess to convey an accurate version of external reality is compromised" (Eagly, Wood, & Chaiken, 1978, p. 424). They suggest that communicators perceived as having a knowledge bias will be perceived as being less competent; those who are perceived as having a reporting bias will be perceived as less trustworthy.

Although social psychologists do not have a comprensive picture of the factors that determine judgments of a knowledge or reporting bias (O'Keefe, 1990), certain factors are of interest in SPS. Status and expertise have frequently been linked to perceptions of credibility and are thus potentially important determinants of attention to communicators of a new idea. Alverez, for example, was a source who, by virtue of perceived general expertise (Nobel prize winner) would be expected to be credible and thus attended to. However, Alverez was no geologist or paleontologist and thus, from this perspective, had low credibility. Raup, who was more of a bystander to the debate, is aware that Alverez has no credentials as a paleontologist but realizes that those who share a theoretical paradigm may not be able to "see" things that an (intelligent) outsider can.

Does credibility matter? If someone with low status in the scientific community had proposed the astroidal impact hypothesis, would as much attention have been paid? Petty, Cacioppo, and Goldman (1981) have shown that, for recipients of a message highly involved with an issue, the credibility of the communicator makes little difference. This finding is consistent with the elaboration likelihood model of persuasion (Petty & Cacioppo, 1986), which holds that credibility and other characteristics of the source are important in determining influence when the involvement of the message recipient is low, but are of little importance when recipients are concerned about the issue. In this situation, influence occurs because of issue-relevant thinking: close examination of the arguments, consideration of other issue-relevant material, and so on (see O'Keefe, 1990; Petty & Cacioppo, 1981). This formulation may explain why Nemeth (1986) found that, even in situations in which the minority was wrong in a problem-solving task (presumably therefore being perceived as having low credibility), the effects were as great on majority problem-solving strategies and problem solution as when they were right (see also Boms & Avermet, 1980). According to Raup, Alverez's status as a Nobel prize winner was helpful, although there was significant resentment of an "outsider" operating in an area in which he had no credentials. This ambivalence is well described by Raup (1986):

> One could argue that Luis Alverez, as a high-energy physicist, has no business making such pronouncements about geology. But people outside a

> discipline can sometimes be very insightful, in part because they are not steeped in conventional ways of interpreting data in that discipline. (p. 79)

The problem of getting attention and credibility goes even deeper at both an individual and a group level. Alverez's paper was published as the lead article in *Science*. Such high-prestige outlets are known to have stiff requirements for publication. Geologists and paleontologists, seeing the title of the paper, might give it the benefit of the doubt based on attributions of credibility to the review process. Moreover, since the mode of self-presentation was that of a serious scientist, working from empirical data, the article would be more difficult to ignore. Even more profound are the implications for group consensus. Alverez, Alverez, Asaro, and Michel were not just posing a minority view on some aspect of the uniformitarian paradigm; they were attacking the very foundations of the paradigm. The social structure of science grows around careers based on a particular scientific consensus, so the threat to that consensus is not just intellectual.

Capturing and maintaining the attention of those who support the majority view may depend on the nature of the problem. There are two approaches to this issue. Laughlin (1988) has distinguished between tasks that have greater or lesser demonstratability, that is, where the data are either intellective or attitudinal. In science, an attitudinal situation might hold where various hypotheses had been put forward to account for some phenomenon but where little data had been generated (e.g., the early years of the debate on whether the universe was expanding or contracting). An intellective situation might obtain where the accumulation of data appears to constrain the plausible hypotheses. In the intellective situation, Laughlin (1988), using hypothesis-testing and evaluation task in which subjects were required to guess the nature of an experimenter-generated rule governing sequences of playing cards, found that majorities tended to ignore the input of minorities. Rosenwein and Koenemund (1991), using a similar task, created a situation in which the degree of ambiguity in the data was very high, a situation approximating an attitudinal condition. They found that majority members were more likely to attend to minority hypotheses, particularly where the majority was least confident of its own hypotheses, unlike a comparison condition in which the data were more clear.

Another approach to considering the effects of the nature of the problem on majority attention to the minority is to focus on the importance of the challenge raised by the minority view. In the astroidal impact example, the challenge is to a fundamental set of assumptions about how physical changes on the earth have occurred. If the majority position is well established, as in this case, it could be argued that scientists who acquired status through work based on the majority paradigm would have a strong vested interest in attending to the challenge, particularly if it was perceived as credible. Those directly in the field and deeply invested in the current paradigm (particularly those who held strong feelings about catastrophism) were much more scornful, hostile, and impugning of the motives of Alverez and his colleagues, in particular, calling the credibility of the authors into question. It would be interesting to see if challenges to paradigmatic assumptions in other cases elicit similar responses and whether responses are qualitatively different if the challenge threatens less fundamental aspects of majority consensuses.

Consequences to the Minority of Majority Attention

The essential tension implies that the concentrated attention of the majority will affect the actions of both the minority and the majority. One factor of great interest has been the consistency with which the minority expresses its views. Although minority consistency has been advanced as a significant factor in predicting minority influence, recent research (Levine, 1989) indicates that issue is far from simple. Minority consistency alone does not seem to be a necessary condition for influencing the majority. Instead, the most influential minority style is one in which the minority displays a patterned consistency (i.e., the consistent behavior is responsive to changes or differences in the situation). Levine, Saxe, and Harns (1976) and Levine and Ranelli (1978) found that the most influential minority was one that began by agreeing with the majority but then deviated from the majority position. Nonrepetitive behavior that preserves the appearance of consistency may be more effective than repetition (Nemeth & Brilmeyer, 1987). Why should this be so in science? I believe it reflects the fact that the minority is willing to "play by the rules." What are these rules? As noted above, one rule is at least to remain within the community, rather than splitting off and creating a new group (my guess is that Pearson and Weldon weakened their position when they broke away to form their own group). Playing by the rules means being willing to answer one's critics, to respond "appropriately" to criticism within the linguistic and behavioral practices of the community. But even more important, it is to work *within a common epistemological frame* in which data representing the real world are collected, suggested by a theoretical formulation or from which such a formulation is deduced.

Gerard (1985) has discussed what responses minorities might make when challenging the majority view. Those espousing minority views are frequently under threat; just like a majority that feels threatened, a greater degree of cohesiveness and esprit is created. One consequence is cross-pressure to maintain a common front and to play down differences among minority members (Janis, 1982). Moreover, espousing the minority view cuts one off from majority group membership and reduces options for other group membership; so a minority group member will become more normatively dependent on the minority group. This process is accompanied by self-justification and other forms of cognitive bolstering (Gerard, 1985). Further bolstering may come from the recruitment of new members to the minority position. Additions to a minority reduce conformity. Social support for a minority both reduces a person's fear of majority retaliation and reduces informational uncertainty (Allen, 1975).

A person in the minority is likely to feel distinctive, to feel that he/she is in the same social category as fellow minority group members, to see the issues as important, and to project his/her own feelings about him/herself on the issues onto other minority members. Minority status may lead to the growth of ingroup feelings, and the perception of the majority as the outgroup. This leads to a greater sense of what Gerard calls "staunchness" (i.e., a mixture of courage and confidence). Tajfel and Turner (1979) have suggested that the issue at stake is social identity. Ingroup bias enables the ingroup to compare itself favorably with the outgroup and "consequently enables the group members to safeguard or otherwise enhance their social identity" (Ng & Cram, 1988, p. 749, see also Mackie, 1986).

A staunch minority is likely to be consistent in the face of majority pressure; and the majority will perceive a staunch minority to be committed and convinced of its position (Levine, 1989). An augmenting principle is at work, the attribution of behavior to a facilitative cause (commitment and conviction) when an inhibitory force (the pressure of the majority) is present (Maass & Clark, 1984; Maass et al., 1987). A corollary discounting principle may be at work with the majority, the commitment to the majority position may be attributed to commitment or to social pressure. The curious paradox is that the greater the pressure, the more seriously the minority may have to be taken and the weaker the commitment of majority members to their positions.

Consequences to the Majority of a Minority Challenge

There is disagreement about whether the consequences to those who espouse a majority position may be governed by the same processes described for minority responses (Stasser, Kerr, & Davis, 1980; Latane & Wolf, 1981; Kruglanski & Mackie, 1990). It has been suggested that the same cognitive and intragroup processes described for the minority may occur when the motives in both groups are the same (Chaiken & Stangor, 1987). Where the majority perceives a significant and credible challenge from the minority (as in the astroidal impact example), a similar process of mobilization of resources, cognitive and otherwise, may occur.

Curiously, while the minority may have an initial advantage because of its high issue involvement (more elaborated, well-worked-out arguments), a significant threat may lead the majority to mobilize what are its greater resources. In particular, the majority has more members by definition. It therefore has a greater pool of ideas from which to draw, material resources are likely to be superior, the probability of one or more articulate spokespersons being available is greater, and so on (see Laughlin, 1988, on social combination processes). Moreover, under conditions of threat, the same ingroup-outgroup dynamics as described for minorities may obtain. Thus, a closing of ranks among members of the majority, greater pressures for uniformity and consensus, an increase in ingroup feelings, and an enhanced sense of social identity may occur. In the astroidal impact case, these processes seemed to have occurred. The geological and paleontological communities most immediately threatened by the new hypothesis mobilized their resources on a number of fronts, in the process reviving and strengthening an alternative hypothesis. As a result, the scientific community interested in this problem is split down the middle on the hypotheses now offered for species extinction. Part of the issue here is also demonstratability. Both hypotheses are clearly underjustified by whatever data are claimed to support them, and the argument is driven by appeals to theory as much as by appeals to data.

Further, in the service of collective action, majorities tend to amplify their power (Rosenwein & Campbell, 1992). These tendencies are illustrated in studies of concurrence-seeking tendencies in groups (Janis, 1982; t'Hart, 1991), group polarization and sampling (Vinokur & Burnstein, 1978), and the management of discontent (Gamson, 1968). As part of the process of collective mobilization, minority voices are silenced or ignored.

Where the issue is not as pressing, when the degree of ego involvement on the part of the participants is less, resource mobilization may not occur. In these situations, Chaiken and Stangor (1987) argue that differences in motivation create qualitatively different processes. If the minority position has dedicated and committed proponents, and if other conditions described in this chapter are satisified, this motivational energy may compensate for the resource advantages the majority is likely to have. On the other hand, minority staunchness may not be effective in creating change where the minority's power is small relative to the majority or where the majority can exercise forms of power assertion (Raven & Kruglanski, 1970). If the majority is also equally committed, the minority is at a great disadvantage. No matter how strong or "staunch" the minority, the majority may reject or ignore the minority.

A case illustrating majority attention to minority challenge is the history of the acceptance of the theory of plate tectonics (and the idea of continental drift) to account for geologic change in the earth's crust. In 1915 Alfred Wegener, described by Giere as "an exemplar of the innovative scientist who begins at the margin of his field, since his doctoral degree was in astronomy" (1922, p. 229), articulated the first well worked out defense of continental drift, in opposition to the more widely accepted notion at the time of fixed continents. Although there were geologists and geophysicists sympathetic to Wegener's ideas, it was a distinctly minority view. These ideas occasioned strong response and the mobilization of significant research effort on the part of majority "fixists." Intellectually, the focus of criticism was the lack of a plausible *mechanism* for explaining how continents could drift. Only in the 1960s, when the great oceanic rifts were discovered, was a plausible mechanism enunciated, followed by general acceptance of plate tectonic theory. Thus, a minority may be most persuasive when it can provide a mechanism which is acceptable to the community to account for a phenomenon. Of course, this is a complex episode. Several versions that are rich in the kind of detail that might interest students of SPS are available (Frankel, 1988; Giere, 1992; LeGrand, 1988).

Concept of Consensus and Nature of Scientific Belief Change

Kuhn (1970), in his descrtiption of paradigmatic change in science, suggests an important hypothesis about the conditions under which dissenting views become influential. His description of "normal science" is that it is based on assumptions about the world that determine the direction in which scientists seek answers to questions that are partly determined by paradigmatic assumptions. But over time, anomalies in the data begin to accumulate. At first, these are marginal and easily rationalized. At some point, however, the anomalies produce confusion and uncertainty, like a Sherif experiment where ambiguity is maximized. In this context, scientists search for a way out. During such crises, there may be greater receptivity to the minority view, particularly if it is asserted "staunchly" and satisfies proper epistemological criteria. The majority becomes informationally dependent on the minority to redefine the paradigm or replace it (Paichler, 1976). Another way of conceptualizing this point is to characterize the situation in which anomalies have built up as making a *norm of originality* more salient as opposed to other norms (Moscovici & Lage, 1978; Vinokur & Burnstein, 1978), and thus minority originality as more acceptable.

Kuhn's hypothesis assumes uniformity in commitment to a majority position. Such unifority is assumed in much of the social influence literature to define consensus. Gilbert and Mulkay (1984) and Fuller (1988; Chapter 3, this volume) have been critical of this assumption. They have suggested that there may be much less agreement on scientific beliefs at any point in history than is generally assumed. In social influence research and theory, there has been renewed interest in individual differences in social influence and a focus on majority members' motives (Chaiken & Stangor, 1987; Levine, 1989; Kruglanski & Mackie, 1990).

Based on the assumption that there are different levels of commitment in those persons expressing agreement with a majority position, I suggest there are at least four logical subgroups within the majority. The first, the dedicated majority (DM), have personal, professional, or ideological commitments to the majority position. Weldon and Pearson and the professional uniformitarian geologists and paleontologists of the two case studies are examples. Members of this subgroup resist converting to a new position the longest and would find ways to rationalize a change on their part. The second subgroup, the committed majority (CM), are those who believe in the consensus position but who will entertain other ideas. Derbishire and Schuster and Raup are examples. A third subgroup, the noncommitted majority (NM), are yielders who are normatively dependent on the majority. The majority rewards or punishes these individuals such that they subscribe to the majority position in order to survive. The logical fourth group are independents, as in Wolf and Latane's formulation (1985). They are neither members of the majority nor members of the minority. Like the breeders and doctors, they function as a jury, judging positions based on their own self-interest or other criteria.

What would we expect the "conversion history" of these four subgroups to be? Minority influence, when it produces conversion, does not do so in ways that the converted want made public. Members of the majority may privately be converted to the new position, but they comply publicly with the original majority position (Allen, 1975; Nemeth, 1986; Nemeth & Chiles, 1988).

Why should this happen? Self-presentational processes play a part. It is embarrassing for a member who has espoused a majority position to suddenly admit a change of heart if the majority felt threatened. It might be dangerous to admit conversion, depending on one's reliance on the majority. Subjects are more likely to convert to a minority position when the minority has left the setting (Moscovici, 1985). People avoid being identified too openly with the minority position in front of the majority, especially when the minority is a derogated outgroup.

We can speculate about the "conversion histories" of majority subgroups. If the scientific community converts to a new position, the DMs would be the last to change, would engage in the greatest self-bolstering and rationalization, would be most defensive, and would most actively defend the old position. The CMs would be most likely to engage in research that might prove decisive in changing people's positions, would be less defensive than DMs, and would move to public acceptance earlier than would NMs. The latter are likely to delay making a private conversion public before a critical mass of relevant individuals had gone public on the assumption that they feel they have the most to lose in breaking with the current consensus. Finally, independents are likely to go public the earliest with a conversion experience because they have no direct investment in a relationship with either the majority or the minority. Little data exist on such conversion histories as they develop over time.

Conclusion

What directions should the study of social influence in science take? First, social influence should be studied rather than artificially distinguishing between majority and minority influence. In this chapter, I did distinguish the two in order to highlight the essential tension between consensus formation and dissent as important in understanding how new beliefs come to be shared in a scientific community.

> [However], rather than debating in general the issue of process uniformity or process distinctiveness, investigators of minority and majority influence might do well to focus their research efforts on defining the contexts in which differences between majorities and minorities will have significant implications for influence and those conditions under which they will not. (Kruglanski & Mackie, 1990, p. 255)

For SPS, it "may make more sense to examine the specific factors that produce (underlie) compliance and persuasion and assess the relevance of those variables to scientific decision-making" (Anonymous, 1991).

Second, what is needed now in SPS is the *real-world, case-study data base* (see also Donovan, Laudan, & Laudan, 1988). Social psychologists of science need to specify a range of well-documented scientific controversies in which minority views did or did not prevail, studying those factors that affected the temporal course of the demise or acceptance of the viewpoint. They also need to debate the qualities or characteristics that make for appropriate case studies. Such an enterprise might help test current hypotheses, help suggest new ones for the laboratory, and help social psychologists to think harder about what science is as an enterprise and how they can best understand it.

References

Allen, V. L. (1975). Situational factors in conformity. In L. Berkowitz (Ed.), *Advances in experimental social psychology* (pp. 1–43). New York: Academic Press.

Allen, V. L., & Wilder, D. A. (1977). Social comparison, self-evaluation, and conformity to the group. In J. M. Suls & R. L. Miller (Eds.), *Social comparison processes: Theoretical and empirical practices* (pp. 42–83). Washington, DC: Hemisphere.

Allen, V. L., & Wilder, D. A. (1980). Impact of group consensus and social support on stimlus meaning: Mediation of conformity by cognitive restructuring. *Journal of Personality and Social Psychology, 39,* 1116–1124.

Alverez, L. W., Alverez, W., Asaro, F., & Michel, H. M. (1980). Extraterrestrial cause for the Cretaceous–Tertiary extinction. *Science, 208,* 1095–1108.

Alverez, W., & Asaro, F. (1990). An extraterrestrial impact. *Scientific American, 263*(4), 61–84.

Anonymous. (1991). [Reviewer's comments on first draft of this chapter.]

Asch, S. (1951). Effects of group pressure upon the modification and distortion of judgments. In H. Guetzkow (Ed.), *Groups, leadership and men* (pp. 177–190). Pittsburgh, PA: Carnegie Press.

Asch, S. E. (1955). Opinions and social pressure. *Scientific American, 193,* 31–35.

Asch, S. E. (1956). Studies of independence and conformity: A minority of one against a unanimous majority. *Psychological Monographs, 70*(9, Whole No. 416).

Campbell, D. T. (1961). Conformity in psychology's theories of acquired behavioral dispositions. In I. A. Berg & B. M. Bass (Eds.), *Conformity and deviation* (pp. 125–142). New York: Harper & Row.

Campbell, D. T. (1988a). A tribal model of the social system vehicle carrying scientific knowledge. In E. S. Overman (Ed.), *Methodology and epistemology for social science: Selected papers* (pp. 489–503). Chicago: University of Chicago Press. (Original work published 1979)

Campbell, D. T. (1988b). The experimenting society. In E. S. Overman (Ed.), *Methodology and epistemology for social science: Selected papers* (pp. 290–314). Chicago: University of Chicago Press.

Campbell, D. T. (1990). Asch's moral epistemology for socially shared knowledge. In I. Rock (Ed.), *The legacy of Solomon Asch: Essays in cognition and social psychology* (pp. 39–55). Hillsdale, NJ: Erlbaum.

Chaiken, S., & Stangor, S. (1987). Attitudes and attitude change. *Annual Review of Psychology, 38,* 575–630.

Chubin, D., & Hackett, S. (1990). *Peerless science: Peer review and American science policy*. Albany: State University of New York Press.

Collins, H. M. (1985). *Changing order: Replication and induction in scientific practice*. Newbury Park, CA: Sage.

Courtillot, V. E. (1990). A volcanic eruption. *Scientific American, 263*, 84–92.

Deconchy, J. (1985). The paradox of "orthodox minorities": When orthodoxy infallibly fails. In S. Moscovici, G. Mugny, & E. Van Avermaet (Eds.), *Perspectives on minority influence* (pp. 187–200). New York: Cambridge University Press.

Deutsch, M., & Gerard, H. B. (1955). A study of normative and informational social influences upon individual judgment. *Journal of Abnormal and Social Psychology, 51*, 629–636.

Doms, M., & Avermaet, E. (1980). The minority influence effect: An alternative approach. In W. Doise & S. Moscovici (Eds.), *Current issues in European social psychology* (pp. 73–97). New York: Cambridge University Press.

Donovan, A., Laudan, L., & Laudan, R. (Eds.). (1988). *Scrutinizing science: Empirical studies of scientific change*. Dordrecht, The Netherlands: Kluwer.

Eagly, A. H., Wood, W., & Chaiken, S. (1978). Causal inferences about communicators and their effect on opinion change. *Journal of Personality and Social Psychology, 36*, 424–435.

Festinger, L. (1954). A theory of social comparison processes. *Human Relations, 7*, 117–140.

Frankel, H. (1988). Plate tectonics and inter-theory relations. In A. Donovan, L. Laudan, & R. Laudan (Eds.), *Scrutinizing science: Empirical studies of scientific change* (pp. 269–288). Dordrecht, The Netherlands: Kluwer.

Fuller, S. (1988). *Social epistemology*. Bloomington: Indiana University Press.

Gamson, W. (1968). *Power and discontent*. Homewood: Illinois University Press.

Gerard, H. B. (1985). When and how the minority prevails. In S. Moscovici, G. Mugny, & E. Van Avermaet (Eds.), *Perspectives on minority influence* (pp. 171–186). New York: Cambridge University Press.

Giere, R. (1992). *Explaining science: A cognitive approach*. Chicago: University of Chicago Press.

Gilbert, G. N., & Mulkay, M. (1984). *Opening pandora's box: A sociological analysis of scientists' discourse*. New York: Cambridge University Press.

Gleick, J. (1987). *Chaos: A new science*. New York: Penguin.

Janis, I. L. (1982). *Victims of groupthink* (2nd ed.). Boston: Houghton Mifflin.

Kelman, H. C. (1958). Compliance, identification and internalization: Three processes of opinion change. *Journal of Conflict Resolution, 2*, 51–60.

Kim, K. (in press). *Explaining scientific consensus: A sociological analysis of the reception of Mendelian genetics*. New York: Guilford Press.

Kruglanski, A., & Mackie, D. M. (1990). Majority and minority influence: A judgemental process analysis. In W. Stroebe & M. Hewstone (Eds.), *European review of social psychology* (Vol. I, pp. 93–122). New York: Wiley.

Kuhn, T. (1970). *The structure of scientific revolutions* (2nd ed.). Chicago: University of Chicago Press.

Latane, B., & Wolf, S. (1981). The social impact of majorities and minorities. *Psychological Review, 88*, 438–453.

Laughlin, P. (1988). Collective induction: Group performance, social combination processes, and mutual majority and minority influence. *Journal of Personality and Social Psychology, 54*, 254–267.

LeGrand, H. E. (1988). *Drifting continents and shifting theories.* New York: Cambridge University Press.

Levine, J. M. (1989). Reactions to opinion deviance in small groups. In P. B. Paulus (Ed.), *Psychology of group influence* (2nd ed.) (pp. 187–232). Hillsdale, NJ: Erlbaum.

Levine, J. M., & Moreland, R. I. (1985). Innovation and socialization in small groups. In S. Moscovici, G. Mugny, & E. Van Avermaet (Eds.), *Perspectives on minority influence* (pp. 143–169). New York: Cambridge University Press.

Levine, J. M., & Ranelli, C.J. (1978). Majority reactions to stable and shifting attitudinal deviates. *European Journal of Social Psychology, 8*, 55–70.

Levine, J. M., & Russo, E. M. (1987). Majority and minority influence. In C. Hendrick (Ed.), *Group processes* (pp. 13–54). Newbury Park, CA: Sage.

Levine, J. M., Saxe, L., & Harris, H. J. (1976). Reactions to attitudinal deviance: Impact of deviate's direction and distance of movement. *Sociometry, 39*, 97–107.

Maass, A., & Clark, R. D. (1984). Hidden impact of minorities: Fifteen years of minority influence research. *Psychological Bulletin, 95*(3), 428–450.

Maass, A., West, S. G., & Cialdini, R. B. (1987). Minority influence and conversion. In C. Hendrick (Ed.), *Group processes* (pp. 55–79). Newbury Park, CA: Sage.

Mackie, D. M. (1986). Social identification effects in group polarization. *Journal of Personality and Social Psychology, 40*, 720–728.

Merton, R. (1973). Priorities in scientific discovery. In N. Storer (Ed.), *The sociology of science* (pp. 286–324). Chicago: University of Chicago Press.

Mitroff, I. (1974). Norms and counter-norms in a select group of Apollo moon scientists: A case study of the ambivalence of scientists. *American Sociological Review, 39*, 576–595.

Moreland, R. L. (1987). The formation of small groups. *Review of Personality and Social Psychology, 8*, 80–110.

Moscovici, S. (1974). Social influence I: Conformity and social control. In C. J. Nemeth (Ed.), *Social psychology: Classic and contemporary integrations* (pp. 179–216). Chicago: Rand-McNally.

Moscovici, S. (1985). Innovation and minority influence. In S. Moscovici, G. Mugny, & E. Van Avermaet (Eds.), *Perspectives on minority influence* (pp. 9–51). New York: Cambridge University Press.

Moscovici, S., & Lage, E. (1978). Studies in social influence: IV. Minority influence in the context of original judgments. *European Journal of Social Psychology, 8*, 349–365.

Moscovici, S., & Nemeth, C. J. (1974) Social influence II: Minority influence. In C. J. Nemeth (Eds.), *Social psychology: Classic and contemporary integrations* (pp. 217–249). Chicago: Rand-McNally.

Nemeth, C. J. (1986). Differential contributions of majority and minority influence. *Psychological Review, 91*, 23–32.

Nemeth, C. J., & Brilmeyer, A. J. (1987). Negotiation vs. influence. *European Journal of Social Psychology, 17*, 45–56.

Nemeth, C. J., & Chiles, C. (1988). Modelling courage: The role of dissent in fostering independence. *European Journal of Social Psychology, 18*, 275–280.

Nemeth, C. J., & Kwan, J. L. (1987). Minority influence, divergent thinking and detection of correction solutions. *Journal of Applied Social Psychology, 17*, 788–799.

Nemeth, C. J., & Staw, B. (1989). The trade-offs of social control and innovation in groups and organizations. *Advances in Experimental Social Psychology, 22*, 175–210.

Ng, S., & Cram, F. (1988). Intergroup bias by defensive and offensive group in majority and minority conditions. *Journal of Personality and Social Psychology, 55*, 749–757.

O'Keefe, D. J. (1990). *Peruasion: Theory and research*. Newbury Park, CA: Sage.

Paichler, G. (1976). Norms and attitude change I: Polarization and styles of behavior. *European Journal of Social Psychology, 7*, 405–427.

Papastamou, S., & Mugny, G. (1985). Rigidity and minority influence: The influence of the social in social influence. In S. Moscovici, G. Mugny, & E. Van Avermaet (Eds.), *Perspectives on minority influence* (pp. 113–136). New York: Cambridge University Press.

Petty, R. E., & Cacioppo, J. T. (1981). *Attitudes and persuasion: Classic and contemporary approaches*. Dubuque, IA: W. C. Brown.

Petty, R. E., & Cacioppo, J. T. (1986). The elaboration likelihood model of persuasion. In L. Berkowitz (Ed.), *Advances in experimental social psychology*, (Vol. 19, pp. 123–205). New York: Academic Press.

Petty, R. E., Cacioppo, J. T., & Goldman, R. (1981). Personal involvement as a determinant of argument-based persuasion. *Journal of Personality and Social Psychology, 41*, 847–855.

Raup, D. M. (1986). *The nemesis affair: A story of the death of dinosaurs and the ways of science*. New York: W. W. Norton.

Raven, B., & Kruglanski, A. (1970). Conflict and power. In P. Swingle (Ed.), *The structure of conflict* (pp. 69–109). New York: Academic Press.

Rosenwein, R. E., & Campbell, D. T. (1992). Mobilization to achieve collective action and democratic majority/plurality amplification. *Journal of Social Issues, 48*, 125–138.

Rosenwein, R. E., & Koenemund, K. (1991). [Mutual majority and minority influence as a function of task ambiguity in science: An experimental analogue]. Unpublished research.

Ross, L., Bierbauer, G., & Hoffman, S. (1976). The role of attribution processes in conformity and dissent: Revisiting the Asch situation. *American Psychology, 31*, 148–157.

Schachter, S. (1951). Deviation, rejection and communication. *Journal of Abnormal and Social Psychology, 46*, 190–207.

Sherif, M. (1936). *The psychology of social norms*. New York: Harper & Row.

Stasser, G., Kerr, N. L., & Davis, J. H. (1980). Influence processes in decision-making groups. In P. B. Paulus (Ed.), *Psychology of group influence* (2nd ed.) (pp. 279–326). Hillsdale, NJ: Erlbaum.

Tajfel, H., & Turner, J. C. (1979). An integrative theory of intergroup conflict. In W. G. Austin & S. Worchel (Eds.), *The social psychology of intergroup relations* (pp. 33–47). Belmont, CA: Wadsworth.

Tanford, S., & Penrod, S. (1984). Social influence model: A formal integration of research on minority and majority influence processes. *Psychological Bulletin, 95*, 189–225.

t'Hart, P. (1991). Irving L. Janis' *Victims of groupthink* [Review of *Victims of groupthink*]. *Political Psychology, 12*, 247–278.

Tuckman, B. W. (1965). Developmental sequences in small groups. *Psychological Bulletin, 63*, 384–399.

Vinokur, A., & Burnstein, E. (1978). Novel argumentation and attitude change: The case of polarization following group discussion. *European Journal of Social Psychology, 8*, 335–348.

Wolf, S., & Latane, B. (1985). Conformity, innovation and the psychosocial law. In S. Moscovici, G. Mugny, & E. Van Avermaet (Eds.), *Perspectives on Minority influence* (pp. 201–215). New York: Cambridge University Press.

CHAPTER 9

Scientists' Responses to Ethical Issues in Research

JOAN E. SIEBER

In the last few decades, there has been a growing awareness that science is not value free and may cause harm (e.g., Katz, 1972; Kelman, 1968; Ryan, 1971; Vinacke, 1954). Significant changes have occurred that bear on the ethics of science. Some of these are social and institutional in nature, such as the establishment of institutional review boards (IRBs) (e.g., Gray, Cook, & Tannenbaum, 1978). Other changes are perceptual and attitudinal, and are manifested in a greater tendency of special-interest groups, government, and scientists themselves to raise ethical issues pertaining to science; some examples of such issues are (1) whether human subjects may be deceived about research in which they participate, (2) whether investigators must share their raw data, (3) whether substantial payment for research participation by poor people is coercive, and (4) whether individual informed consent is appropriate in non-western cultures in which all authority is vested in a leader. Decisions about what constitutes ethical behavior in such matters is no longer left to the individual scientist. Scientists, individually and collectively, must now comprehend and respond to many ethical issues raised by others. The process of collective and individual response to ethical issues in science is the topic of this chapter.

When an ethical issue is first raised, scientists typically deny that the issue is serious or that scientists should alter their behavior. By the time the issue is settled, scientists respond according to a new norm of behavior to resolve the ethical issue—perhaps not as the issue was initially defined but rather as it ultimately came to be defined. What is the psychology underlying such change in perspective and behavior?

Two processes of problem solving are involved: (1) the ethical issue is defined and rules to guide scientists are formulated; and (2) scientists learn to understand and apply the rules. Both processes cause conflict between those emphasizing the magnitude of the concern and those emphasizing the undesirability of constraints on science. Although little is known about the dynamics of these two interrelated processes, some relevant variables readily come to mind. For example, participants may be reasonable and competent problem solvers, or acrimonious and unyielding extremists; the rules they forge may be wise and creative or misguided and procrustean. Those involved in defining the issue and formulating the rules may have enlisted the talents of competent scientists, may have rebuffed scientists' efforts to assist, or may have found no scientists interested in assisting. Thus, ethical analysis and socialization occur in many ways, cause varying amounts of conflict, and achieve varying degrees of ethical socialization and benefit to science and society.

These processes can be studied to help science and society achieve better solutions. However, as a research topic, the ethical socialization of scientists is fraught with problems: The subjects—scientists, representatives of scientific societies, and their critics—are relatively powerful people and may successfully resist serving as subjects of research (Sieber, 1989). Behavioral, cognitive, and attitudinal processes that occur as an ethical issue moves from recognition to behavioral change of scientists are complex and not easily measured. At what entry points and with what conceptual and methodological tools might researchers study these processes? Although the ethical socialization of scientists is a difficult and politically sensitive research topic, there are compelling reasons to research it:

1. It is defensive and shortsighted of science not to understand its own responses to value conflicts.
2. Many resources and opportunities are squandered in science through unenlightened conflict about ethical issues.
3. Research on scientists' own ethical socialization may lead to the development of important new professional insights, research skills and methods.

What Is an Ethical Issue in Science?

Matters such as informed consent, confidentiality of data, and data sharing involve ethics because they involve the perception of some form of good in science or society. They become issues when they cause conflicts among values or interests. For example, informed consent might mean informing subjects of circumstances under which confidentiality could not be assured (e.g., if the data were subpoenaed, if the subject disclosed details of child abuse, or if one intended to share data including unique identifiers with colleagues). Such full informing might diminish response rate or candor. But not all ethical issues arise out of conflicts. For example, some investigators refuse to share their data because they do not wish to invite warranted criticism of their work.

Almost weekly, articles in *Science* or *Chronicle of Higher Education* discuss ethical issues in science, such as scientific fraud, unwillingness of scientists to share data, and dual role relationships. Typically, (1) a scientist is alleged to have behaved unethically; (2) someone within government calls for federal investigation, claiming that universities are ill-equipped or unwilling to investigate or punish misbehaving senior scientists; and (3) university officials claim that the misbehavior, if any, is an isolated case and best investigated by the scientist's institution. In some cases, (4) a government agency, scientific or professional society, or funder commands the scientific community to effect some changes. However, not all ethical issues unfold this way:

- In the late 1960s, biomedical research practices that were heedless of human dignity and welfare (Katz, 1972; Sieber, 1982a) attracted congressional attention, stimulating development of bioethical analysis and resulting in laws governing biomedical, social, and behavioral research on humans and massive changes in the control and conduct of *all* human research.
- In the mid 1970s, animal rights activists protested cruel and nonessential animal research. They gathered vocal popular support, and then gained allies in government and achieved regulation of animal research and many changes in research practices (Zola, Sechzer, & Sieber, 1984). However, some "fringe" animal rights activists now call for eliminating all animal research, and are damaging laboratories, stealing research animals, and terrorizing researchers. These acts resulted in some loss of popular, judicial and legislative support for the animal rights movement.
- Since 1970, five social scientists have used deception research methods to study the editorial practices of key social science journal editors, for which they have been harassed, tried for professional misconduct, and professionally blackballed. The hypocrisy of their treatment has been discussed in various public forums (the entire May 1982 issue of *Behavioral and Brain Sciences*, the Winter 1990 issue of *Science, Technology and Human Values*, and Sieber, 1989).

Efforts to create guidelines or acceptable methods for studying

> powerful people have stalled for lack of institutional sponsors. Institutions, such as the professional associations that would be involved in establishing guidelines for research on the powerful, have a built-in resistance to examining themselves. At most, a professional association may appoint a committee to discuss the issue, but such committee activities tend to amount to window dressing; no real institutional change is likely to result. However, as the cases at the end of this chapter suggest, overarching scientific organizations such as the National Academy of Sciences are more permeable to new values, less vulnerable to complaints from particular scientists who seek to avoid scrutiny, and more likely to foster institutional change.

Thus, some ethical issues are raised by those in the mainstream of society, some are powerful fringe concerns, and some are raised by scientists. Some concerns eventuate in massive or minor changes in science, and some concerns go nowhere. But apart from simple descriptions, there is little understanding of how such issues evolve, or of interventions that would render the process more beneficial. A model of the process is prerequisite to research on the psychology of responses to ethical issues in science.

The Search for a Suitable Model

Research on ethical issues tends to focus on an isolated aspect of the process (Stanley, Sieber, & Melton, in press). Without a model, researchers are like the mythical blind men studying the elephant. How does the part being examined relate to other parts? How does the whole process work? A rudimentary model is offered here as a starting point for efforts to fit together what is known about the process of ethical socialization in science, to improve the model, and to design interventions that test and extend the model.

Kuhn's (1962) characterization of scientific revolutions is not an apt model. Unlike scientific revolutions, most ethical changes receive their initial impetus from outside science, and do not result in radical changes in world view. Nor are psychological theories of moral socialization helpful, for unlike research on moral socialization, the study of ethical socialization in science begins when some matter becomes an ethical issue, not later when individual scientists begin to learn the resulting norm of scientific behavior. The early struggle between science and other institutions over whether an issue exists probably affects the ensuing ethical socialization of the individual scientist.

The Proposed Model

The most pertinent model is a composite of many processes. This model arises out of my research that documents elements of ethical issues, and tries to invent solutions that scientists will accept when strong forces require that *something* be done.

The model is based primarily on events surrounding the 1974 federal mandate that established human subjects review committees, or IRBs as they are variously called, and that required that informed consent be obtained for most human research. Nine stages of this case suggested the stages of the proposed model. After the discussion of each stage, possible research or intervention is suggested.

Stage 1: Initial Confrontation

Scientists are (perhaps rudely) informed that there is an ethical issue to resolve. Scientists typically accuse the news bearers of ignorance of science or irrelevance to their discipline.

For example, in 1974 the Federal Government mandated that scientists studying human subjects must obtain informed consent and that IRBs must be established at research institutions expecting to receive federal funds. Most social scientists considered this mandate an appropriate, overdue response to biomedical research but a most inappropriate and uninformed response to social and behavioral research. Social and behavioral research cannot harm people, they argued; moreover, informed consent could destroy research that requires deception. (Alternatives to deception and consent to deception were not considered.) IRBs had been required for all research sponsored by the Public Health Service as of February 8, 1966. However, the initial confrontation did not occur until 1974 when informed consent and IRB review requirements were extended by the National Research Act to all research involving human subjects at institutions that receive federal funds.

Most new ethical issues in science have a longer history than is generally acknowledged. The basic concept of informed consent was set forth at Nuremberg in 1949 (Katz, 1972). Vinacke's (1954) critique of deception research, Kelman's *A Time to Speak* (1968), and Katz's *Experimentation with Human Beings* (1972) are excellent resources for understanding why Congress investigated the conduct of human research, and what differences between biomedical and social science must be understood before sensible research norms can be formulated. Yet, criticisms such as those of Vinacke and Kelman were ignored even though these psychologists are highly respected scientists.

Intelligently organized funding would have dignified, integrated, and institutionalized scholarship such as Vinacke's and Kelman's, making it impossible to ignore. In addition, Stage 1 is the point at which to begin study of the ensuing evolution of the behavior of the scientists involved.

Stage 2: Insistence, from outside Science, on the Problem

The issue is repeatedly restated by persons or institutions with power to threaten scientists, or to attract their attention. In response, groups of scientists meet and take action to defend their prerogatives.

For example, in 1974, the National Commission for the Protection of Human Subjects of Biomedical and Behavioral Research was established by the National Research Act to formulate the ethical principles that should underlie biomedical and behavioral research on human subjects and to make recommendations to the Secretary of the Department of Health, Education and Welfare and to Congress. The twice-monthly hearings of the commission often made headlines. Scientists were invited to testify and to help Congress to craft reasonable regulations. By 1975, some scientists had defended their methods at hearings of the commission (e.g., Berkowitz, 1976).

As debate develops, critics sharpen their arguments about why current practice is wrong but are insensitive to the costs of proposed changes to science and society. Scientists sharpen their arguments about why they cannot use the suggested reform without harm to science and remain deeply embedded in the norms and folkways of their discipline. One can be both impressed by the logic of both sides and dismayed at their intransigence.

At this stage, the polemic between investigators and scientists may take on the character of a trial in which the prosecutor presents the worst possible scenario and the defense paints a picture of utter innocence. Scientists, when under powerful attack, may see no room for compromise or problem solving. It would benefit both sides if they would devote some resources to developing viable solutions. Examples of such successful problem solving appear at the end of this chapter.

Stage 3: The Debate Within

Scientists explore and debate the issue. Some embrace the new ethic and find themselves in the minority (Warwick, 1975). Scientists who publish ethical criticism of current research practice tend to be senior, because this is a poor strategy for tenure and promotion. Those who promulgate moderate ideas are often attacked by both sides—as in my research on students' and scientists' attitudes toward animal research laboratory courses (Sieber, 1986).

If the psychology of science were to develop as an area of respectable inquiry, greater legitamization might be given to the scientific study of ethical change and socialization in science. Thus, young scientists might have a respectable context in which to examine these issues. Respectable journals might publish their findings, helping them avoid the suspicion of peers that they are useless gadflies or incipient whistle blowers.

Stage 4: Science Protects Its Turf

Scientists speak out against the new ethical requirements. Professional publications are typical avenues for this polemic. For example, Berscheid, Dermer, and Libman (1973) published an article in *American Psychologist* showing that college students are not interested in reading (incredibly long, tedious) informed consent statements.

Incentives for more creative and constructive responses to new ethical issues should be offered. An attractive, inexpensive approach would be for professional societies to offer prizes for research in which small studies of the effects of variations in ethical procedures are grafted onto larger studies that raise ethical/procedural questions. An example of such research within research is Singer's (1978) study of the effects of variations in informed consent and promises of confidentiality on response rate and candor, nested within a larger survey research. Unfortunately, Singer received no special reward for her pioneering work. The fostering of such change by professional associations might incite membership revolt. Perhaps such enlightened programs and contests are better left to organizations such as the National Academy of Science and the American Association for the Advancement of Science (AAAS).

Stage 5: Scientists as Ethical Problem Solvers

A few social scientists recognize that they can use the theories and methods of their discipline to study ethical issues and devise solutions.

For example, Boruch and Cecil (1979) present sophisticated procedural, methodological, statistical, and legal solutions to risks to confidentiality in social research, and Stanley, Sieber, and Melton (in press) summarize the sizable literature of empirical research on social science ethics.

This is the beginning of systematic scientific development of knowledge, methods, and procedures in response to ethical issues. Because these efforts are not supported by any professional association, any funding agency, or most journal editors, these innovators must be highly ingenious in creating methods that work and will be seen as important contributions to the methodology of science. One hopes eventually for a cadre of peer reviewers to evaluate and support good ideas in this realm. The National Science Foundation (NSF) program, Ethics and Values Studies, is a step in this direction.

Stage 6: Gatekeepers Enforce the New Rules

Critical gatekeepers (government, funders, professional associations, university administrators) enforce the new norm. This is the first time many scientists have confronted the new rule or thought about how to implement it. This is the beginning of their ethical socialization.

For example, by 1980, most universities had established IRBs. IRBs were respected by some scientists but scorned by others who perceived them as unreasonable and costly to society and science; some scientists learned to be ethical while others learned to skirt IRB requirements.

Regulation is now a major element in the ethical socialization of scientists. The structure and function of regulatory mechanisms and their effect on scientists should be studied. Research on the functioning of IRBs was undertaken shortly after their inception to improve them (Gray, Cook, & Tannenbaum, 1978). Additional intervention research should be undertaken as the regulatory process matures, and it should focus on both the regulators and the resultant effects on scientists.

Stage 7: Perfunctory Socialization of Scientists and Students

At this point, "Sunday Christians" abound—individuals who neither understand nor practice the ethical principles they profess. Some scientists propose courses on ethics and research, but turf battles within and between departments prevent such courses from counting for credit in a major; they die from underenrollment. Textbook authors mention the new ethical norm without endorsing it so much as to offend those faculty who consider it useless. For example, a page may be devoted to ethics or informed consent, but the topics are oversimplified and students are not told how to implement these new norms.

In Stages 5 and 6, procedural and methodological research is not accorded much prestige in scientific funding or publication. We need a better understanding of the attitudes within the scientific establishment that account for this lack of interest. Such knowledge might hasten change, producing a "respectable literature" in prestigious journals, allowing textbooks to cite examples of methodological research driven by ethical concerns, and encouraging students to devise their own research on such problems.

Stage 8: Professional Socialization Begins

Eventually, the ethical issues become clearly defined and largely agreed on. Procedural, methodological, statistical, and legal solutions to ethical problems find their way into the literature. Professional workshops, codes of ethics, and university courses present the (now rather old) new ideas in a positive light, creating new norms of science. Some funding for course development, materials and workshops may become available. Conferences, and symposia are held.

For example, many books now deal with ethics in social science. The notion that science is value free and that value issues have no place in science seems antiquated. The journal *IRB: A Review of Human Subjects Research* has been in circulation since 1978, and *Ethics and Behavior* was launched in 1991. Two organizations give workshops on research ethics for those who administer and serve on human subjects review committees: Public Responsibility in Medicine and Research focuses primarily on biomedical research; its sister organization, Applied Research Ethics National Association, focuses on social and behavioral research. Socialization is under way, and a literature is being generated, but it is perceived as a literature on applied ethics and research administration, not on research methodology. A self-fulfilling prophecy is at work here. For example, in *IRB: A Review of Human Subjects Research*, I (Sieber, 1982b), published a taxonomy of kinds of deception, kinds of harm each might cause (Sieber, 1983), and approaches to choosing the least harmful deception methods that would satisfy the research objective. I did not publish this paper in a journal that would reach most deception researchers, such as *Journal of Personality and Social Psychology*, because I believed it would be subjected to a long review process followed by rejection. Consequently, my ideas were not cited and discussed in social psychology methods texts for several years.

With increased understanding of ethical issues having methodological implications, we need a concerted effort to incorporate the resulting ideas into mainstream research methodology. Perhaps I was correct in writing *Planning Ethically Responsible Research* as part of the Applied Social Research Methods Series, not as a special ethics book. Methodologists should publish reviews of methods emphasizing ethics and their relationship to validity and use of results.

Stage 9: The New Norm Becomes Established

Within a few decades, students and professionals do not recall the old norm. Even ethically insensitive scientists routinely accept the new norm.

For example, IRBs are now accepted by researchers and administrators as a convenient way to control students and colleagues who exercise poor judgment in the design and conduct of human research. In these litigious times, scientists and administrators are concerned about liability. The notion that research, especially funded human research, should occur unfettered by ethical review seems outdated.

Now we have a new norm. But history spirals onward, especially where topics and environments of research are changing and highly value laden. Ethical issues in science rarely arise because scientists are immoral, but rather because they are insensitive to changing work contexts or methods. A new ethical issue may arise at any time. An institution such as science can neither adapt to every change nor flinch at every social sensitivity. It should, however, strive to adapt to significant changes in the surrounding culture.

If the psychology of ethical socialization in science were to become a serious topic of theory and inquiry, emerging ethical issues might be viewed, at least by students of ethical socialization, as topics for scholarship and research and not as accusations to be denied. The options within governmental and scientific institutions for responding to emerging ethical issues would be increased and clarified. Objective assessment of attitudes and sensitivities to emerging ethical issues might occur, and interventions might be sought that enhance the process of change from the perspectives of science and society. While no one is prescient enough to devise the best course of action without sufficient analysis, it may be possible to improve scientists' skill at ethical analysis.

Research on and Intervention in the Overall Process

Many psychologists have studied isolated parts of scientists' responses to ethical issues. The question is whether research can capture or intervene in the larger dynamic of this process. To examine this possibility, consider briefly three cases. Two are efforts to require data sharing—one successful and one not—based on my own experience with these cases. The third is an example of a highly successful intervention in the area of animal welfare and research regulation which contrasts sharply with some of the more acrimonious efforts in that area.

> *Case 1*: In 1985, the NSF initiated a pilot project to require data sharing in the social sciences, and developed flexible policies for administering this requirement. Scientists were expected to make their documented raw data available on request to other scientists at no more than the marginal cost of reproducing and delivering the

requested data. (Documentation means describing the data so that others can use the data file correctly, for example, describing the sampling and data cleaning procedures and the handling of missing values and outliers.) Following the success of this pilot program, the director of the NSF announced to university presidents in 1989 that all scientists who receive NSF funds were henceforth to share their documented raw data and other research materials (e.g., software and samples). The most remarkable aspect of the NSF data-sharing requirement is its quiet success.

In the course of this experiment, the NSF encouraged Stage 5 scientific activities. I received two such grants. In 1987, I held a conference to explore social scientists' concerns about data sharing. This resulted in a January 1989 AAAS symposium and a book (Sieber, 1991). Since 1990, my statistician colleague, Bruce Trumbo, and I have been teaching faculty to use shared data sets in computer-based statistics and methodology courses, and as a convenient and economical foundation for their own research programs. We are also contacting scientists and requesting data sets that our workshop participants wish to use, and studying the outcome of each request. We will study the effects of our workshops and follow-up support on the subsequent curriculum development, teaching, and research of the workshop participants.

Case 2: The Public Health Service (PHS) hoped to require data sharing of its grantees. I addressed a PHS data-sharing workshop for biomedical scientists in April 1990 on scientists' concerns about data sharing. Many scientists who had asked to be invited seemed to have come only to protest the whole idea, and they treated my comments as irrelevant to biomedical science. The PHS had not done the careful groundwork that characterized the NSF data-sharing program, and it had no concrete examples of workable data sharing in biomedical sciences to offer. Basic biomedical research is also competitive; sharing might mean losing one's competitive edge. Moreover, many biomedical scientists suspected the PHS of plotting to require voluminous documentation and data storage so that auditors could discover scientific or fiscal fraud. The conference ended in heated debate between the conveners and some science administrators who denounced data sharing.

Case 3: When the PHS strengthened its policy governing animal experimentation in 1985, the focus was on captive laboratory animals, not on free-living wild vertebrate animals. The provisions of this captive laboratory animal policy did not take into account the special problems of zoological studies conducted by field researchers. In fact, application of laboratory guidelines to field studies led to some inappropriate decisions; not only are the species likely to differ, but the purposes and conditions surrounding field and laboratory research are fundamentally different (Orlans, 1990).

Bruce Umminger of the Cellular Biosciences Division and James Edwards of the Environmental Biology Division of the NSF urged the various professional associations of field biologists to develop guidelines for research on wild species in field and laboratory settings. The relevant societies for the study of free-living mammals, birds, amphibians, reptiles, and fish developed guidelines, examining such issues as capture methods, marking for identification, tissue sampling, housing and feeding needs of these diverse species when brought into the laboratory or when studied in a natural habitat, pain perception and how to recognize signs of stress and pain in these various species, how the stages of development of each species relate to its needs, and so on. Indeed, the needs of some of these species defy conventional wisdom about the care of laboratory animals. For example, *wild* mice mark their territory, hence frequent changes of bedding as required for laboratory mice will cause them to constantly re-mark their territory; anesthesia so distresses birds (including laboratory birds such as pigeons) that brief surgical procedures should be without deep anesthesia; chilling amphibian larvae before immersing them in a preservative may seem humane, but rather than numb them, chilling lowers metabolism and delays death. On issues such as these, needless acrimonious debate might have ensued had not accurate knowledge about specific species been used to develop standards of care. In addition, a field scientist, working with minimal equipment that must be carried to the research site, needs more latitude with respect to such activities as surgical procedures and euthanasia than would be permissible in the laboratory. A consensus based on humane and scientifically sound practice needed to be developed for field-based procedures.

On October 8, 1987, a workshop titled "Field Research Standards" was held at the Rockefeller University, sponsored by the NSF and organized by the Scientists' Center for Animal Welfare under the direction of Barbara Orlans. The new draft guidelines for mammalogy, ornithology, herpetology, and ichthyology were reviewed and discussed. Final versions of those guidelines are now available (Orlans, 1988).

The contrast between these three cases suggests that thoughtful early investment in Stage 5 activities may help avoid acrimonious conflict. However, apart from the success or failure of such an innovation, we know little about the underlying psychology of these phenomena. For example, what would interviews reveal about the effects on scientists' attitudes and behavior of early interventions in the process of debate and socialization?

As these three cases suggest, different areas of science have different experiences of ethical issues and ethical socialization. Taken as different treatment conditions, they are part of a natural quasi-experiment (Cook

& Campbell, 1979) within which one may ask such questions as: How are scientists' attitudes and behavior influenced by the socialization to which they are exposed? What arguments do people learn about whether they should engage in a new ethical norm? What difficulties does the new norm create? What negative or positive attitudes are scientists likely to teach their students about the new norm? What factors foster sensitivity in scientists and students to emerging ethical issues? What factors foster development of wise and foolish rules and procedures intended to make scientists more ethical? Research on the process of the ethical socialization of scientists promises to be interesting, challenging—and controversial.

Acknowledgments

I owe much to those who commented on earlier drafts of this chapter, expecially Rachelle Hollander, Bill Keogh, Barbara Orlans, Will Shadish, Judith Swazey, and Bruce Umminger.

References

Berkowitz, L. (1976). *Some complexities and uncertainties regarding the ethicality of deception in research with human subjects.* Background paper prepared for the National Commission for the Protection of Human Subjects of Biomedical and Behavioral Research. Washington, DC: Department of Health, Education and Welfare.

Berscheid, E. R. S., Dermer, M., & Libman, M. (1973). Anticipating informed consent: An empirical approach. *American Psychologist, 28*, 913–925.

Boruch, R. F., & Cecil, J. S. (1979). *Assuring confidentiality of social research data*. Philadelphia: University of Pennsylvania.

Cook, T., & Campbell, D. (1979). *Quasi-experimentation: Design and analysis issues for field settings*. Chicago: Rand-McNally.

Gray, B., Cook, R., & Tannenbaum, A. (1978). Research involving human subjects. *Science, 201*, 1094–1101.

Katz, J. (1972). *Experimentation with human beings*. New York: Russell Sage.

Kelman, H. (1968). *A time to speak: On human values and social research*. San Francisco: Jossey-Bass.

Kuhn, T. S. (1962). *The structure of scientific revolutions*. Chicago: University of Chicago Press.

National Research Act. (1974). Tit. II, Pub. L. No. 93–348.

Orlans, F. B. (1988). *Field research guidelines: Impact on animal care and use committees*. Bethesda, MD: Scientists Center for Animal Welfare.

Orlans, F. B. (1990). Fundamental differences between field and laboratory investigations. *Laboratory Animal, 19*(3), 43–44.

Ryan, W. (1971). *Blaming the victim*. New York: Pantheon.

Sieber, J. E. (Ed.). (1982a). *The ethics of social research: Surveys and experiments*. New York: Springer-Verlag.

Sieber, J. E. (1982b). Deception in social research I: Kinds of deception and the wrongs they may involve. *IRB: A Review of Human Subjects Research, 3* (1–2), 12.

Sieber, J. E. (1983). Deception in social research II: Factors influencing the magnitude of potential for harm or wrong. *IRB: A Review of Human Subjects Research, 4*(1–3), 12.

Sieber, J. E. (1986). Scientists' and students' attitudes on animal research. *American Biology Teacher, 48*(4), 85–91.

Sieber, J. E. (1989). On studying powerful people (or fearing to do so): A vital role for IRBs. *IRB: A Review of Human Subjects Research, 11*(5), 1–6.

Sieber, J. E. (Ed.). (1991). *Sharing social science data: Advantages and challenges*. Newbury Park, CA: Sage.

Singer, E. (1978). Informed consent: Consequences for response rate and response quality in social surveys. *American Sociological Review, 43*, 144–162.

Stanley, B., Sieber, J. E., & Melton, G. (Eds.). (in press). *Psychology and research ethics*. Lincoln: University of Nebraska Press.

Vinacke, W. E. (1954). Deceiving experimental subjects. *American Psychologist, 9*, 155.

Warwick, D. (1975, October). Deceptive research: Social scientists ought to stop lying. *Psychology Today*, 38–40.

Zola, J., Sechzer, J., & Sieber, J. (1984). Animal experimentation: Issues for the 1980's. *Science, Technology and Human Values, 9*(2), 40–50.

CHAPTER 10

Characterizing Niches and Strata in Science by Tracing Differences in Cognitive Styles Distribution

JOHN M. WILKES

The Emergence of the Cognitive Styles Research Tradition

The cognitive styles perspective on science took shape amid the 1960s organizational behavior studies of innovation by industrial scientists in research and development (R&D). Both organizational structure and characteristics of personnel are legitimate objects of study in this research tradition, as both are subject to managerial manipulation.

It was in this setting that Gordon and Morse (1969) solved a key problem in the creativity literature in psychology. They developed a simple two-step model of the creative process, distinguishing problem formulation from problem solving. (Wilkes, 1979) This reconceptualization subsumed ongoing debates about how many different types of creativity there were and how many traits would be needed to account for creativity (Roe, 1953; Guilford, 1959; Ghiselin, 1963; Barron, 1969).

Existing measures of creativity were based on Guilford's (1959) notion of divergent thinking, best measured by ideational fluency. The latter measures (Wallach, 1971) were often combined with IQ measures (convergent thinking in Guilford's schema) to assess people's ability to handle the problem-solving stage of creativity (Hudson, 1966). No indicator of the problem-formulation stage was available, despite the heavy stress on this by noted theorists of the self-actualization school of thought (Rogers, 1959; Maslow, 1959; Fromm, 1959).

Gordon filled this empirical gap by developing a measure of "social differentiation" ability and showing that the measure had value in predicting performance in the problem-formulation stage of creativity (Gordon & Morse, 1969). Gordon also adopted Mednick's Remote Associates Test as his indicator of giftedness at the second (problem-solving) step of the creative process. Mednick's measure was controversial because it had right and wrong answers, which contradicted the prevailing theory that creativity is divergent rather than convergent thinking (Hudson, 1966; Wallach, 1971). Mednick (1963) claimed that the Remote Associates Test got at both because a remote associator must generate a range of possibilities and then recognize one that fits the specifications for an acceptable answer.

Both differentiation and remote association (RA) predicted performance in industrial R&D but were not highly correlated with each other ($r = .21$). Hence, Gordon and Morse (1969) developed a typology of four cognitive styles (see Table 10.1). I have based my 20-year exploration of the relationship between cognitive styles and scientific research accomplishment on this typology.

The Cognitive Styles Perspective

In the cognitive styles literature, people are not creative but processes and solutions may be. People who are gifted in different ways encounter situations with different demands. Where gifts and situations match, people are unusually successful or excel. Creativity is then attributed to them by peers or superiors. The trick is to predict what types will match a given situation.

Gordon and Morse (1969) portrayed Differentiators as perceptive and discriminating, just as an artist might be said to have a discriminating eye for significant details. High Differentiators are said to perceive the environment as distinct parts, attending to unique qualities of things and people, and drawing subtle distinctions based on nebulous or subjective criteria. Hence, early in a research project the High Differentiator is open to diversity, sensitive to nuance, and attuned to subtle discrepancies in theory or data that are often cues as to where investigation might fruitfully begin.

TABLE 10.1. Gordon's Cognitive Styles Typology

Differentiation	Remote association	Type
High	High	Integrator
High	Low	Problem finder
Low	High	Problem solver
Low	Low	Implementor

In contrast, Low Differentiators are more bound by preconceptions and see the environment as homogeneous, tending to suppress discrepancies. When diagnosing a problem or an unexpected result, this cognitive quality is detrimental, although at other times it can be advantageous since high differentiators can more easily get hung up on details and lose track of the overall trend. In the case of diagnostic or critical evaluation tasks, it is the low differentiators that are more vulnerable to selective perception, organizational mind-sets, and paradigm-induced blindness. They overlook or reject the unanticipated, so they have less information to work with at the outset and less ability to navigate in uncharted waters when prevailing theory fails. When generalizing from a proven case, however, they are less likely to be "hung up" on details.

Gordon was interested in RA because it is most evocative of what scientists call the Eureka experience. Answers to the RA items are non-logical. When the answers come, they come suddenly, completely, and with certainty. Gordon viewed RA as an information storage and retrieval quality, allowing people to generate and review alternatives to solve a problem—a directed mental flexibility often described as an intuitive grasp or ingenuity. RA involves recombining available materials while bringing in elements not normally connected with the materials, to fashion fitting, elegant, or optimal solutions.

During the ensuing 2 decades, the value of Gordon and Morse's (1969) typology has been proven in studies of career choice, problem selection, reception of innovation, patterns of productivity, relative success of academic and industrial scientists, interpersonal conflict, group dynamics, leadership styles, and peer review. (Wilkes, 1976; Boynton, Cleary and Plummer, 1989; Feeny, 1980; Charanian, 1964; Wilkes, 1980; Gordon and Morse, 1969; Wilkes and Neumann, 1978) These studies suggest that different kinds of scientists play different roles in science. The fact that scientists stereotype each other in terms that are evocative of these cognitive differences (Mitroff, 1974) and conflict with their cognitive opposites (Boynton et al., 1989) lends credence to the hypothesis. The cognitive styles approach has proven useful in examining issues as diverse as the experience of women in science (Bar-Haim & Wilkes, 1989), the hacker debate in computer science (Wilkes, Mangu, Harriman, & Gewirtz, 1984), and student response to alternative science curricula (Reagan & Hogan, 1991).

The central task of this chapter is to document the tendency of cognitive types to concentrate in various fields, specialties, roles, and problems as they match their cognitive styles to prevailing task environments. One type dominates the structured and mature problem domains, while their cognitive opposites predominate when the nature of the problem is still at issue. In addition, the findings that cognitive style affects peer review and that men and women in the same field can differ greatly in their cognitive style, patterns of relative career success, and group dynamics in small teams are also discussed briefly.

Overall Methodological Orientation

The cognitive styles research tradition is best viewed as a psychological sociology of science rather than a social psychology of science because context is key and sociological questions predominate in shaping the analysis. One must control institutional and situational variables to see the more subtle individual variations. Pelz and Andrews' (1966) study of 1,311 industrial scientists and engineers engaged in R&D demonstrates this point:

> Okay, let's summarize. First, creative ability as we measured it (with the [Remote Association Test]) did not relate to any measure of performance for any group of our scientists. The reason, however, was that some scientists were in situations where creative ability "paid off" for them but others were in situations where creative ability seemed to hurt their performance. (p. 171)

Further, the sociologist's obsession with stratification processes in science is evident in this work. Achievement, advancement, and peer review all received early attention. The stress is on the migrations and performance of individuals that reveal the intellectual complexion of the community and the impact of institutional settings in which science is embedded. Social psychologists explain individual differences in terms of group phenomena. By contrast, the psychological sociologist breaks from the "tabula rasa" assumption typical of sociology (that everyone is basically alike until socialized differently) and examines individual differences related to performance or success in a given organizational environment.

The cognitive styles research program does not try to control for the effects of personality and ability in order to focus on organizational influences, but rather uses the reverse strategy. People with known differences are like marker dyes a medical doctor runs through the complex circulatory system to locate its channels, pools, choke points and blockages. Ideally, one wants to follow the cognitive style groupings over time, see how they fare in different fields, see where each type concentrates and contributes the most, and especially the cognitive mix in the group that rises to the elite leadership positions. In practice, most studies are cross-sectional rather than longitudinal and merely describe the existing distribution. However, cross sections are taken to test process-oriented theories that are basically sociological. (Gordon & Marquis, 1966) In short, the goal is to understand the sorting and self-selection processes that shape and reshape social organizations and technical communities.

Cognitive Styles and the Stratification of Science

The effective use of Gordon's cognitive style typology to illuminate academic science was first demonstrated by me (Wilkes, 1976), in a dissertation study advised by Gerald Gordon. The initial plan was to draw five respondents at random from the regular faculty of the 15 strongest graduate departments (Roose & Anderson, 1970) located in the northeastern United States in each of four fields. The final sample included 194 academic chemists, economists, physicists, and sociologists for whom there was complete cognitive information.

These four fields represented sciences at different Kuhnian paradigm states (or stages of theoretical codification)—the preparadigm, transitional, paradigm, and postparadigm (or "crisis") stages. Differences between these fields were presumed to reflect differences in each field's consensus about the phenomena under study rather than their subject matter per se. Lodahl and Gordon's (1972) study of consensus levels in three of these four fields bolstered this interpretation.

Fields in a high-paradigm state, such as chemistry (with a strong consensus-based conceptual structure), should present an optimal environment for both the problem solver and the implementor, because demands on the individual for problem identification and conceptualization are greatly reduced. In contrast, the preparadigm fields, such as economics and sociology, place strong demands for differentiation skills on their devotees. Emphasis is on the development, critique and testing of conceptual frameworks to support analysis of phenomena amid conflict between schools of thought. These conceptualization and diagnostic demands correspond to the strengths of those high in differentiation ability, although the value of RA is also evident under these research conditions.

The need for a combination of differentiation and RA ability is greatest in postparadigm fields (Masterman, 1970), such as physics in the early 1970s. Things were not very codified at the time, as most of the field's specialties were "showing promise," were in a "tool-sharpening phase" or were "reeling from the impact of new instrumentation" (National Academy of Sciences, 1972). However, the physicists of the early 1970s were not really starting anew. They were reconceptualizing the relationship between elements of a received empirical tradition with considerable predictive power. Under such conditions, one must pierce the blinders of the existing paradigm to deal with anomalous findings and demonstrate the problem-solving efficacy of new approaches. Then, if the new conceptualization is to gain adherents, one must also explain or reinterpret the phenomena the old paradigm dealt with effectively. Hence, the dual cognitive demands of postparadigm research.

This "paradigm state" conceptualization leads one to predict disproportionate numbers of problem solvers among successful academic chemists, because it is the highest paradigm field. Integrators would be disproportionately represented among physicists, because most specialties in physics were in postparadigm turmoil at the time of the study. Problem finders would be overrepresented in both preparadigm fields, economics and sociology, but more so in sociology. Kuhn (1970) viewed economics as having the widest consensus in the social sciences. I construed it as a "transitional" field.

These predictions were consistent with the data (Wilkes, 1977). The problem solvers (Low Diff, High RA) among the chemists outnumbered each of the other cognitive types 2:1. In contrast, the problem finders (High Diff, Low RA) outnumbered problem solvers 2:1 in economics and 4:1 in sociology, although there were unexpectedly large contingents of implementors (Low Diff, Low RA) in both preparadigm fields. The pattern in physics is what one would expect where both qualities are needed but RA is stressed more consistently than is differentiation. Physics had few implementors (13%) and a high (39%) proportion of integrators (High Diff, High RA). Overall, an impressive 68% of the physics sample were high on RA, whereas only 40% of the total sample scored high on this measure.

These data support but do not prove the theoretical point that the state of the field, not differences in subject matter, accounts for these findings. The hypothesis that the pattern is due to disproportionate success by those best matched to field conditions is also still subject to doubt. One might also ask whether Table 10.2 reflects a screening, self-selection, or socialization process? Further analysis sheds light on most of these issues, although the socialization explanation was dismissed more on logical grounds than empirical grounds.

Socialization seems an unlikely explanation given that cognitive styles measures are among the most stable psychological measures over time (Mischel, 1968). There is no evidence that RA can be taught. The nature and origins of differentiation ability are still subject to dispute but seem to be stable by the teenage years. Hence, this pattern is probably the result of a self-selection process or the screening of a system that winnows out the less worthy by using increasingly stringent evaluation processes to sort and assess aspiring scientists over time.

TABLE 10.2. Standardized Distribution of Cognitive Types by Field

	High D, High R	High D, Low R	Low D, High R	Low D, Low R	Total (*N*)
Chemistry (paradigm)	18%	19%	42%	21%	(58)
Economics (transitional)	21%	29%	16%	35%	(51)
Physics (postparadigm)	39%	19%	29%	13%	(44)
Sociology (preparadigm)	26%	32%	8%	34%	(41)
Adjusted frequency	(48)	(49)	(48)	(49)	(194)
Adjusted distribution	25%	25%	25%	25%	(194)
Original distribution	19%	33%	22%	27%	(194)

Note. From Wilkes (1976). Copyright 1976 by Cornell University. Reprinted by permission.

Relevant evidence contradicting the subject matter interpretation of the field differences comes from subanalyses of the economics and physics samples. First, I compared the economists in the most codified specialties (econometrics and public finance) with those in the more substantive areas of the field (such as political economy and the economics of development). The codified regions of economics were as dominated by problem solvers as was the chemistry sample, while the rest of economics had the same 4:1 problem-finder to problem-solver ratio found in sociology.

The comparison of experimental and theoretical physicists yielded a less clear picture. The experimental contingent was disproportionately high on RA ability (55% integrators and 26% problem solvers), but the theorists were more diverse than expected. They split about 50/50 on differentiation ability and were dominated by similar-size contingents of the cognitively opposite problem finders and problem solvers. However, physicists assured me that there are two very different ways of doing theoretical physics. Their descriptions of the two types of theorists were consistent with the distribution of cognitive style groupings found among the theorists.

Alonzi, Comey, and Ducharme (1989) studied a sex-stratified sample of 50 members of the Central Massachusetts American Chemical Society (ACS). Their data set provided an opportunity to convene a panel of chemists on the faculty of Worcester Polytechnic Institute (WPI) to rate the subfields recognized by the ACS in terms of paradigm state. Approximately 40% of the sample was judged to be working in the less structured (often interdisciplinary) specialties. The contingent of chemists in less structured subfields was dominated by problem finders (41% were of this one type). The remainder of the sample working in more mature and structured subfields was disproportionately populated by problem solvers (38% were of this one type). Further, the problem finders, especially among the men, tended to be in academia. The industry based contingent was composed disproportionately of High Remote Associators.

Returning to further discuss the results of the 1976 four-fields study (which was the basis of Table 10.2), the relative career success of the sample members is revealing. There are two ways to view the relative success of academics. One is to look at their rank and ability to achieve tenure. The second is to attribute to them the prestige of their department affliation. The second approach proved to be more interesting in terms of cognitive style.

Approximately equal proportions of problem finders and problem solvers exist in second-rank senior faculty groups of both the social and physical sciences. This is not so in the first-rank departments (those rated 1 to 25 nationally). The most prestigious departments are the ones that exhibit the 2:1 ratio of problem solvers (Low Diff, High RA) to problem finders (High Diff, Low RA) in the senior faculty of the physical sciences and a 1:4 ratio favoring problem finders in the senior social sciences faculty. Hence, the skews found in cognitive distributions by field derive primarily from the pattern found in the elite faculty of each field.

Comparison of the junior and senior faculty distributions in these departments suggests that promotions were rarer for "implementors" (Low Diff, Low RA) than for the other types. Hence, both invitations to first-rank institutions and promotion patterns have sculpted the contours of the research elite in each fields. These findings make the case for screening, rather than self-selection, quite strong, but one body of evidence raised serious doubts about that interpretation—This was a comparison of the relative success of the sample members taking into account the social class of their fathers.

If screening is the key, the upwardly mobile should be the elite in the field since scholarship evaluations and other reviews will have been more continuous during their education. Conversely, if those who were middle to upper middle class in the last generation have an advantage, it might come from their better access to established professionals against whom they could measure themselves and take role models.

The available data suggest that the cognitive skews found in each field are either muted or not present at all in the upwardly mobile half of the sample. By comparison, the cognitive styles skews are found in an even more exaggerated form among respondents who were middle class in the last generation. These data revive the self-selection interpretation. The self-selection hypothesis gets an added boost from related data on the concentration of cognitive types among students by major.

In 1983–1984 Bar-Haim and I characterized the cognitive styles of a sex-stratified sample of 1,006 native born undergraduate and graduate students drawn from most of the departments used in my previous, 1974–1975, faculty study (Wilkes, 1976). The study was designed to examine the process by which the faculty pattern had emerged, and secondarily to compare the distributions of men and women. In this study, the sex difference proved to be less interesting than the comparison of the graduate and undergraduate students, although sex differences in other student samples have been dramatic.

The undergraduates started to show a skew pattern reminiscent of the senior faculty in most fields, perhaps because they had direct access to, and were being evaluated by, the elite faculty. However, the graduate student cohorts ranged from random to diametrically opposite skews in distribution when compared to the faculty distribution in Wilkes's study. This finding was both disconcerting and unexpected. No explanation was apparent, and the first publication based on these data went to press without noting it. (Bar-Haim and Wilkes, 1988) Later, the academic origins of graduate students and the professional aspirations of undergraduates were examined and these two student groups proved to be disjoint sets.

The first-rank department undergraduates in chemistry were typically interested in other (often medical) careers, economics majors were typically interested in business, and few sociology majors planned to go on in the field. The physics majors typically aspired to be scientists and were an exception. Hence, the graduate students were rarely drawn from the ranks of undergraduates trained by the elite research faculty. By default, the top-ranked graduate schools were populated by the best undergraduate students trained by the teaching faculty from small colleges and less-research-oriented universities. Their prior role models were therefore likely to differ considerably from the research faculty they would encounter in graduate school. Of course, many of them probably aspired to become college teachers rather than scientists per se.

Bar-Haim and I went on to examine the cognitive distribution of the science graduate students recruited from prestigious small teaching colleges, where the best teaching faculty in the physical and social sciences might well be concentrated. That student distribution was almost diametrically opposed to the one dominating the elite research faculty, and the contrasting pattern was particularly clear in the physical sciences. Graduate students with undergraduate training in the second-rank graduate departments were the only ones distributed similarly to senior faculty in the top-ranked graduate departments they were entering. Interestingly, the junior faculty at these same second-ranked departments tended to have the same distribution as the elite senior faculty as well.

Considering all this evidence, self-selection is probably the primary mechanism underlying the sorting by cognitive styles that has produced skews in the cognitive distribution of the elite academic scientists by field. However, a constant stream of feedback, via grades and comments from faculty, is influencing the sorting process and could be construed as adding an element of screening to the dynamics of stratification in academic science.

Group Dynamics and Peer Review from a Cognitive Styles Perspective

Evidence that cognitive style related to career choice and success in science led to an interest in the microprocesses that might underlie these patterns. Clearly the peer-review process was responding in some way to cognitive styles differences. Did the research of certain types stand out as qualitatively superior? Was there a research or presentation style that allowed cognitive types to recognize and respond to each other's work through the peer-review process? Different types of scientists would be expected to bring different emphases to their work as peer reviewers.

Neumann's (1976) secondary analysis of data in Gordon's possession demonstrated that the prime peer-review evaluation criteria of the cognitive types were different. He identified the highest and lowest quintile of differentiators from among 148 administrators, physicians, and medical sociologists convened to evaluate uniform summaries of results from all medical sociology projects funded by a program in the National Institute of Mental Health over a period of 2 years. Regression analysis was used in this study to relate three subratings—"importance [of the question]," "innovativeness [of the approach]," or "productivity" [solid work within the state of the art or extending it modestly]—to the reviewer's overall rating of the project.

Neumann's analysis revealed two interesting things. First, High and Low Differentiators systematically responded to different evaluation criteria. The High Differentiators emphasized the importance of the question and procedural innovation, which combined to explain 51% of their total variance with only another 9% explained by the productivity variable. The Low Differentiators emphasized the productivity variable, which explained 59% of their variance. Importance and innovation accounted for only 10% of the variance in the ratings of the Low Differentiators.

The degree of consensus among the two groups also varied greatly. High Differentiators largely agreed as to which projects were innovative, productive, and important. The ratings of Low Differentiators revealed much less consensus. They agreed only on the productivity criterion.

When Gordon (1966) examined the differentiation scores of the principal investigators of the projects under review in Neumann's study, he found that the High Differentiators were doing the projects rated higher on innovation than on productivity, and vice versa for the Low Differentiating principal investigators. Hence, the evaluation panelists were identifying and rewarding the work of their like types through the peer-review process.

Among other things, this finding has implications for the ability of cognitive opposites to work together. Boynton et al. (1989) used the project-based curriculum at WPI to study problem choice, degree of conflict, and complementarity of cognitive types. At WPI, each graduating engineer conducts a major qualifying project (MQP) worth three courses, essentially an undergraduate thesis. MQPs are typically done in groups run as design teams or R&D groups under faculty supervision.

Since WPI is on a quarter system, a full load is three courses per quarter, for a total of 12 in a year. A typical project is worth three courses, and many of them occupy one course slot each term for a full year. Boynton et al.'s (1989) study involved 53 student engineers organized into 14 groups. Seven of the MQP students were not in groups but were working alone. Most of the singleton projects had a specific assignment as part of a larger team building a solar car. In short, the setting approximated R&D teams reporting to a few common supervisors, with some people on special assignment.

Boynton et al. (1989) report three interesting findings. First, 63% of the problem finders gravitated to the less defined projects, while 77% of the problem solvers elected to work on more defined projects. Second, students with cognitive styles matched to the level of structure in the task environment performed especially well. Over half the problem solvers and problem finders got top ratings from their advisers. However, 70% of problem solvers in well-defined projects got top ratings compared to 30%–40% of the students with one of the other three cognitive styles who were working on the more structured projects. The problem finders proved themselves masters of the ill-structured projects by an even more dramatic margin. Ten of 22 students undertaking less structured projects were problem finders, and top ratings were garnered by 60% of these problem finders. Only 17% of the students of other cognitive styles received top ratings.

Third, the mix of cognitive types on each team had a considerable impact on performance. Far more conflict was reported between persons of the opposite cognitive type (i.e., integrator–implementor and problem-finder–problem-solver combinations) than with anyone else. Some of this conflict was fruitful and some of it destructive.

Half of the people working alone got "very good" project ratings. So did half of the four groups in which all team members were of the same cognitive type. The groups that did better than this average performance level had a moderate mix of types—that is, four of six groups that had similarity on one cognitive dimension and diversity on the other achieved "very good" ratings. None of the three groups with one person of divergent cognitive style got a top rating, and only one of five groups in which everyone had a different cognitive type got a "very good" rating. Hence, there may be an optimum level of cognitive diversity in small working groups. A curvilinear relationship is suggested by these data.

Gender Differences and Learning Styles: Areas under Examination

Bar-Haim and I first predicted sex differences in cognitive distribution of a high-paradigm field that is male dominated in a National Science Foundation Proposal submitted in 1981. We theorized that the early waves of women entering technical fields would be disproportionately of the "pioneering" problem-finder type, yet they would often be entering fields dominated by problem solvers (see Bar-Haim & Wilkes, 1989, for details). Shablin (1985) confirmed this prediction in her study of 120 WPI engineering students (Table 10.3). Forty percent of the WPI women in her stratified sample were "problem finders," compared to only 23% of the WPI men.

This difference in distribution by sex was expected to be a passing phenomenon. Data gathered at WPI 5 years later with a similar sample (Francis & Pietras, 1990) suggest that while the male and female distributions still differed (the women were still more likely to be high on differentiation ability), the next wave of women had more integrators and fewer problem finders. Indeed, the 1990 female distribution closely approximated the male distribution of 1985. The male distribution of 1990 had also changed, so the sex difference was persisting although changing in nature. Also, one of the mechanisms involved in this change had

TABLE 10.3. Cognitive Styles by Sex: 1984–85—Worcester Polytechnic Institute

	High D, High R	High D, Low R	Low D, High R	Low D, Low R	Total (*N*)
Males	32%	23%	19%	26%	(72)
Females	21%	40%	16%	24%	(48)

Note. Data from Shablin (1985).

ominous implications. Female recruitment eroded over the 5-year period from an entering class that was 24% female in 1985 to an entering class that was 18% female in 1990. Under these circumstances, efforts to reach and hold more of the female problem finders might be in order.

The coincidence of a highly visible social difference with an underlying shift in cognitive distribution is a potentially important finding I have come to call the "shadow" variable phenomenon. In this instance of a male and female divergence the main concern is that attempts to construe cognitive conflict in terms of gender theory would greatly exacerbate an already tense situation. Given a pattern of heavy female attrition (mostly major changing) that makes sense in cognitive terms (since men of the same type are behaving similarly) but is unusually visible in the case of women and considered much more problematic, the potential for confusion and conflict is great. This may be a case in which intervention to control perceived gender discrimination has little effect, since it is a cognitive type, not women per se, that is not valued.

Concurrent studies of high school students suggest that the way in which science and technology are presented matters greatly in attracting the different cognitive types to the technical professions. Student reactions to the controversial Chemistry in the Community (ChemCom) curriculum, which presents technical issues in their social context, are dramatically different from their reactions to the more abstract traditional texts. ChemCom was favored by 90% of problem finders in four high school classes exposed to one module from the curriculum. In contrast, 75% of problem solvers urged a return to the traditional text (Reagan & Hogan, 1991).

A concurrent study of a WPI introductory biology class for nonmajors also experimented with a nontraditional text and replicated the ChemCom findings (Sabin, 1991). A unit from the alternative text under study, *From Gaia To Selfish Genes* (Barlow, 1991), had extraordinary appeal to problem finders but turned off problem solvers, who experienced it as more complex and difficult to comprehend than the traditional text. Hence, the traditional approach to science teaching caters to problem solvers at the expense of other cognitive types. Shifts in curricula designed to broaden the appeal of science may reduce the appeal of science courses to the problem solvers, in a classic trade-off situation.

Summary and Conclusion

The cognitive styles approach has developed slowly on the margins of the social study of science for 30 years. The latest wave of research demonstrates considerable theoretical promise and practical utility. However, much current research is being done by science and engineering students who do not publish in social sciences outlets, and seminal pieces have appeared in the management and organizational behavior literatures. Hence, much of this research is unknown to psychologists and some key evidence is being handed down orally and in unpublished reports. Further, the basic cognitive measures need attention from developmental and cognitive psychologists, both to explore their origins and to see how well they correlate with other, better-known personality and cognitive measures.

The study of cognitive styles can reveal both strategic microprocesses and major structural changes in technical fields. The key is to connect the degree of structure and predictability in a task environment with the capacity of different cognitive types to operate effectively in more or less codified areas. The range of promising applications starts at the intake of the science manpower pipeline, with learning styles and student response to alternative science curricula, and goes on to illuminate the career choices of mature scientists by field and the stratification processes that sort scientists by prestige of institution. As such, the cognitive styles perspective helps integrate diverse processes shaping the scientific community.

Studies have been done of major choice, women in science, problem choice, small-group dynamics, peer review, productivity, innovation, and relative accomplishment. These studies portray the contribution and likely fate of different kinds of scientists amid moving cyclical processes of development and change in the state of the relevant bodies of knowledge. This connection between cognitive distribution and state of field closes a feedback loop—as the state of a field impacts the mix of talent in science and vice versa. The capacity to accommodate both cross-sectional descriptive study and dynamic longitudinal process studies, while connecting individual behavior, community composition and group social structure, is an appealing characteristic of this research tradition. The cognitive styles approach to the study of science seems ripe for rapid development.

References

Alonzi, A., Comey, K., & Ducharme, D. (1989). *Men, women and the American Chemical Society*. Worcester, MA: Worcester Polytechnic Institute, Interactive Qualifying Project Report.

Bar-Haim, G., & Wilkes, J. (1988). Comparisons of male and female student aspirants to a scientific career: Perceptions of promising science talents. *International Journal of Comparative Sociology*, xxix, 3–4.

Bar-Haim, G., & Wilkes, J. (1989). A cognitive interpretation of the marginality and underrepresentation of women in science. *Journal of Higher Education, 60*(4), 371–387.

Barlow, C. (Ed.). (1991). *From Gaia to selfish genes: Selected writings in the life sciences.* Cambridge, MA: MIT Press.

Barron, F. (1969). *Creative person and creative process.* New York: Holt, Rinehart & Winston.

Boynton, D., Cleary, R., & Plummer, J. (1989). *Personal factors in group organization.* Worcester, MA: Worcester Polytechnic Institute, Interactive Qualifying Project Report.

Charanian, T. (1964). Creativity and the research and development proposal. Unpublished master's thesis, University of Chicago, Chicago, IL.

Crowther, B. (1970). Differentiating ability in the use of personality scales. *Perceptual and Motor Skills, 333*, 325–326.

Feeny, J. (1980). *Scientific acceptance: the psychological basis.* Worcester, MA: Worcester Polytechnic Institute, Interactive Qualifying Project Report.

Francis, P., & Pietras, M. (1990). *Cognitive style and imagery of women engineers.* Worcester, MA: Worcester Polytechnic Institute, Interactive Qualifying Project Report.

Fromm, E. (1959). The creative attitude. In H. Anderson (Ed.), *Creativity and its Cultivation* (pp. 44–54). New York: Harper & Row.

Ghiselin, B. (1963). Ultimate criteria of two levels of creativity. In C. Taylor & F. Barron (Eds.), *Scientific creativity* (p. 30). New York: Wiley.

Gordon, G. (1966). *The identification and use of creative abilities in scientific organizations.* Paper presented at the 7th National Research Conference on Creativity, Greensboro, NC.

Gordon, G., & Marquis, S. (1966). Freedom, visibility of consequences and scientific innovation. *American Journal of Science, 72*(2), 195–202.

Gordon, G., & Morse, E. (1969, Spring). Creative potential and organizational structure. *Journal of the Academy of Management, XI*(3), 37–49.

Guilford, J. (1959). Traits of creativity. In H. Anderson (Ed.), *Creativity and its Cultivation* (pp. 142–161). New York: Harper & Row.

Hudson, L. (1966). *Contrary imaginations.* Edinburgh, Great Britain: Methuen.

Kuhn, T. S. (1970). *The structure of scientific revolutions* (2nd ed.). Chicago: University of Chicago Press.

Lodahl, J., & Gordon, G. (1972). The structure of scientific fields and the functioning of graduate departments. *American Sociological Review, 37*, 57.

Maslow, A. (1959). Creativity in self actualizing people. In H. Anderson (Ed.), *Creativity and its cultivation* (pp. 83–95). New York: Harper & Row.

Masterman, M. (1970). The nature of a paradigm. In I. Lakatos & A. Musgrave (Eds.), *Criticism and the growth of knowledge* (pp. 59–91). Cambridge, England: Cambridge University Press.

Mednick, S. (1963). The associative basis of the creative process. In S. Mednick & M. Mednick (Eds.), *Research in personality* (p. 583). New York: Holt, Rinehart, & Winston.

Mischel, W. (1968). *Personality and assessment.* New York: Wiley.

Mitroff, I. (1974). *The subjective side of science.* Amsterdam: Elsevier.

Mitroff, I. & Inez, F. (1977). On the psychology of the Apollo moon scientists. *Human Relations, 30,* 657–674.

National Academy of Sciences. (1972). *Physics in perspective* (Vol. 1, pp. 94–380). Washington, DC: National Research Council.

Neumann, Y. (1976). *Social differentiation and the evaluation process.* Paper presented at the Conference on Cognitive Style, Boston University, Boston, MA.

Pelz, D., & Andrews, F. (1966). *Scientists in organizations: productive climates for research and development.* New York: Wiley.

Reagan, M., & Hogan, R. (1991). *Response to S–STS curricula: A study of Chem-Com.* Worcester, MA: Worcester Polytechnic Institute, Interactive Qualifying Project Report.

Roe, A. (1953). A psychological study of eminent psychologists and anthropologists and a comparison with biological and physical scientists. *Psychological Monographs, 69*(2), 1–67.

Rogers, C. (1959). Toward a theory of creativity. In H. Anderson (Ed.), *Creativity and its cultivation* (pp. 69–82). New York: Harper & Row.

Roose, K., & Andersen, C. (1970). *A rating of graduate programs.* Washington, DC: American Council on Education.

Sabin, D. (1991). *Response to an S–STS curriculum: A study of from Gaia to selfish genes.* Worcester, MA: Worcester Polytechnic Institute, Interactive Qualifying Project Report.

Shablin, E. (1985). *Sex discrimination in scientific imagery.* Worcester, MA: Worcester Polytechnic Institute, Interactive Qualifying Project Report.

Wallach, M. (1971). *The intelligence/creativity distinction.* Morristown, NJ: General Learning Press.

Wilkes, J. (1976). *Cognitive research styles in paradigm and preparadigm field.* Unpublished doctoral dissertation, Cornell University.

Wilkes, J. (1977). *Scientific styles and disciplinary emphases.* Paper presented at the Annual Meeting of the Society for Social Study of Science, Boston, MA.

Wilkes, J. (1979). *Cognitive issues arising from study in the sociology of science.* Paper presented at the Annual Meeting of the American Psychological Association, New York.

Wilkes, J. (1980). *Cognitive styles and scientific activities.* Paper presented at the Annual Meeting of the American Psychological Association, Montreal.

Wilkes, J., Mangu, R., Harriman, P., & Gewirtz, D. (1984). *The hacker challenge: artistry, addiction or subversion?* Paper presented at the Eastern Sociological Society Annual Meeting, Boston, MA.

Wilkes, J., & Neumann, Y. (1978). *The interaction of style, values, and evaluative criteria in scientific publication and peer review.* Paper presented at the Annual Meeting of Midwestern Sociology Society, Omaha, NB.

CHAPTER 11

The "Atmosphere of Pure Work": Creativity in Research and Development

TERESA M. AMABILE

The labor of love aspect is important. The most successful scientists often are not the most talented, but the ones who are just impelled by curiosity . . . . They've got to know what the answer is.

—ARTHUR SCHAWLOW (Nobel prize-winning physicist)

Over the past three or four decades, science studies have dealt with a fascinating array of important topics: the development and change of paradigms, the influence of philosophical movements in science, and the place of science within the larger society. Amazingly, very little of the science studies literature has explicitly mentioned the ultimate driving force behind all of scientific progress: human creativity. To be sure, considerations of creativity are just beneath the surface of many science studies writings. With this volume's clear focus on the social psychology of science, creativity can emerge for closer examination—because it is social psychological research that has made some of the most startling discoveries about scientific creativity in recent years.

The approach to creativity that I present in this chapter is different from most psychological work on scientific creativity since 1950 (which is widely considered to be the starting point for serious psychological research on the topic). Most simply, I focus on the effect of *social factors on motivation* and the effect of *motivation on creativity*—in essence, a *social psychology of creativity,* which I have applied to scientific creativity as well as to other domains. By contrast, the majority of psychological research on scientific creativity takes a personality approach, examining the special personal qualities that set successfully creative scientists apart from ordinary people or from less creative scientists. Most often, the methodologies involve self-report personality or biographical questionnaires. For example, some studies have found that Nobel prize-winning scientists are more likely to come from Jewish families than from gentile families (e.g., Berry, 1981), or that scientists who hold a large number of patents are more likely to have had permissive parents than are less successful scientists (Buel, 1965). Others found, for example, that productive researchers are likely to be ambitious, unsociable, nonanxious, defensive, and approval seeking (Rushton, Murray, & Paunonen, 1983) or that creativity ratings of scientists correlated negatively with sociability, good impression, and communality but positively with capacity for status and empathy (Weiss, 1981), or that women scientists were more serious, radical, confident, dominant, intelligent, and adventurous than women in general but less sociable, group dependent, and sensitive (Bachtold & Werner, 1972).

Mostly, then, psychological studies of scientific creativity have endeavored to relate how scientists (or creative scientists) are different from other people. Occasionally, motivation has been mentioned, but, again, as a stable personality characteristic. One study found that highly productive scientists were more likely to have strong motivation, autonomy, and self-direction (Fox, 1983). Another described the quality of motivation experienced by scientific researchers: It is marked by intellectual stimulation and desire for an original contribution and novelty, and by an absence of concern with financial reward (Maini & Nordbeck, 1973).

This latter finding begins to interface with research on the social environment by leading us to ask: If researchers are not motivated strongly by desire for financial reward what is the effectiveness of reward in motivating scientific creativity? It turns out that this is a central concern of my own research program, a concern that my students and I have attacked with an arsenal of methodologies including (principally) experimental studies and interview/questionnaire studies as well as (to a minor degree) autobiographical reports and case studies. Most of our experimental work has utilized verbal or artistic creativity measures, but some has examined problem-solving creativity. The interview/questionnaire studies have focused primarily on scientific creativity. As a body, this work has produced generally consistent results—whether the subjects were children telling stories or adults grappling with scientific problems. It has led us to the intrinsic motivation principle of creativity: People will be most creative when they feel motivated primarily by the interest, enjoyment, satisfaction, and personal challenge of the work itself—and not by extrinsic motivators such as tangible reward, evaluation concern, deadlines, and external dictates. It is that "labor of love aspect" that Arthur Schawlow deemed so important.

Definitions

Because the definition of creativity is often in dispute, it is necessary to clarify the way in which we define and assess this ephemeral entity in our research. In our experimental studies, we have subjects make products in response to clearly defined tasks; the tasks are designed so as not to depend heavily on special skills in drawing, verbal fluency, mathematical ability, and so on. Some such tasks that my students and I have used include solving Luchins's water jar problems (a mathematical/scientific task), making paper collages, and writing haiku poems. Once these products have been made in the context of our studies, experts—people who are familiar with the domain—rate them on creativity and other dimensions. For example, we ask studio artists to rate the paper collages. This approach to creativity assessment was used by a few previous researchers (e.g., Getzels & Csikszentmihalyi, 1976; Kruglanski, Friedman, & Zeevi, 1971) with apparent success. We call it the consensual assessment technique.

The consensual assessment technique for creativity is based on this operational definition: A product or response is creative to the extent that appropriate observers independently agree it is creative. Appropriate observers are those familiar with the domain in which the product was created or the response articulated (Amabile, 1982a). The conceptual definition of creativity that my students and I use in theory building makes assumptions about what judges are responding to when they rate products as more or less creative: A product or response will be judged as creative to the extent that (1) it is a novel and appropriate, useful, correct, or valuable response to the task at hand, and (2) the task is heuristic (open ended) rather than algorithmic (having only one clear, obvious path to solution) (Amabile, 1983a).

A Working Model of Creativity

The componential model of creativity, which we have been using as a theoretical guide in our research, attempts to include all factors that contribute to creativity—person factors as well as social–environmental variables. The model includes three major components of creativity, each of which is necessary for creativity in any given domain. (It is described in more detail elsewhere [Amabile, 1983a, 1983b, 1988a, 1988b].)

Domain-Relevant Skills

These skills are the basis from which any performance must proceed. Domain-relevant skills include memory for factual knowledge, technical proficiency, and special talents in the domain in question. For example, a bioengineer's domain-relevant skills include his innate talent for imagining and thinking about complex scientific problems as well as sensing out the important problems in that domain, his factual knowledge of biochemistry and the techniques of genetic engineering, his familiarity with past and current work in the area, and the technical laboratory skill he has acquired.

The domain-relevant-skills component can be viewed as the set of cognitive pathways that may be followed for solving a given problem or doing a given task. As Newell and Simon (1972) poetically describe it, this component can be considered the problem solver's "network of possible wanderings" (p. 82).

Creativity-Relevant Skills

Herein lies the "something extra" of creative performance. Assuming that an individual has some incentive to perform an activity, performance will be "technically good" or "adequate" or "acceptable" if the requisite domain-relevant skills are in place. However, even with these skills at an extraordinarily high level, an individual will not produce creative work if creativity-relevant skills are lacking. Creativity-relevant skills include a cognitive style favorable to taking new perspectives on problems, an application of heuristics for the exploration of new cognitive pathways, and a working style conducive to persistent, energetic pursuit of one's work.

Creativity-relevant skills depend to some extent on personality characteristics related to independence, self-discipline, orientation toward risk taking, tolerance for ambiguity, perseverance in the face of frustration, and a relative unconcern for social approval (Barron, 1955; Feldman, 1980; Golann, 1963; Hogarth, 1980; MacKinnon, 1962; Stein, 1974).

Our bioengineer's arsenal of creativity skills might include his ability to break perceptual set when observing experimental results, his tolerance for ambiguity in the process of deciding on the appropriate interpretation for puzzling data, his ability to suspend judgment as he considers different approaches, and his ability to break out of strict algorithms for attacking a problem. The bioengineer might also have learned to employ some of the creativity heuristics described by theorists: "When all else fails, try something counterintuitive" (Newell, Shaw, & Simon, 1962); "Make the familiar strange" (Gordon, 1961). Finally, if the bioengineer is productively creative, his work style is probably marked by an ability to concentrate effort for long periods (Campbell, 1960; Hogarth, 1980) and an ability to abandon unproductive strategies, temporarily putting aside stubborn problems (Simon, 1966).

Intrinsic Task Motivation

We define intrinsic motivation as the motivation to engage in a task primarily for the sake of task engagement—because the activity itself is interesting, enjoyable, or personally challenging to the individual. Extrinsic motivation is the motivation to engage in a task primarily in order to achieve some goal extrinsic to the task itself—such as meeting a deadline or winning a prize. An individual can have no motivation for doing a task, a primarily intrinsic motivation, or a primarily extrinsic motivation; obviously, intrinsic and extrinsic motivations for the same task may coexist. However, one is likely to be primary. On the basis of our empirical results, we argue that a primarily intrinsic motivation will be more conducive to creativity than a primarily extrinsic motivation. We propose that, when intrinsic motivation is primary, the individual will be able to work more creatively because *task involvement* will be higher.

For practical purposes, there are two ways in which motivation can be considered the most important of the three creativity components. First, it may be the easiest to affect in a straightforward way because intrinsic–extrinsic motivation is strongly subject to social influences, even subtle ones (see, e.g., Lepper & Greene, 1978; Deci & Ryan, 1985). Second, no amount of skill in the domain or in methods of creative thinking can compensate for a lack of intrinsic motivation to perform an activity. Without intrinsic motivation, an individual either will not perform the activity at all or will do it in a way that simply satisfies the extrinsic goals. But, to some extent, a high degree of intrinsic motivation can make up for a deficiency of domain-relevant skills or creativity-relevant skills. A highly intrinsically motivated individual is likely to draw skills from other domains, or to apply great effort to acquiring necessary skills in the target domain (see, e.g., Harter, 1978; Dweck, 1986). Task motivation makes the difference between what our bioengineer can do and what he *will* do. The former depends on his levels of domain-relevant skills and creativity-relevant skills. But it is the bioengineer's task motivation that determines the extent to which he will fully engage his domain-relevant skills and creativity-relevant skills in the service of creative performance.

Within the componential model, task motivation includes two elements: the individual's baseline attitude toward the task and the individual's perceptions of his/her reasons for undertaking the task in a given instance. For example, the bioengineer approaches each task with a baseline level of interest—probably quite high for most of his research tasks, but perhaps quite low for one particular set of problems. For any given research task, however, the bioengineer's interest will vary from the baseline as a function of any extrinsic constraints imposed on him (such as evaluative pressure or constrained choice) and his own strategies for dealing with those constraints. Our scientist may be highly intrinsically motivated to undertake a new project of his own design, but he may be singularly uninterested in a project handed to him by the director of the laboratory.

Experimental Evidence

In our experimental research, my students and I have used the basic overjustification paradigm (cf. Lepper, Greene, & Nisbett, 1973): Subjects work on an interesting creativity task in either the presence or the absence of a specific extrinsic constraint. Subsequently, their products are rated on creativity by several independent experts. We have carried out such experiments with a wide range of independent variables (extrinsic constraints), subject groups ranging in age from preschool children to working adults, and a variety of artistic, verbal, and problem-solving creativity tasks. Although they are not without their complexities, our results reveal consistent patterns that strongly support the intrinsic motivation principle of creativity. Indeed, some of our findings have been replicated by other researchers:

- *Evaluation*. Expected evaluation has a detrimental effect on creativity (Amabile, 1979; Amabile, Goldfarb, & Brackfield, 1990; Hennessey, 1989; Bartis, Szymanski, & Harkins, 1988). Actual prior positive evaluation has a detrimental effect on subsequent creativity (Berglas, Amabile, & Handel, 1979).
- *Surveillance*. Being watched while working has a detrimental effect on creativity (Amabile et al., 1990).
- *Reward*. Contracted-for reward has a detrimental effect on creativity (Amabile, Hennessey, & Grossman, 1986; Hennessey, 1989; Kruglanski, et al., 1971). "Bonus" reward (not contracted for) has a positive effect on creativity (Amabile et al., 1986).
- *Competition*. Competing for prizes has a detrimental effect on creativity (Amabile, 1982b, 1987).

- *Restricted choice.* Restricted choice in how to do an activity has a detrimental effect on creativity (Amabile & Gitomer, 1984; Hennessey, 1989).
- *Extrinsic orientation.* Simply thinking about extrinsic motivators can lead to lower levels of creativity (Amabile, 1985).

Rather than attempting to describe each of these experimental studies, I present one as an illustration of our basic paradigm (Amabile, 1987). A group of 49 corporate executives, lower-level managers, educators, and researchers participating in a creativity conference served as subjects in this study of the effects of competition on problem-solving creativity. All the participants received a booklet containing Luchins's (1942) water jar problems. There were five set-making problems (in which the solution was always the same equation) and one set-breaking problem at the end. This last problem could be solved either by the familiar algorithmic solution or by a more elegant (simple) solution. At the end of the booklet, there were a few riddles to occupy people who finished before the others. Half the participants (randomly assigned) received noncompetitive instructions at the front of their booklet, stating simply that the problems were provided as a means to illustrate problem solving. The rest of the participants, however, received competitive instructions, stating that a contest would be held, with the winner being the person who correctly completed all of the problems the fastest. These competition subjects were asked to write down their starting and stopping times.

On a number of measures, the noncompetition group did better. More subjects in this group solved all six problems correctly. More important, for the set-breaking problem, subjects in the noncompetition group discovered the elegant solution significantly more often than did subjects in the competition group. Interestingly, although we had intended the riddles to serve simply as a filler task, we found that there were differences between the two groups on that task, too. The noncompetition participants were much more likely to try the riddles, suggesting that they were indeed more intrinsically motivated toward this activity than was the competition group.

Observational Data

In an effort to investigate the influence of social–psychological and motivational variables on real-world scientific creativity, I worked with colleagues at the Center for Creative Leadership to interview a large number of research and development (R&D) scientists about creative and noncreative events in their work experience (Amabile & S. S. Gryskiewicz, 1988). We asked the participants, who were 120 R&D scientists from over 20 corporations, to describe for us an example of high creativity and an example of low creativity from their work experience (defining creativity as they saw fit). We told them that we were particularly interested in anything about the events that stood out in their mind—anything about the person or persons involved and anything about the work environment. We felt that by using this critical incident technique, we would be more likely to avoid the interjection of personal beliefs about creativity than if we simply asked interviewees what they thought was important for supporting or undermining creativity in research organizations.

In our search for information about the major influences on creativity and innovation, we did a detailed content analysis of verbatim transcripts of these tape-recorded interviews. The incidents that our interviewees described covered the range of R&D activities. In decreasing order of frequency, they are the development of a new product, the development of a new process, the improvement of an existing product, and the improvement of an existing process. The prominent features of these events also fell into four major categories. Rank-ordered by frequency, they are qualities of environments that promote creativity, qualities of environments that inhibit creativity, qualities of problem solvers that promote creativity, and qualities of problem solvers that inhibit creativity. In our system, "qualities of environments" are any factors outside the problem solvers themselves (including other people) that appeared to consistently influence creativity positively, as in the high-creativity stories, or negatively, as in the low-creativity stories. "Qualities of problem solvers" are any factors of ability, personality, or mood within the problem solvers themselves that seemed to consistently influence creativity either positively or negatively. We found that environmental factors were mentioned much more frequently than were personal qualities. Because this finding appeared in both the high- and the low-creativity stories, and because a large percentage of the stories did not involve the interviewee as a central character (problem solver), we feel that this preponderance of environmental factors cannot be dismissed as a simple attributional bias.

The prominence of the environment in these interviews is an important finding for the social psychology of creativity. It suggests that our laboratory research on the effects of social constraints has produced not only statistically significant findings but ecologically significant ones as well. The *environment* was a much more salient factor than the *individual* for these R&D scientists in their experience of specific creative and uncreative events. This does not mean that, in an absolute sense, environmental factors account for more of the variance in creative output than do individual difference factors. Certainly, at a macroscopic level, personal factors such as general intelligence, experience in the field, and ability to think creatively are the major influences on output of creative ideas by R&D scientists. But, assuming that hiring practices at major corporations favor scientists who exhibit relatively high levels of these personal qualities, the variance above this baseline may well be accounted for primarily by factors in the work environment. Social factors may be responsible for only a small part of the total variance in creative behavior, but they may account for the lion's share of the variance that anyone can do anything about. It is almost always easier to change the social environment (or one's perception of it) than to change traits and abilities.

Our detailed content analysis of the interviews, done by independent coders, revealed several environmental factors that inhibit creativity. Among these are many that we had already studied in experimental paradigms, but several new inhibiting factors were also revealed. In decreasing order of the frequency with which they were mentioned, the environmental obstacles to creativity were (1) various organizational characteristics having to do with poor communication, infighting, and excessive red tape; (2) constraint or restriction of choice in how to do one's work; (3) organizational disinterest or apathy toward the project; (4) poor project management in the form of unclear goals or overcontrolled work assignments; (5) evaluation pressure; (6) insufficient resources; (7) insufficient time; (8) emphasis on the status quo or the maintenance of standard procedures; and (9) competition, especially within the organization itself.

Perhaps as important, several environmental *stimulants* to creativity were revealed in these interviews. In decreasing order of frequency, they were (1) freedom in deciding how to do one's work; (2) good project management in the setting of work assignments; (3) sufficient resources; (4) encouragement; (5) various organizational characteristics having to do with communication, cooperation, and collaboration; (6) recognition and feedback; (7) sufficient time; (8) challenging work; and (9) pressure arising from the urgent need for a solution.

Although the personal qualities mentioned in these creativity stories largely replicate the findings of the personality literature on creativity cited earlier, two of our findings bear mentioning here because of their relevance to the intrinsic motivation principle of creativity. Among the unfavorable personal qualities that appeared in the low-creativity stories, "unmotivated" was the most frequently mentioned and "externally motivated" was the fourth most frequently mentioned. Among the favorable personal qualities that appeared in the high-creativity stories, "self-motivation" was the second most frequently mentioned.

More recent research on creativity in organizations has included both R&D scientists and nonscientists, and has utilized an environmental assessment questionnaire (the Work Environment Inventory) in place of the more laborious interviews on which the questionnaire was based. This research has confirmed the central role of freedom, challenging work, and a number of organizational–environment variables in influencing creativity (Amabile & N. Gryskiewicz, 1989).

Of the case studies describing scientific creativity in recent years, one of the most interesting was the report of Data General Corporation's invention of the "Eagle" superminicomputer in the late 1970s (Kidder, 1981). In that relatively small, entrepreneurial, high-tech company, an elite band of scientists and engineers was chosen to work in a highly protected environment under the management of a "maverick" engineer. In this description of the reward system driving them, notice the emphasis on the intrinsic value of the work itself:

> They didn't have to name the bigger game. Everyone who had been on the team for a while knew what it was called. It didn't involve stock options. Rasala and Alsing and many of the team had long since decided that they would never see more than token rewards of a material sort. The bigger game was "pinball." . . . "You win one game, you get to play another. You win with this machine, you get to build the next." Pinball was what counted. (Kidder, 1981, p. 228)

It is perhaps not surprising that one of the scientists on the Eagle project described the environment as "an atmosphere of pure work." This may very well describe the optimal environment for all scientific creativity.

Conclusion

Clearly, the picture that emerges from our experimental and observational research is not a simple one. We cannot simply say that all evaluation, reward, and constraint will undermine creativity. We found scientists in organizations reporting that their creativity was undermined when they felt extreme evaluation pressure *and* when they felt that no one paid any attention to their work. They reported low creativity when tangible reward was overemphasized *or* underemphasized, when goals were set so tightly that they felt no room to maneuver, *or* when they had no sense of what the overall goals for a project *were*. Combining this evidence with evidence on the personal qualities of successful scientists, and with our experimental evidence on environment, motivation, and creativity, we can fashion a tentative explanation. It appears that any factors leading the individual scientist to focus on the extrinsic motivators in a situation, rather than the intrinsic motivation of involvement in the work itself, will undermine creativity. Thus, for example, organizational apathy toward a project or lack of any clear goal for work may convey to the scientist that the work itself is unimportant. Lack of equitable pay and reward for good effort can lead individuals to focus on those extrinsics just as surely as the frequent dangling of reward "carrots" can. Moreover, scientists who start out with a low level of intrinsic interest in the work may be motivated by nothing but extrinsics. On the other hand, scientists who start with a very high level of intrinsic motivation may be able to retain that motivation even in the face of strong extrinsic pressures. We are currently pursuing such questions in our research.

In sum, it may be said that scientific creativity can best be fostered when the social environment respects that "labor of love aspect," when there is a clear understanding of Einstein's (1949) statement that "it is a very grave mistake to think that the enjoyment of seeing and searching can be promoted by means of coercion and a sense of duty" (p. 19).

References

Amabile, T. M. (1979). Effects of external evaluation on artistic creativity. *Journal of Personality and Social Psychology, 37*, 221–233.

Amabile, T. M. (1982a). Social psychology of creativity: A consensual assessment technique. *Journal of Personality and Social Psychology, 43*, 997–1013.

Amabile, T. M. (1982b). Children's artistic creativity: Detrimental effects of competition in a field setting. *Personality and Social Psychology Bulletin, 8*, 573–578.

Amabile, T. M. (1983a). Social psychology of creativity: A componential conceptualization. *Journal of Personality and Social Psychology, 45*, 357–377.

Amabile, T. M. (1983b). *The social psychology of creativity*. New York: Springer-Verlag.

Amabile, T. M. (1985). Motivation and creativity: Effects of motivational orientation on creative writers. *Journal of Personality and Social Psychology, 48*, 393–399.

Amabile, T. M. (1987). The motivation to be creative. In S. Isaksen (Ed.), *Frontiers in creativity: Beyond the basics* (pp. 223–254). Buffalo, NY: Bearly Limited.

Amabile, T. M. (1988a). A model of organizational innovation. In B. M. Staw & L. L. Cummings (Eds.), *Research in organizational behavior* (Vol. 10, pp. 123–167). Greenwich, CT: JAI Press.

Amabile, T. M. (1988b). From individual creativity to organizational innovation. In K. Gronhaug & G. Kaufman (Eds.), *Innovation: A cross-disciplinary perspective* (pp. 139–166). New York: Cambridge University Press.

Amabile, T. M., & Gitomer, J. (1984). Children's artistic creativity: Effects of choice in task materials. *Personality and Social Psychology Bulletin, 10,* 209–215.

Amabile, T. M., Goldfarb, P., & Brackfield, S. (1990). Social influences on creativity: Evaluation, coaction, and surveillance. *Creativity Research Journal, 3,* 6–21.

Amabile, T. M., & Gryskiewicz, N. (1989). The creative environment scales: The work environment inventory. *Creativity Research Journal, 2,* 231–254.

Amabile, T. M., & Gryskiewicz, S. S. (1988). Creative human resources in the R&D laboratory: How environment and personality impact innovation. In R. L. Kuhn (Ed.), *Handbook for creative and innovative managers* (pp. 501–524). New York: McGraw-Hill.

Amabile, T. M., Hennessey, B., & Grossman, B. (1986). Social influences on creativity: The effects of contracted-for reward. *Journal of Personality and Social Psychology, 50,* 14–23.

Bachtold, L. M., & Werner, E. E. (1972). Personality characteristics of women scientists. *Psychological Reports, 31,* 391–396.

Barron, F. (1955). The disposition toward originality. *Journal of Abnormal and Social Psychology, 51,* 478–485.

Bartis, S., Szymanski, K., & Harkins, S. (1988). Evaluation and performance: A two-edged knife. *Personality and Social Psychology Bulletin, 14,* 242–251.

Berglas, S., Amabile, T. M. & Handel, M. (1979). *An examination of the effects of verbal reinforcement on creativity.* Paper presented at the meeting of the American Psychological Association, New York.

Berry, C. (1981). The Nobel scientists and the origins of scientific achievement. *British Journal of Sociology, 32,* 381–391.

Buel, W. D. (1965). Biographical data and the identification of creative research personnel. *Journal of Applied Psychology, 49,* 318–321.

Campbell, D. T. (1960). Blind variation and selective retention in creative thought as in other knowledge processes. *Psychological Review, 67,* 380–400.

Deci, E., & Ryan, R. (1985). *Intrinsic motivation and self-determination in human behavior.* New York: Plenum Press.

Dweck, C. S. (1986). Motivational processes affecting learning. *American Psychologist, 41,* 1040–1048.

Einstein, A. (1949). Autobiography. In P. Schilpp (Ed.), *Albert Einstein: Philosopher–scientist* (p. 19). Evanston, IL: Library of Living Philosophers.

Feldman, D. (1980). *Beyond universals in cognitive development.* Norwood, NJ: Ablex.

Fox, M. F. (1983). Publication productivity among scientists: A critical review. *Social Studies of Science, 13,* 285–305.

Getzels, J., & Csikszentmihalyi, M. (1976). *The creative vision: A longitudinal study of problem-finding in art.* New York: Wiley.

Golann, S. (1963). Psychological study of creativity. *Psychological Bulletin, 60,* 548–565.

Gordon, W. (1961). *Synectics: The development of creative capacity.* New York: Harper & Row.

Harter, S. (1978). Effectance motivation reconsidered: Toward a developmental model. *Human Development, 21*, 34–64.

Hennessey, B. (1989). The effect of extrinsic constraints on children's creativity while using a computer. *Creativity Research Journal, 2*, 151–168.

Hogarth, R. (1980). *Judgement and choice*. Chichester, England: Wiley.

Kidder, T. (1981). *The soul of a new machine*. Boston: Atlantic–Little, Brown.

Kruglanski, A. W., Friedman, I., & Zeevi, G. (1971). The effects of extrinsic incentive on some qualitative aspects of task performance. *Journal of Personality, 39*, 606–617.

Lepper, M., & Greene, D.(1978). *The hidden costs of reward*. Hillsdale, NJ: Erlbaum.

Lepper, M., Greene, D., Nisbett, R. (1973). Undermining children's intrinsic interest with extrinsic rewards: A test of the "overjustification" hypothesis. *Journal of Personality and Social Psychology, 28*, 129–137.

Luchins, A. (1942). Mechanization in problem solving: The effect of Einstellung. *Psychological Monographs, 54*(6, Whole No. 248).

MacKinnon, D. (1962). The nature and nurture of creative talent. *American Psychologist, 17*, 484–495.

Maini, S. M., & Nordbeck, B. (1973). Critical moments, the creative process and research motivation. *International Social Science Journal, 25*, 190–201.

Newell, A., Shaw, J., & Simon, H. (1962). The processes of creative thinking. In H. Gruber, G. Terrell, & M. Wertheimer (Eds.), *Contemporary approaches to creative thinking* (pp. 63–119). New York: Atherton Press.

Newell, A., & Simon, H. (1972). *Human problem solving*. Englewood Cliffs, NJ: Prentice-Hall.

Rushton, J. P., Murray, H. G., & Paunonen, S. V. (1983). Personality, research creativity, and teaching effectiveness in university professors. *Scientometrics, 5*, 93–116.

Simon, H. A. (1966). Scientific discovery and the psychology of problem solving. In R. G. Colodny (Ed.), *Mind and cosmos: Essays in contemporary science and philosophy*. Pittsburgh, PA: University of Pittsburgh Press.

Stein, M. (1974). *Stimulating creativity* (Vol. 1). New York: Academic Press.

Weiss, D. S. (1981). A multigroup study of personality patterns in creativity. *Perceptual and Motor Skills, 52*, 735–746.

CHAPTER 12

Thinking by Groups, Organizations, and Networks: A Sociologist's View of the Social Psychology of Science and Technology

RON WESTRUM

In this chapter, I consider two broad aspects of thinking by social ensembles and then present a short case study. First I examine how groups, organizations, and networks possess intellectual systems for understanding the world, and how these systems are developed and maintained. Second I consider how thinking might be done with more or less efficiency and show how social structure affects this success. Finally, I describe research on the discovery of the "battered child syndrome" and apply the above perspectives to this particular instance.

Communities of Thought

The basic idea of communities of thought is that groups develop *schemas* to interpret their experiences. Kuhn's (1972) discussion of these schemas as *paradigms* is familiar. A schema lays out the concepts and ideas that the community will use and the connections between them. Included in such systems is a set of statements that the community believes to be true and a set of procedures to test current knowledge and yield new knowledge.

Consciousness is selective, and so are schemas. The system influences its possessors to attend to some things and to ignore others (Fleck,

1979). Therefore, just as systems bring some things to light, other matters are kept in darkness as "hidden events" (Westrum, 1982). Although consciousness is socially shaped (Campbell, 1979/1988), many of the activities and maneuvers that produce it are not themselves conscious. Thus, many assumptions that guide what people do and say are unconscious or unexamined; many of them in retrospect seem false or wrongheaded. Nonetheless, these assumptions are essential for thinking, because productive work involves an ability to forge ahead without examining the epistemological basis of everything we do.

Thinking by social ensembles involves the development, dissemination, and integration of observations, information, and theories. This is done by familiar social processes such as public relations, moral entrepreneurship, and education. A key concept is consciousness—how events and ideas become "ideas in good currency," in the words of Schon (1971). For Karl Marx, the relevant unit of analysis was the social class, but evidently we can study consciousness in ensembles of whatever size and composition we choose. The concept of "organizational learning," for instance, is already common in economics (e.g., Argote & Epple, 1990).

English sociologists of science were among the first to examine the consequences of the "social construction of reality" posed by this view. They also linked it to "interests" of certain groups, including social classes. In other words, views of reality exist not only to make sense of things, but also implicitly to validate or enfranchise certain groups, and to invalidate or disenfranchise certain others. For instance, Donald MacKenzie, in a series of outstanding studies in fields as disparate as statistics and missile guidance, has argued that these interests play a tacit but important role in shaping scientific views (e.g., MacKenzie, 1990).

Recent sociology of science has examined the processes by which new thought systems come to dominate an intellectual scene. Michel Callon, Bruno Latour, and others set forth an entrepreneurial theory of change in which proponents of particular systems try to overcome resistance through strategic maneuvers so that others must use their system (Callon, Law, & Rip, 1986; Latour 1987). This theory has called attention to the manner in which intellectual enterprises are built up by their proponents and deconstructed by their opponents.

Efficiency of Thinking

My second focus is, "What makes for useful thought?" The development of scientific or technological communities has vital consequences for quality of life and human survival. The viability of the human species may well depend on the kinds of thinking institutions we use. Accordingly, these institutions can be studied for their effectiveness and their efficiency.

Early experiments on group structure and problem solving by Bavelas (1951) and others drew attention to the relationship between structure and performance. If this is true for small systems, could it be true for large ones? For instance, Ben-David suggested that the success of the American university system is partly due to its decentralized structure (Ben-David & Zloczower, 1962). The same relationship between decentralization and complex problem solving was found in some of the small-group experiments. Other parallels might well exist.

Consider also the "groupthink" studies of Janis (1982). He drew attention to how groups can fail to use their full cognitive resources, both to maintain consensus, and to avoid anxiety. He examined political fiascos in which self-censorship, group consensus, and premature closure interfered with information processing. Studies of interaction in airline cockpits has shown that similar problems exist there (Foushee, 1984). Information does not flow freely under captains who feel, for example, that outside advice might threaten their authority. Work to combat these systemic problems is called "cockpit resource management," but the basic principles are widely applicable elsewhere as well.

Many phenomena explored in the groupthink research are relevant both to less cohesive groups and to large-scale systems. "Mindguarding," self-censorship, and other symptoms of groupthink can be found in the Challenger tragedy, in the failure to spot the South Pole ozone hole earlier, and elsewhere (Westrum, 1988). The key variable is information flow, the readiness of elements of the system to share information, which is associated with the system's reaction to "messengers" with good or bad tidings. I (Westrum, 1988) have suggested considering the general variable as "willingness to think."

We might use the following scale to measure this willingness to think:

1. *Suppression.* Signals buried, messengers shot.
2. *Encapsulation.* Signals refused, messengers stalled.
3. *Public relations.* Signals watered down, messengers drowned out.
4. *Local fix.* Signals accepted, messengers politely listened to, presenting problem fixed.
5. *Global fix.* Signals accepted, solution widely proclaimed and disseminated.

6. *Thoughtful redesign.* Signals lead to reflection, the system not only responds to the presenting problem but also reforms basic policies.

A key aspect of the willingness to think is an ability to use ideas and information regardless of their point of origin in the system, including the ability to make use of information that orginates in the "wrong" subsystem. How able is the organization to make use of answers from unexpected sources?

An example is the artificial fertilization of fish eggs in the *departement* of the Vosges in France, in 1841. Two "simple fishermen," Gehin and Remy, guessed that mixing the male "milt" with female eggs would increase the yield of new fish, and they tried it. Their experiment greatly increased the yield of fish. The Society of Emulation of the Vosges gave them a bronze medal and a sum of money.

> They were subsequently employed to exercise their system in the different rivers and streams of the department and of those in the adjacent departments. In the course of a short time, they succeeded in stocking these waters with *millions* of trout. (Anonymous, 1855)

In March 1849, Dr. Haxo, perpetual secretary of the Society of Emulation, wrote to the Academy of Sciences at Paris. The government of France then rapidly applied the system to all of the French rivers. Gehin and Remy were given government jobs at good salaries.

This tale illustrates how a "technological frame" (Bijker, 1987) shapes the application of ideas. The process was discovered in 1758 by Count Von Golstein, a German naturalist, and was repeated on several occasions by others. None of the savants, however, saw the economic potential in this process. It was simply not part of their "frame," or paradigm. For the fishermen, increasing the number of fish, not simply artificial fertilization, was a primary goal. Fortunately, in the case of Gehin and Remy, the system responded to and used the knowledge developed by nonsavants.

The intellectual efficiency of the small research and development teams called "skunkworks" is common knowledge in the business world. Yet very few attempts have been made to understand why these small but intense teams succeed while larger teams fail. Similarly, certain large organizations have been intentionally structured to generate creativity, such as the Naval Weapons Center in China Lake, California. We might learn a great deal by studying the success and failure of such organizational designs (Westrum & Wilcox, 1989).

In the remainder of this chapter, I will apply these concepts to a case study, the social construction of the battered child syndrome. I demonstrate how consciousness of child abuse began with uncorrelated observations, proceeded through a stage of controversy, and finally ended in scientific acceptance (see also Westrum, 1982).

Uncorrelated Observations

For many centuries it was recognized that some parents and other caretakers beat and killed their children. In modern times, however, violence toward children was assumed to be rare, the outcome of insanity or drug abuse. Thus, when physicians observed what appeared to be traces of parental beatings, the signs were often rationalized or simply ignored. Stories about abuse seldom appeared in the medical literature or in newspapers. For the medical profession child abuse was a rare event. However it was not rare, only hidden. How it remained hidden from view and was later exposed is what I explain here.

A well-known paper by Pfohl (1977) claimed that "the 'discovery' of child abuse offered pediatric radiologists an alternative to their marginal medical status. By linking themselves to problems of abuse, radiologists became indirectly tied into the crucial clinical task of patient diagnosis" (p. 318). This simplistic explanation is quite misleading. Among the many facts it conceals is that a pediatrician, not a radiologist, transformed child abuse into an important social problem. The real story is more complex and interesting.

The first "moral entrepreneur" to take up this cause was John Caffey, a pediatrician and radiologist. When he published *Pediatric X-ray Diagnosis* in 1945, the book virtually established pediatric radiology as a medical specialty. In the course of his subsequent career, Caffey garnered a large share of honors and published several more editions of his book, which grew in size and authority (e.g., Silverman, 1965).

Caffey also became the first radiologist to describe what he called "multiple unsuspected traumata." Caffey's x-ray films allowed him to see what many other pediatricians could not: In certain children, the "long bones" of the arms and legs had been broken and were in different stages of healing, suggesting repeated trauma. Although Caffey suspected the existence of a child abuse problem *in the 1920s*, he kept hoping that some unknown medical condition would exonerate the parents. No such condition appeared, however, and in 1946, Caffey finally published "Multiple Fractures in the Long Bones of Infants Suffering from Chronic Subdural Hematoma."

Although subdural hematoma ("brain bleeds") was usually associated with breach births in infants, Caffey suspected that here the damage was intentionally inflicted by another human being. The paper's obscure title effectively concealed this. Even in a second paper (Caffey, 1957) he was still diffident. An even more serious failure was the lack of discussion of inflicted trauma in Caffey's *Pediatric X-ray Diagnosis* until the 1967 edition. About 1% of the first edition (1945) of the book is devoted to "trauma." Although trauma as a cause of radiologic appearances is also included in the 1956, 1961, and 1967 editions, the first mention of abuse as a cause of this trauma is in the following lines: "Occasionally infants and children are under the control of psychotics who go into sudden rages or alcoholics who cause severe traumatic lesions in their victims and later have no memory of the episode, or intentionally deny it" (Caffey, 1967, p. 881).

Although "multiple unsuspected traumata" was apparently used by Caffey for many years to describe the battered child syndrome (B. Girdany, personal communication, October 19, 1981), a formal description is absent from all editions of *Pediatric X-ray Diagnosis* until 1972. This lack suggests the relatively minor importance of child abuse to pediatric radiology, contrary to Pfohl's thesis (Cf. Pfohl, 1977). So much was being discovered in the late 1940s and early 1950s that pediatric radiology had no need of child abuse to bolster its professional stature (J. Holt, personal communication, October 29, 1981).

While Caffey tiptoed around the issue of child abuse in print, he was much more direct in person, and especially with his two resident physicians, Bertram Girdany and Frederic Silverman. For Caffey, child abuse was not part of the worldview of pediatrics, but a severe anomaly. For his residents, Caffey's views were a starting point. When they later saw apparent child abuse, they described it as such. Both were to become moral entrepreneurs themselves on the issue.

For instance, soon after Silverman arrived in Cincinnati to practice, he became aware of local cases that matched those Caffey had described to him as a resident. In September 1951, he presented a paper to the American Roentgen Ray Society, which convinced many radiologists that when x-rays show signs of trauma, the physician must not trust the parents' silence as to relevant history (Silverman, 1953).

Others now entered the scene. Paul Woolley and William Evans (1955), a pediatrician and a radiologist, respectively, wrote a particularly important paper. This paper, a model of scientific exposition, was a study of 25 infants whose provisional diagnosis was subdural hematoma. They separated the children into three groups: (1) those whose injury was an isolated accident, (2) those whose parents created an "injury-prone environment" through negligence, and (3) those who had been intentionally injured. This pathbreaking paper still did not create the public furor that was to follow Kempe's efforts in 1961. By keeping the issue within the bounds of medicine, the authors had kept the problem encapsulated, but this was shortly to change.

Controversy

The papers by Silverman (1953) and by Woolley and Evans (1955) moved the issue from the "uncorrelated observations" to the "controversy" stage. And there was controversy, for some doubted the reality of child abuse. One social worker, Elizabeth Elmer, who made her own discovery of incidences of child abuse, related: "I can remember some of the tumultuous meetings in which, for example, a pediatrician would say, 'If I believed the parent could abuse the child, I would leave pediatrics immediately'" (E. Elmer, personal communication, November 18, 1981). One physician, Roy Astley (1953), wrote a paper suggesting that the problem was merely bone fragility.

For the majority of physicians who dealt with children, awareness grew slowly. Radiologists who had relevant evidence usually found themselves in a state of what Floyd Allport called "pluralistic ignorance." Each knew only his/her own experience, and was unaware of the experiences of others. The suspicions were strengthened, however, when outsiders suggested that child abuse was a generic problem. Sometimes the opening came after a lecture by a physician who was willing to speak publicly about child abuse. Silverman and Woolley both found that giving talks on child abuse led to subsequent questions and contacts. Paul Woolley related:

> The first three or four times I went out in the boondocks to talk over this—I remember one trip down in San Antonio—usually everybody gathered, and then some of the older, experienced pediatricians would sneak up to you at cocktail hour: "You know I've seen a number of these and wondered about it. There's nothing written on it." (P. V. Woolley, Jr., personal communication, November 13, 1981)

Some reports and x-ray films came in by mail from physicians who needed help in diagnosis or simply moral support for opinions with which they felt uncomfortable. Just as the individual relieved doubts and concerns by relating such observations, the experts became fortified in their belief that they were on to something; the data base increased. Social consciousness also increased through contacts between idea generators and disseminators and providers of data, who had experiences that needed interpretation. The system was beginning to think about child abuse.

Yet the problem would not yield to an easy solution. Radiologists had a restricted role: They were to diagnose, not to treat. Pediatricians had a broader role, but they were used to operating in a clinical context, with little control over what went on beyond the clinic. Many physicians, including pediatricians, had difficulty seeing themselves as the advocate of a child *against* the parents, even when the child had been savagely beaten (Boardman, 1962). Thus there was no immediate way to help the children, and it is tempting to hypothesize that the problem would not be recognized until somebody could see a solution. This is a particular application of a more general law: *The system cannot think about that over which it has no control.* In the late 1950s the helping professions began to realize that they were ineffective in dealing with child abuse—a problem more widespread than they had imagined. But the ability to think well about the problem came in tandem with a solution for handling it: the interdisciplinary team.

Scientific Acceptance of Child Abuse

The key role in the recognition of child abuse was played by a remarkable pediatrician named C. Henry Kempe, then at the University of Colorado School of Medicine. Kempe was an innovator. In 1957, at age 34, he had been invited to come to Denver to chair the Department of Pediatrics. Within a few years, he had brought about dramatic changes, getting grants, increasing staff, and generally making the hospital a better place for children to stay. He encouraged mothers to stay with their children in the hospital and to bring ethnic foods to cook for their children and themselves, and he made early use of foster grandmothers and grandfathers. Meanwhile, Kempe was playing a major role in the worldwide eradication of smallpox, a task he was later to describe as his lifework. His intellectual contributions included coauthorship of the *Handbook of Pediatrics* and of the many editions of *Current Pediatric Diagnosis and Treatment*. Like Caffey, Kempe was an intellectual leader of considerable gifts (Silver, 1980).

The University of Colorado Medical School at this time was an important intellectual center for the West. Its faculty frequently consulted for physicians in sparsely populated neighboring states that did not have their own medical schools. These outreach activities, along with his work in public health, taught Kempe the value of public arousal and interdisiciplinary cooperation in coping with medical problems. Thus, both Kempe's skills and his university were ideally suited for a problem that required moral entrepreneurship.

Kempe's involvement began with concerns about his residents' misdiagnoses:

> I was *intellectually* offended at first, before my better instincts took over, by the simply silly diagnoses made by bright house staff in situations where nothing but child abuse *could* be the diagnosis. For example, I was shown in those first few years cases of "spontaneous subdural hematomas" (a condition then described in neurosurgical texts; no more) in children who had thrived from birth and now, at age 6 months, had these serious brain bleeds. Some of the children had other findings of abuse such as bruising. I was presented with cases of "non-specific bleeding disorders" with a family history given of "being easily bruised." These children did not bruise in hospital! Their lab test showed no bleeding problems whatsoever. Children were shown to me who were in coma and we showed them to be poisoned by barbiturates or vodka administered by somebody. We saw burns of palms of the hands that looked to me like cigarette burns, I saw burns which could come only from dunking a child's bottom in hot water with symmetrical burns, often, of the feet as well. I saw children whose tongue frenulum was torn from what I thought had to be forceful bottle push, and thus trauma by someone. The common denominator was the denial of child abuse by these fine young doctors who simply could not imagine the facts of life. (C. H. Kempe, personal communication, April 10, 1983)

What could be done about this problem? Kempe tried to inform his colleagues in pediatrics, but a paper that he and his colleague Henry K. Silver sent to the Society for Pediatric Research in 1959 was "read only by title," and not chosen for delivery. Similar lack of interest greeted Bertram Girdany, then at the University of Pittsburgh, when he and Elizabeth Elmer approached the American Academy of Pediatrics about the issue.

Meanwhile, Kempe and his colleagues had discovered the importance of an interdisciplinary team. Bringing the necessary elements (radiology, pediatrics, and social work) together allowed an effective approach to a problem that was both medical and social.

> I was angry that we ourselves [pediatricians] were not doing the job and also that we were not helping the social work departments who were supposed to be doing it, had been doing it for a hundred years, being the protective services right along. Doctors didn't help them . . . . There were social workers over there, the taxpayers would support them, and doctors were over here . . . social workers and doctors didn't talk until the mid-50's. (C. H. Kempe, personal communication, February 21, 1984)

Team approaches developed, not only in Denver but also at the Children's hospital at the University of Pittsburgh, where Elmer (social work) collaborated with Girdany (radiology) and Thomas McHenry (pediatrics), and at Children's Hospital of Los Angeles, under Helen Boardman, director of Social Services. These teams were prototypes for similar efforts elsewhere, and from them came the impetus for the intellectual, legal, and institutional changes that would greatly enlarge society's protection of abused and neglected children.

Team formation aided the perception of child abuse by those involved in treating it. When social workers were part of the team, there was no need to feel helpless in the face of manifest signs of child abuse. Physicians now became more willing to *see* child abuse, to talk about it, and to prevent it. Social workers, in turn, could be taken on to the wards to see the gravity of the problem. Child abuse could thus be better seen when there was an explicit social organization to deal with it.

But these were local fixes. How could the problem be brought before a wider audience? The breakthrough came when Kempe was made program chairman of the American Academy of Pediatrics in 1960. Kempe used his prerogative as chairman to choose one symposium on a key morning, and Kempe suggested "child abuse" as the topic (C. H. Kempe, personal communication, April 10, 1983). On Tuesday morning, October 3, 1961, in the Grand Ballroom of the Palmer House in Chicago, over 1,000 people gathered to hear a symposium on the battered child syndrome. The panel was multidisciplinary: Frederic Silverman, M.D. (Pediatric Radiology, University of Cincinnati); Chair Henry Kempe (Pediatrics, University of Colorado); Elizabeth Elmer (Social Work, Pittsburgh); John Horty (Director of Health Law Center, University of Pittsburgh); Brandt Steele (Psychiatry, University of Colorado); and Honorable Benjamin S. Schwartz (Judge of the Juvenile Court, Cincinnati).

Afterwards, Kempe and the other panelists were besieged by physicians who came to them for advice and catharsis:

> Many were moved and came to tell us all day, long after the morning meetings, about their cases and how troubled they had been, mail poured in and the Newsletter of that Academy meeting reported the symposium well before it was published. (C. H. Kempe, personal communication, April 10, 1983)

Using a multidisciplinary panel meant that the various interest groups represented by the panel were all put on notice that their involvement might be necessary. Soon after, an article was published on the battered child syndrome in the *Journal of the American Medical Association* (Kempe et al., 1962). The enormous impact of this article was partly a result of the changed awareness. But two internal features of the article are noteworthy. The first was a survey of hospitals and district attorneys, carried out by William Droegemueller, then a third-year medical student. Droegemueller remembers that he "looked upon the survey as a very gross analysis of the incidence and, secondly, as a method to persuade others" (W. Droegemueller, personal communication, March 26, 1984). The survey established that 749 cases had come to the attention of the respondents in the last year, only the tip, it was suspected, of a large iceberg. Today, an estimate of reported cases of child abuse usually runs into hundreds of thousands.

The second important feature of this article was the name. Caffey orally used the term *multiple unsuspected traumata* and Silverman, in his unpublished papers, used the term *unsuspected trauma syndrome*. Neither of these terms was in common use. Kempe's short, graphic term seized the imagination. Now anyone could understand what was involved. Between 1962 and 1965, articles on child abuse appeared in *Time, Newsweek, Life, Good Housekeeping*, and the *Saturday Evening Post*. During the previous decade, virtually no mention of the problem appeared in these publications. It is more difficult to gauge the reaction of the newspapers. During the 1950s, one or two cases per year typically appeared in *The New York Times*. Coverage only gradually increased during the 1960s, but by the 1970s dozens of cases were printed in a single year. The single most influential media event, however, was a television episode of *Ben Casey*. This single program was only one of several that dealt with child abuse (Paulsen, Parker, & Adelman, 1966). The battered child syndrome had become a social problem.

During the decades that followed, legal changes followed the changes in medical knowledge. Reporting of child abuse cases was mandated by law for certain personnel and protected for other categories. The number of protective service workers dealing with child abuse increased. As they did, awareness of cases of child abuse rose enormously. In other words, the creation of institutions to detect and treat caused the production of data on the events to be detected and reacted to. Organs of action and consciousness grew apace. Furthermore, recognition of physical child abuse led through analogy to recognition of sexual abuse of children and physical abuse of spouses and the elderly.

Conclusion

We have seen how the idea of the battered child syndrome was developed and how institutional change came with it. The case study illustrates how a hidden event can become a consciously addressed social problem. But change in a social system's schema requires action, and one can speculate that systems may differ enormously in their encouragement of such action. Fortunately, American medicine, through its decentralized structure, encouraged Kempe and others to act. Yet the same system discouraged Caffey from putting his conclusions into print. Schemas are hard to change, as the slow recognition of AIDS was to show (Shilts, 1988). The study also suggests that a system's willingness to become aware of problems is associated with its ability to act on them, a hypothesis that might be further pursued in later studies.

Our society depends, in a myriad of ways, on the quality of the thought processes of groups, organizations, and networks. The issues raised by the battered child syndrome case are evident in many other cases of "hidden events" (Shilts, 1988; Westrum, 1988). But more broadly, the study of thinking beyond the individual level is essential for us to improve the quality of decisions made in science and elsewhere. It is heartening to see an increasing interest in organizational learning, organizational memory, and organizational culture. Organizations, however, are not the whole picture. Much thinking goes on in dispersed networks, industries, and scientific fields. We can only hope that the work of Janis is elaborated on and extended by other scholars to cover a much larger range of activities.

References

Anonymous. (1855, January 6). The artificial production of fish. *Littell's living age, 44*(554), 24–30.

Argote, L., & Epple, D. (1990, February 23). Learning curves in manufacturing. *Science 247*, 920–924.

Astley, R. (1953). Multiple metaphyseal fractures associated with subdural hematoma in small children (Metaphyseal fragility of bone). *British Journal of Radiology, 26*, pp. 577–583.

Bavelas, A. (1951). Communication patterns in task–oriented groups. In D. Lerner & H. Lasswell (Eds.), *The policy sciences* (pp. 193–202). Stanford, CA: Stanford University Press.

Ben-David, J., & Zloczower, A. (1962). Universities and academic systems in modern societies. *European Journal of Sociology, III*(1), 45–84.

Bijker, W. (1987). The social construction of Bakelite: Toward a theory of invention. In W. Bijker, T. P. Hughes, & T. Pinch (Eds.), *The social construction of technological systems.* pp. 159–187. Cambridge, MA: MIT Press.

Boardman, H. (1962). A project to rescue children from inflicted injuries. *Social Work, 7*, 43–51.

Caffey, J. (1945). *Pediatric x-ray diagnosis.* Chicago: Yearbook Publishers.

Caffey, J. (1946). Multiple fractures in the long bones of infants suffering from chronic subdural hematoma. *American Journal of Roentgenology, 56*(2), 163–173.

Caffey, J. (1957, May). Some traumatic lesions in growing bones other than fractures and dislocations: Clinical and radiological features. *British Journal of Radiology, 30,* 225–238

Caffey, J. (1967). *Pediatric x-ray diagnosis* (4th ed.). Chicago: Yearbook Publishers.

Callon, M., Law, J., & Rip, A. (1986). *Mapping the Dynamics of science and technology.* Dobbs Ferry, NY: Sheridan House.

Campbell, D. (1988). A tribal model of the social system vehicle carrying scientific knowledge. In E. S. Overman (Ed.), *Methodology and epistemology for social science: Selected papers* (pp. 489–503). Chicago: University of Chicago Press. (Original work published 1979)

Fleck, L. (1979). *The genesis and development of a scientific fact.* Chicago: University of Chicago Press.

Foushee, C. (1984). Dyads and triads at 25,000 feet: Factors affecting group processes and aircrew performance. *American Psychologist, 39,* 885–893.

Janis, I. L. (1982). *Groupthink* (2nd ed.). Boston: Houghton Mifflin.

Kempe, C. H., Silverman, F. N., Steele, B. B., Doregenmueller, W., & Silver, H. K. (1962, July 7). The battered child syndrome. *Journal of the American Medical Association, 181*(1), 17–24.

Kuhn, T. S. (1972). *The structure of scientific revolutions* (2nd ed.). Chicago: University of Chicago Press.

Latour, B. (1987). *Science in action.* Cambridge, MA: Harvard University Press.

MacKenzie, D. (1990) *Inventing accuracy: A historical sociology of nuclear missile guidance.* Cambridge, MA: MIT Press.

Paulsen, M., Parker, G., & Adelman, L. (1966, March). Child abuse reporting laws—some legislative history. *George Washington Law Review, 34*(3), pp. 482–506.

Pfohl, S. J. (1977). The discovery of child abuse. *Social Problems, 24*(3), 310–323.

Schon, D.A. (1971). *Beyond the Stable State.* New York: W. W. Norton.

Shilts, R. (1988). *And the band played on: Politics, people, and the AIDS epidemic.* New York: Penguin.

Silver, H. K. (1980). Presentation of the Howland Award: Some observations introducing C. Henry Kempe, M.D. *Pediatric Research, 14,* 1151–1154.

Silverman, F. N. (1953). The Roentgen manifestations of unrecognized skeletal trauma in infants. *American Journal of Roentgenology, 69*(3), 413–427.

Silverman, F. N. (1965). Presentation of the John Howland medal and award of the American Pediatric Society to John Caffey. *Journal of Pediatrics, 67* (5, Part 2), 1000–1007.

Silverman, F. N. (1972). Unrecognized trauma in infants, the battered child syndrome, and the syndrome of Ambroise Tardieu. *Radiology, 194,* 337–353.

Westrum, R. (1982). Social intelligence about hidden events: Its significance for scientific research and social policy. *Knowledge: Creation, Diffusion, Utilization, 3,* 381–400.

Westrum, R. (1988, October). *Organizational and inter-organizational thought.* Paper presented to the World Bank Conference on Safety Control and Risk Management, Washington, DC.

Westrum, R., & Wilcox, H. (1989, Fall). Sidewinder. *American Heritage of Invention and Technology, 5*(2), 56–63.

Woolley, P. V., Jr., & Evans, W. A., Jr., (1955). Significance of skeletal lesions in infants resembling those of traumatic origin. *Journal of the American Medical Association, 158*, 539–543.

PART III

Discussion

CHAPTER 13

Making Scientific Knowledge a Social Psychological Problem

STEPHEN TURNER

The fundamental explanatory strategy of social psychology is "generalization from experiment." It works like this. Experimental subjects, such as undergraduate students, are placed in artificial experimental situations that can be manipulated in ways that produce measurable effects. To explain something in the real world, the psychologist matches up the known experimental conditions to the known real world situation, and applies the mechanisms that explain the experimental outcomes to the real world processes. We may call this "explaining by analogy to experiment." If something goes wrong when we try to apply experimental results to the real world, it is because there are causal conditions in the particular real-world situation that produce the unexpected outcomes. Identifying such causal conditions and their effects can itself be done within the framework of this model of explanation: We can perform experiments that identify mechanisms that correspond to the real-world causal processes that produced the unexpected outcomes.

Identifying conditions and their effects is the business of psychological science. In principle, all the problems that arise from the explanatory strategy of "generalization from experiment," such as problems of external validity and ecological validity, can be solved within the methodological paradigm of experimental-psychological science. In practice, of course, the real world is complex, and complexity may mean that as a practical matter we cannot solve all of these problems. We may be unable to control, measure, manipulate, or identify all of the relevant variables or find good real-world analogues.

Science is an especially complex activity, and the study of the causal processes by which beliefs are established among scientists is especially subject to the methodological problems that complexity produces. So we should expect that predicting real-world scientific opinion formation on the basis of variables of demonstrated importance in the social psychology laboratory may be a hazardous affair. But there is no reason to think that the processes are fundamentally different, and that, therefore, the descriptive vocabulary established in the psychology experiment is somehow inapplicable to real-world episodes in natural science. Science involves both cognitive and social psychological processes such as interpersonal influence. Psychology accounts for these things. Hence, psychological explanation ought to account for a part, indeed a large part, of science.

The reasoning presented in the previous paragraph is compelling. Accounting for scientific belief in psychological terms is a natural extension of the project of psychological science itself. Science, however, presents formidable obstacles to being made into a subject for social psychology. Some kinds of analogizing between the social psychology experiment and the "real" world of science are credible. But some kinds are not. This chapter suggests some differences between the cases where analogizing works and where it does not.

Psychological Science versus Science

The standard explanatory strategy of psychological science, the explanation of real-world action, belief, and conduct on the basis of experiments designed to bring out the general relations that hold between psychological variables and conditions, has recently been examined in a "social constructionist" study by Kurt Danziger (1990). Danziger argues against the model presented in the opening paragraphs of this chapter. Psychology, he suggests, has created through its specific strategy of experimentation an artificial "subject" which has no determinate relation to the "subjects" in the real world to which psychological explanations are applied. The experiments that are central to this strategy, he claims, do not have any determinate "general" significance. The attempt to present such results as uniquely pure or general, Danziger suggests, is purely rhetorical. Much of his book is devoted to debunking the statistical rhetoric on which the impression of generality is based, and to explaining the odd history of this statistical rhetoric—a history that is essentially unique to psychology, but which psychologists wrongly tend to identify with science, or with cognition generally.

One aspect of Danziger's argument that is pertinent to the problem of making science and scientists subjects of psychological science is the problem of conceptualizing "stimuli." Danziger notes that in the era of

Wundt, psychophysical research was performed in a laboratory setting with a social structure quite different from that of the later generalized experimental situation. In Wundt's lab, experimenters manipulated stimuli. But their colleagues were the subjects. Observers and subjects were socially and practically interchangeable: The subjects were trained psychologists who knew what to report and, in general, what the experiments were about. In time, this situation was replaced with the kind illustrated in the present volume, in which undergraduate subjects perform tasks under circumstances manipulated by the observer. This kind of experiment, Danziger suggests, is a novel social structure—unlike that of Wundt's laboratory.

The peculiarity of this structure is not acknowledged by the experimenter. But it has a striking effect in the case of science. In later psychophysical research, it was simply assumed that the actual qualitative reports of persons experiencing the sensations artificially produced by the experimenter could be ignored. Only data produced correctly, in a predetermined "correct" form, were considered. Narrow answers about what was heavier than what, for example, counted; other responses did not. Analogous data-limiting assumptions must be made in other kinds of experiments. The experimenter must assume that the experimenter's own descriptions of the stimuli or conditions are unproblematic—and certainly not in any sense negotiable with the subject. In the case of science, however, the subjects cannot be so easily dismissed.

Whose descriptions of the stimuli or conditions count? The subjects? The experimenters? Or the subjects, but only when the response is in a form the experimenter counts as correct? If the subject is a research scientist and the stimuli or conditions are the scientific facts or data, the answer has to be that the scientist's counts and the experimenter's does not, and that the form of the scientist's answer counts too—if the scientist says, on the basis of his or her training in a particular research community, that something is, for example, a scientific fact, it is not the same as reporting a strong opinion. The correct descriptions of the phenomena that the scientist studies are, after all, the scientist's own. When a scientist explains his or her belief in a given scientific result, the explanation is given in the language of the scientist's particular, and often rather tiny, scientific community. This terminology is unproblematic only for members of this community. For the rest of us, it must be mastered through great effort. For the social psychologist to pretend that this kind of belief is equivalent (and therefore governed by the same psychological mechanisms) to opinions that can be manipulated in laboratory experiments and scored on a Likert scale is disingenuous.

The difficulty here is simple. If we are to characterize the cognitive states of scientists, we must use the specific vocabulary of the scientist as the scientist uses it. We cannot substitute our characterizations of conditions or stimuli, because the scientists whose cognitive states we are characterizing are themselves thinking in scientific terms. What we, as psychologists but not specialists in particular natural science disciplines, can describe is different from what a specialist can describe. Moreover, the psychologist's descriptions of the factual situation will be, according to science, wrong or imprecise. We can ignore these differences in the context of the social psychology experiment itself, because our aim there is to determine general relations between variables. But when we apply social psychological mechanisms to scientists we must also claim that our generalized descriptions of conditions and stimuli correspond to the particular case of science we are attempting to explain. It is this match that is problematic.

Other kinds of commentators on science, such as practitioners of the sociology of scientific knowledge (SSK) and the history and philosophy of science, have responded to the problem of "whose descriptions are correct" by *conceding* the primacy of the scientists' modes of description. These writers do not take it upon themselves to argue with scientists about scientific facts, or presume that they have a superior way of describing them. Of course, they are often concerned to show how scientists' descriptions of the world change, and they often reject elements of the self-understanding of the scientist, such as scientists' beliefs in the "reality" of the objects of their theories or the brute factuality of a given result, which the commentator may show to be a product of a specific rhetoric of "factuality." But neither kind of commentator ever pretends to offer a soup-to-nuts replacement for scientists' reasons for their beliefs.

The psychologist who applies "social psychological" experimental findings to actual episodes in science, however, does, in effect, offer such a replacement. This puts the psychologist in the position of arguing with the scientist about science. In this volume, for example, one chapter assumes that the victory of Mendelian genetics over Pearson's statistical variation approach can be described as a matter of consensus formation. The participants in this episode—the geneticists and plant scientists—thought that they were establishing a scientific fact or mastering a new scientific result. The scientists would reject the social psychological description of their beliefs because they rejected the analogy. The term "consensus formation" is properly used to describe a situation in which *opinions* change. But the scientific facts which the scientists accepted were not matters of opinion, but matters of fact.

If the scientists are right, the application of experimental results about consensus formation is wrong. The experiments are not analogous to the historical episode. They can be regarded as analogous only by someone who does not understand and appreciate the science. Thus a "social psychological science" that proceeds naively in the traditional manner of psychology by generalizing from experiments and accounting for the real world in the language of the laboratory comes into conflict with science itself. But if psychologists accept that this is an absurd position to get into and avoid such conflicts, as sociologists and historians of science do, what sort of role might psychology play in understanding science?

Two Appropriate Uses of Analogies to Experiments

Conflicts arise when the "scientific" descriptions of psychological science come into conflict with "folk" descriptions given by scientists of their science. But conflicts are only a problem when the psychologist's descriptions involve characterizations of the contents of science, as they do when a scientist's trained judgement of a scientific result is characterized as an opinion like the kinds of opinions experimental subjects might be manipulated to hold. When the content of science itself is not at stake, scientists are like other folk, whose descriptions and explanations *are* appropriately challenged and replaced by psychologists.

There are two main kinds of analogical extensions of experimental psychological results to science that do not produce conflicts with the contents of science, which I will call "pragmatic" and "niche" applications.

1. *Pragmatic Applications.* In many situations scientists and other "folk" use vague terms, such as "creativity," that apply both to science and to non-science. It is assumed, probably correctly, that there is some commonality between the underlying pychological facts of creativity in different domains. If one's practical goal is to stimulate creativity, for example by manipulating a working environment, it may be that the lessons of laboratory studies in creativity will be useful in producing the practical results one needs. In these uses there is no conflict with science.

Studies of puzzle solving experiments might be construed in pragmatic terms. Solving puzzles is certainly a part of science, and scientists often work collectively on such problems. But in this case one potentially runs into the same problems encountered above in connection with consensus formation. If one attempts to apply what has been learned in social psychology experiments to actual historical episodes in science, one may well be forced to characterize these episodes in terms that conflict with the science.

One of the achievements of SSK has been the improvement of description of the process of establishing results in science. These fine-grained descriptions of scientific activity tend to make experimental results pertaining to generalized collective problem solving less rather than more relevant—the better one's description of the details of a given discovery in the history of science, the less it will have in common with the standardized experimental case of collective problem solving. This raises the question of whether the application of social psychological experimental results to science depends on the use of poor descriptions of the scientific activity in question.

If our concerns are exclusively pragmatic, we may ignore these considerations, as long as we are successful in producing results, such as improved creativity or problem solving. If our concerns are explanatory, however, the social psychologist and the historian or sociologist will talk past one another, because they will describe the same episode in radically different ways.

2. *Niche explanations.* Social psychology may provide mechanisms to account for processes that are already understood as social psychological (for example as "irrational" and group-induced) in character and thus do not conflict with scientists' descriptions.

Historically, the primary use of the social psychology experiment in the history of social psychology was to provide such mechanisms. Concepts like influence were part of the sociological literature. Psychologists, often at the urging of sociologists who wished to have more rigorous definitions and concepts, provided terms and operationalizations tested in the social psychology laboratory. In these cases, psychological mechanisms did not compete with other explanations, but were used to better specify processes already appealed to by sociologists.

It is sometimes assumed by authors in this volume that the same relationship might hold between present SSK, understood to be about such things as scientific opinion, and a future SPS. Unfortunately, the present forms of analysis popular among sociologists, such as actor-network theory and rhetorical analysis, do not appeal explicitly to psychological mechansisms. Nevertheless, there may be many other areas in science studies in which this relationship might hold—in which social psychological terms are used loosely and in ways that the social psychologist can improve by linking the uses to the social psychological tradition.

Postscript

During the Manhattan Project, the psychologist Ernest Hilgard was invited to Los Alamos by Robert Oppenheimer "to come and do a study of a very interesting illustration of the cooperation of many scientists in solving a problem" (Hilgard, 1974, p. 141). According to Hilgard, he and Oppenheimer "had often discussed such matters (without the details of content) during [Hilgard's] wartime visits to Washington" (p. 141). Hilgard declined the invitation. Had he accepted, perhaps the history of science studies would have been quite different—centered around the social psychological problems of collective problem solving rather than the problem of the social construction of scientific reality (a problem or niche that was itself created by the failure of the philosophical program of logical positivism). Perhaps SSK would be where SPS is now—searching for a place to contribute in an already well-defined domain.

But perhaps Hilgard grasped implicitly the point I have urged here—that there could be no simple extension of existing psychological science to the topic of scientific problem solving. Hilgard would not have asked whether "the details of content" would have been in principle an obstacle to the extension of the experimental paradigm of psychological science. But perhaps it is a question that should be asked.

References

Danzinger, L. (1990). *Constucting the subject: Historical origins of psychological research.* New York: Cambridge University Press.

Hilgard, E. R. (1974). Autobiographical statement. In G. Lindzey (Ed.), *A history of psychology in autobiography* (Vol. 6, pp. 129–160). Englewood Cliffs, NJ: Prentice-Hall.

CHAPTER 14

On The Social Psychology of Science

RYAN D. TWENEY

The contributions to this volume make one thing abundantly clear: There is in fact a rich research tradition of social psychological research that has relevance to science. Shadish, Fuller, and Gorman (Chapter 1, this volume) have done a fine job of providing an overview of the relations of the social psychology of science (SPS) to other fields in science studies. My purpose in this present chapter is to explore the relations that might emerge from a specifically cognitive point of view, and to use these relations as the basis for a critique (and an appreciation) of the other chapters in this book. Along the way, I hope it will become clear that a specifically social–psychological view can help in the project of the cognitive science of science, and that the purely cognitive approach contains much that is relevant and necessary for a true SPS.

The perspective of this chapter is that of an outsider in the sense that my background is in cognitive experimental psychology and, in some ways, in the history of science (although I have no formal training in the latter). My education in psychology included little formal training in social psychology. My most recent work in science studies has had a strong historical focus insofar as I have concerned myself with the cognitive principles that underlie an important scientific diary, that of Michael Faraday. Lately I have been as much historian of science as psychologist. I believe that my stance can shed light on some aspects of the project of SPS that may not be apparent to those more directly engaged (and more firmly socialized within social psychology).

Two big assumptions underlie almost everything that has been said in the chapters in this volume. The first is that the best approach to understanding processes, be they cognitive processes or group processes, is to make a distinction between causes and effects and then to try to identify the causes that produce particular effects. The second is that there are universal laws that are applicable to the study of scientific thinking. In the case of both assumptions, the contributors to this volume appear to assume further that the causal relationships and the universal laws are specifically social–psychological in character In both cases, also, the assumptions are epistemological in character, insofar as each has strong implications for how knowledge about social process is to be had; thus, each assumption has enormous implications for the appropriate methodologies needed to study science. I argue here that neither of these assumptions can be taken to be true in the literal form stated; each must be modified in light of what we know to be true about cognitive systems in general, and about scientists in particular.

Causality and the Social Psychology of Science

Accepting any sort of determinism in our picture of the universe entails the further acceptance of the role of causality; to be a determinist is to acknowledge that events have causes. Put this way, it is hard to challenge the assumption without rejecting the entire point of trying to be scientific, but we need to be careful not to let the causal tail wag the deterministic dog. If we assume that every event has a cause (or causes), we can make a serious error. In fact, it makes sense to speak of one event as the cause of another event only when the two events are independent (as with the oft-used billiard ball example). If we assume that all causality is of this sort, we are mechanists, not just determinists.

Much of the traditional mechanistic approach to causality emerges in discussions of the role of experimentation, an approach that owes much (at least in the American literature on the subject) to John Stuart Mill. Mill (1843/1846) talks about the ways in which we can isolate causes and effects using what he called the method of agreement, the method of difference, and so on. His work stands as a major contribution in the history of epistemology (for all the abuse that historians of logic, per se, sometimes heap on his head), not least for the strong link to specific methodologies that Mill showed to inhere in such positions. In recent decades, Mill's approach has been strongly asserted and extended by American psychologists, an entire generation of whom seems to have read, for example, the wonderful book by Campbell and Stanley (1963).

It may seem strange to link Campbell and Stanley with Mill, especially since, in the first few pages, they disavow a Fisherian background to their work and instead align their approach with a post-Darwinian evolutionary epistemology (Campbell & Stanley, 1963; see also Campbell, Chapter 2, this volume). While a full discussion of the point would require a different chapter, the distance from Mill to evolutionary epistemology is not really very great; Mill's epochal *On Liberty* can be read as an evolutionary epistemology, and was published in the same year—1859—as Darwin's *Origin of Species*. Further, Campbell and Stanley themselves make clear their debt to Mill, first in a reference to Mill (p. 17), as well as by their entire discussion of quasi-experimentation, all of which is couched in Millean terms. The later work by Cook and Campbell (1979) reinforces my point insofar as the first chapter is devoted to an analysis of causation in a Mill-cum-Popper fashion; the entire discussion presumes exactly the two assumptions I have outlined above.

Just as, in Hanson's (1958) famous reversible rabbit–duck figure, it would not make sense to ask whether the rabbit causes the duck, so, too, the separability of the events that are being related in cause-and-effect relationships is crucial, a prior requirement before any cause-and-effect arguments can make sense. Yet the assumption is made a bit too quickly, at least to my taste, in a lot of the work that is described in this book. In effect, not all the phenomena that we need to discuss in the application to scientific processes will yield to the right kind of separation. As Salmon (1984) puts it, perhaps the problem here is that we should be relying on a process-based ontology rather than an event-based ontology; perhaps, in the present case, we should speak of the social psychological processes of science rather than of the social psychological events that cause other social psychological events. My concerns in this regard are aroused even in the proposal that Shadish et al. (Chapter 1, this volume) make. In speaking of the mind of the scientist as the locus of science, they are clearly presuming separability of the sort I am talking about. At first glance, their claim appears obviously true. Clearly, science occurs inside the skin of a scientist, and the skin makes for as distinct a separation between inside and outside as one could wish for. However, the whole point of a social perspective (whether sociological, social psychological, anthropological, etc.) is that what goes on inside is affected by the social context in special ways. The social realm is not just one other external cause (like, e.g., gravity). In particular, the social context is capable of being modified by the psychological context and, in the cases of most interest to social psychology, this modification is a continuous interaction, an ongoing process in which the proper unit is "person in environment," not "person as such."

Clearly there is a degree of separability, of course. Persons are special insofar as they have experiences that leave a long-term residue, act as semiautonomous agents in some causal senses, and manifest phenomena (thoughts, feelings, etc.) that occur invariably and exclusively inside the skin. Cognition makes a difference, which is why most of us believe that a purely Skinnerian analysis fails; a person is more than just a reinforcement history. Shadish et al.'s (Chapter 1, this volume) approach has great heuristic value these days in particular, because science studies disciplines are still reeling from the effects of the "Strong Programme," an attempt, in my view, to take away the autonomy of the person as such. In fact, I believe, the true situation is more complex. Human beings are complex semi-autonomous systems that cannot be understood apart from their context. They are not simply reducible to the effects of those contexts. This view of human functioning has been most clearly elaborated by Simon (1981) and, more recently, Newell (1990). It says, in brief, that any system that reaches a particular degree of complexity can only sustain goal-directed activity if it is organized hierarchically into independent or semi-independent subsystems. While the engineer will strive for fully independent subsystems (e.g., the fuel injection system of a modern automobile is fully independent from the steering system), most complex natural systems are semi-independent. Thus, for human cognition in particular, we can consider the memory system to be semi-independent of the perceptual system; the independence is not total, because of the prevalence of, for example, the effects of cultural and social norms on distortions in memory (Bartlett, 1932, is the classic reference here).

A total "inseparabilist" (to coin a term) would not worry about the problem of where to set the limits between the "inside" of a person and the "outside." If the limits between the inside and the outside of a person are just fictions used for a heuristic purpose, mere convention is enough. However, if the systems are semi-independent, with respect to certain variables, two or more systems may appear to be completely independent in some contexts. Thus, it becomes an empirical question whether and where to draw a limit for a particular case. In my own work on Faraday, I have frequently adopted such a perspective, assuming, for example, that I could trace the course of Faraday's thinking through his diary entries, and that there would be meaningful relationships to be found there (Tweney, 1985; Tweney & Hoffner, 1987). If I want to know, say, whether Faraday used his research on acoustic vibrations as a source of ideas for his electrical researches, I do not need to get into all the details of how the social context provided concepts, cues, reference groups, and so on. I could not, however, maintain that view in discussing the origin of his ideas about acoustic vibrations or electricity, origins that are inseparable from the social context of other scientists' views on the same topics, inseparable from Faraday's social interests as a young and rising professional, and inseparable even from the religious context of his ideas (for the latter, see Cantor, 1991). Faraday's thought is, in these respects, not independent of the larger context. How large a context is again an empirical issue. Note, finally, that my analysis of Faraday's ideas, whether limited to his diaries or concerned with the larger social context as well, did not really ask causal questions; I was not asking what proportion of the variance is due to Faraday's social context or to Faraday's own thoughts. Those are silly questions that only make sense if one acknowledges a pure separability perspective.

Recently, in response to a paper by Fuller (1991), I presented an example of misguided separability which I think sharpens the issue:

> Imagine three Martians trying to determine how a terrestrial automobile functions. One is a peripheralist and says that all the insulated wiring is clearly the principal causal factor. The second, a centralist, says that it must be those funny little cylinders that move up and down, and that the wiring is merely incidental. The third, "Edinburghian," Martian advocates a social account: The key factor is clearly the presence of a human operator acting according to social codes in a social context—the wiring and metal are merely the objectified results of other social processes. The Martians could certainly conduct experiments to resolve the issue, and if they did, each would decide that 100% of the variance was accounted for by his/her/its favored hypothesis. Thus, for example, the peripheralist could easily show that ripping the wiring out of the cars in an experimental group produces different performance compared to that of a control group in which the wiring was not interfered with. Ditto for the pistons and for the drivers, in the experiments conducted by the other two groups.
>
> Obviously something is wrong here, and I think the problem faced by the three Martians is exactly paralleled in much of the discussion about social factors in science. You simply cannot treat such issues as "zero-sum games" and expect to get very far. Like the Martians, we must acknowledge that 100% of the variance in science is "caused by" social factors, and 100% of the variance is also "caused by" cognitive factors, and so on. The mistake, in short, is to assume that a deterministic account of a phenomenon demands that we isolate external and separable causes. The assumption fails when faced with the complexity of a modern automobile precisely because an automobile is a complex system of coacting parts. To account for its behavior requires an account of its structure and of the dynamics of its operation. Experiments can play an important role, of course, but there is simply no point to conducting experiments to try to resolve the "winner-take-all" issues that a simple-minded causal model posits. Nor is it enough to say that we must be looking for interactions among factors, not simply for main effects. Including interactions in this sense still commits us to the zero-sum game of accounting for all the variance. Besides, unless there are only two or three factors involved, hunting for the interactions becomes exponentially burdensome. We should instead look for hows, not whys; otherwise, we will end up by arguing over whether the battery or the fuel or a licensed driver is more important to the operation of a motor vehicle. (Tweney, 1991, pp. 204–205)

Only the pure separabilist needs to worry about setting limits as a *prior* issue to be resolved before research can begin. Such a person had better be able to specify what the unit of analysis is before trying to specify cause-and-effect relationships. Consider, for example, Amabile's studies of creativity in the workplace (Chapter 11, this volume). Her approach is to develop a model that "includes three major components of creativity, each of which is necessary for creativity in any given domain" (Amabile, Chapter 11, pp. 318–319). She is certainly right that domain-relevant skills, creativity-relevant skills, and intrinsic motivational factors are important to the understanding of creativity; what I object to is the causal-like language of the premise, a language that betrays the very assumption that needs investigation. Frankly, I distrust the implication that the four complex processes in question (i.e., creative activity, two kinds of skills, and motivational factors) are separable in any meaningful sense.

Instead, they are interlocking parts of a dynamic set of relationships that call for a much different conceptualization. This is not to deny that meaningful empirical relationships can emerge from such an approach, or that such relationships could be turned to useful practical applications. Empirical relationships are empirical relationships; there either is or is not a lawfulness there that might be useful, but I fail to see that such relationships are leading toward a greater social psychological understanding. Such understanding will involve picking apart the dynamic interrelationships of the variables construed as dimensions of an interlocked system. To show that these variables are causally related to each other requires one first to demonstrate that they are separable and at least semi-independent, and that has not yet been done.

My concerns in this regard are also heightened by some of the uses to which meta-analysis can be put (Rosenthal, Chapter 6, this volume; Miller & Pollock, Chapter 7, this volume). For all the advantages of the procedure (and the two chapters are very persuasive about those), the real danger is that meta-analysis cannot be applied except in those cases in which the separability of variables is already assured. Yet, as we have seen, one can happily presume separability and go on to talk for a long time about the resulting empirical relationships without ever seeing that the real window on the processes involved has been closed. The danger is not that one will fail to find empirical relationships, or even that the relationships found will be spurious. The danger is that one will find such relationships and stop the process of inquiry.

Universal Laws in the Social Psychology Of Science

One of Wilhelm Wundt's many overlooked contributions to psychology is his fine appreciation for the types of laws that can be found in science, and how those laws might be applicable in psychology. Essentially,

Wundt (1986) said, there are two kinds of laws, those that reflect empirical generalizations (the domain of the social sciences as he construed them) and those that reflect the operation of universal principles (the domain of the natural sciences). For the scientist it is important to keep the two separated because the role of exceptions differs in the two cases. If we are seeking empirically true generalizations, we may actually expect exceptions to the rule. In the case of universal principles, however, exceptions must be "explained away" (perhaps as muddy data or the coacting influence of other principles).

Most of us, most of the time, presume that the phenomena we are studying are subject to the second kind of law—we would rather be Galileo than Aristotle. Thus we make an assumption, one that is related to the first one about the separability of cause and effect; we believe that lurking somewhere under the rich complexity of scientific behavior, scientific thought, and scientific action are some universal laws that are going to be applicable to an understanding of science. I am enough of a realist and enough of a determinist to agree with the idea that there probably are universal principles that underlie social interaction, social effects on cognition, and so on. The alternative view, that there are no such principles, seems to power much of the recent discussion of hermeneutics but strikes me as self-defeating for science. Nonetheless, caution is necessary: Premature formulation of universal principles is not likely to help.

As before, attending to this issue as epistemological demands certain methodological reorientations. In particular, given any presumptive universal principle of functioning on whatever level, the real test of whether it can be used to understand science is whether the principle can be mapped into a highly detailed process description of the real-world process of science. In other words, the operation of universal principles must be shown to be instantiable in a microstructural account of the moment-by-moment changes in scientific behavior. In effect, postdiction is as important as prediction.

Note that Wundt's distinction, when applied to SPS, implies that we may not be able to carry out such mappings for all of our presumed universal principles. We might, in any given case, have to settle for the empirical generalization that says, in effect, that what we are observing is the aggregated result of lots of different processes; we might, in short, have to reacknowledge history as a necessary component of social psychological accounts. For Wundt, of course, this led to a distinction between two kinds of psychology, one allied to the natural sciences and one more like the social sciences or humanities (see Bringmann & Tweney, 1980; or the recent reiteration of Wundt's point by Cahan & White, 1992). Any SPS will necessarily need to cope with this issue, however it is resolved.

The assumption that universal laws exist invokes other presumptions about the nature of what it is we are studying and how it is we are going to measure it. Consider the following anecdote. Suppose that I have developed a gadget that I claim is a gravity machine. It has a dial on it that has varying kinds of numbers. Suppose I take this gravity machine and say, "Look, this measures gravity." The first question is going to be, "How?" As a good psychologist, I would probably start by demonstrating that the readings are highly reliable. To establish the machine's validity, I could demonstrate that the numbers get larger as I move it down to the floor and smaller as I pick it up to the ceiling. If I move it toward a massive mountain, the readings go up very slightly, again correlating perfectly with the way in which we expect the gravitational field to behave. Furthermore, I can show that the readings vary as the square of the distance from the center of the earth. At this point I claim to have proven that the machine measures the gravitational field; it possesses both reliability and validity. We might agree that we have really stumbled onto something, and we might agree that I could take this over to the faculty of the physics department and try to sell it to them. Now, would they buy it in the face of this kind of evidence? The answer to that is that they absolutely would not buy it. They might send me off to geology and say, "Those folks might find a use for that gadget but we in physics are not going to buy it unless or until you show us how the machine works. Never mind all those correlations, show us how the damn thing works. Show us the mechanisms." Unfortunately, many of social psychology's most used variables are measurable in much the same way that gravity is measurable by the gravity machine; although the measurement techniques possess reliability and validity, the techniques do not carry with them closely specified mechanisms or clear demonstrations that these mechanisms are plausible explanations of the empirical reliability and validity of the technique. In this category is virtually everything done using questionnaires. Questionnaires supposedly measure an inner state, usually in the absence of a detailed theory of how the inner state produces the behavior in the paper-and-pencil task; there is a mystery between the thing to be studied and the numbers to study it with. Such measurement may in fact have practical utility (that is what reliability and validity show us), but it lacks the minimum requirements for a measure that has theoretical utility.

Alternative Approaches

What are the alternatives? If my claims are correct, how do we approach the problem of analyzing the complex social processes manifest in science? In a nutshell, the main answer is to focus on the characterization of social psychological processes in science as interconnected dynamic systems. Within cognitive science, the need for such approaches does not constitute a new problem, and it has in fact been resolved with at least some partial success.

The first clear statement of the need for such characterization occurred in Simon's *Sciences of the Artificial* (1981). He uses an analogy to make his point:

> We watch an ant make his laborious way across a wind- and wave-molded beach. He moves ahead, angles to the right to ease his climb up a steep dunelet, detours around a pebble, stops for a moment to exchange information with a compatriot. Thus he makes his weaving, halting way back to his home. So as not to anthropomorphize about his purposes, I sketch the path on a piece of paper. It is a sequence of irregular, angular segments—not quite a random walk, for it has an underlying sense of direction, of aiming toward a goal.

Simon's supposition is that the complexity of the irregular path reflects the complexity of the ant's environment, not of the underlying laws that govern its behavior. So, too, with human thinking; the apparent complexity of human problem solving reflects the complexity of the information environment in which it occurs, even though the underlying properties of the human information-processing system that explain the performance are quite simple in and of themselves. The trick to understanding either ants or people, then, is to make sure that the relevant complex features of their respective environments are not lost in the process of analysis, and to make sure that the laws invoked in the explanations are indeed laws of the type Wundt described as natural scientific in character, rather than statistical in character.

This supposition leads directly to some fairly profound methodological points, namely, first, one must include a finely grained temporal dimension as part of the characterization (sampling the ant's position every 30 seconds, for example, would be insufficient), and, second, one must pay close attention to the details of individual behavior. Imagine what would happen to the complexity of the ant pathways if one simply averaged lots of ant paths and tried to understand that data. In the realm of cognitive analysis, Simon's own work in problem solving is the best exemplar of the approach, and it is the reason why single-subject analysis of think-aloud protocols is a cornerstone of cognitive methodologies (see Newell & Simon, 1972).

Interesting cognitive systems must be studied as dynamic systems, that is, as continually changing systems. Campbell recognizes this, to be sure, as evidenced by his long-standing focus on evolutionary models of epistemological change in science (see Campbell, Chapter 2, this volume), although the point must be extended to the short-term dynamics as well, to the second-by-second articulation of how change occurs, and to the explanation of that change. Such explanation is common in the physical sciences (hence the prevalence of differential equations in those disciplines) but is still too rare in the social sciences. For SPS in particular, it is going to be necessary to ask about such dynamic processes. Rudwick's magisterial book on the Devonian controversy in Victorian geology (Rudwick, 1985) is, as Shadish et al. (Chapter 1, this volume) note, a big step in the right direction. Rudwick has gone a long way toward showing us how it is possible to reduce such accounts to visual representations to capture some of the dynamic properties across time. It remains for us to extend the range to the microstructural dynamics of such accounts. An important part of such extension must involve the detailed analysis of case studies (Tweney, 1989).

A possible avenue for explanatory extensions has opened recently because of the rise of connectionist models, "neural networks," which have great promise in the understanding of the dynamics of social psychological processes. Thus, a small group engaged in the process of solving a problem is a parallel distributed processing system. Since the formal machinery for modeling such processes is now available commercially in the form of low-cost and powerful neural network software, it is a direction that should be tried. Could, for example, Rudwick's graphic summaries of the change of ideas across time be captured in the behavior of a neural net? Thagard's (1992) recent work using a connectionist approach to historical case studies of science is relevant here, as will be our current effort to capture the dynamic flow of the entries in Faraday's diary (Chitwood & Tweney, 1992).

Information processing in a social community of scientists thus can be seen as involving a set of interconnected dynamic parallel systems. The interesting questions involve the emergent properties that might be found in connectionist models of such systems. Within cognitive psychology as such, even the simplest associative memory network models proved to have emergent properties that were not programmed in at the outset. Thus, these models generalize across similar classes of stimuli in an almost magical fashion, without the need to build in a generalization function. Instead, generalization just happens as a natural outgrowth of the architecture. In modeling a social system, we could expect such emergent properties to be an important part of the value of the approach. Thus, what Fuller refers to as "transactive" memory (i.e., the extended memory that a group possesses) can be modeled directly in a connectionist architecture. Could we then expect to see "coalitions" form? Would such coalition formation mirror the observed properties of coalitions in the real world? These strike me as exciting, untested possibilities.

In sum, I must end on an optimistic note. There are untried prospects for theoretical exploration, room to immerse ourselves in empirical problems of a fresh sort, challenging meta-theoretical issues that must be argued and that will force us to reexamine the presuppositions of science itself, psychology in general, and the claims of SPS. For the next several decades, there is nowhere to go but forward.

Acknowledgment

Grateful acknowledgement is made to Will Shadish and Susan Chitwood for their comments on an earlier draft of this paper.

References

Bartlett, F. C. (1932). *Remembering: A study in experimental and social psychology*. Cambridge, England: Cambridge University Press.

Bringmann, W. G., & Tweney, R. D. (Eds.). (1980). *Wundt studies: A centennial collection*. Toronto: C. J. Hogrefe.

Cahan, E. D., & White, S. H. (1992). Proposals for a second psychology. *American Psychologist, 47*, 224–235.

Campbell, D. T., & Stanley, J. C. (1963). *Experimental and quasi-experimental designs for research*. Chicago: Rand-McNally.

Cantor, G. (1991). *Michael Faraday, Sandemanian and scientist: A study of science and religion in the nineteenth century*. London: Macmillan.

Chitwood, S. T., & Tweney, R. D. (1992, November). *Real world scientific inferences may require confirmation bias*. Paper presented at 33rd Annual Meeting of the Psychonomic Society, St. Louis, MO.

Cook, T. D., & Campbell, D. T. (1979). *Quasi-experimentation: Design and analysis issues for field settings*. Chicago: Rand-McNally.

Darwin, C. (1859). *Origin of species*. London: John Murray.

Fuller, S. (1991). Is history and philosophy of science withering on the vine? *Philosophy of the Social Sciences, 21*, 149–174.

Hanson, N. R. (1958). *Patterns of discovery: An inquiry into the conceptual foundations of science*. Cambridge, England: Cambridge University Press.

Mill, J. S. (1846). *A System of logic, ratiocinative and inductive*. New York: Harper & Bros. (Original work published 1843)

Mill, J. S. (1908). *On Liberty*. London: Longmans, Green. (Original work published 1859)

Newell, A. (1990). *Unified theories of cognition*. Cambridge, MA: Harvard University Press.

Newell, A., & Simon, H. A. (1972). *Human problem solving*. Englewood Cliffs, NJ: Prentice-Hall.

Rudwick, M. J. S. (1985). *The great Devonian controversy: The shaping of scientific knowledge among gentlemanly specialists*. Chicago: University of Chicago Press.

Salmon, W. C. (1984). *Scientific explanation and the causal structure of the world*. Princeton, NJ: Princeton University Press.

Simon, H. A. (1981). *The sciences of the artificial*. (2nd ed.). Cambridge, MA: MIT Press.

Thagard, P. (1992). *Conceptual revolutions*. Princeton, NJ: Princeton University Press.

Tweney, R. D. (1985). Faraday's discovery of induction: A cognitive approach. In D. Gooding & F. James (Eds.), *Faraday rediscovered: Essays on the life and work of Michael Faraday, 1791–1867* (pp. 189–210). New York: Stockton Press.

Tweney, R. D. (1989). A framework for the cognitive psychology of science. In B. Gholson, W. R. Shadish, Jr., R. A. Neimeyer, & A. C. Houts (Eds.), *Psychology of science: Contributions to metascience* (pp. 342–366). Cambridge, England: Cambridge University Press.

Tweney, R. D. (1991). On bureaucracy and science: A response to Fuller. *Philosophy of the Social Sciences, 21*, 203–213.

Tweney, R. D., & Hoffner, C. E. (1987). Understanding the microstructure of science: An example. *Proceedings of the Ninth Annual Meeting of the Cognitive Science Society* (pp. 677–681). Hillsdale, NJ: Erlbaum.

Wundt, W. (1886). Ueber den Begriff des Gesetzes, mit Ruecksicht auf die Frage der Ausnahmslosigkeit der Lautgesetze. *Philosphisches Studien, 3*, 195–215.

CHAPTER 15

Social Psychology and Science Studies: More Commonality of Purpose than Metatheory

THOMAS D. COOK

Introduction

Although a comprehensive descriptive theory of science is not yet possible, we can identify some elements likely to be important in such a theory. The first section of this chapter does this, arguing that science should be conceived in terms of interrelationships between individual scientists and the social contexts in which they are embedded. Social psychology should have a major role to play in such a descriptive theory of science because its central task is also to understand how individuals and social contexts relate to each other. Moreover, the relevance of other academic disciplines to a comprehensive theory of science is less obvious. Sociology's predominant interest in larger aggregates makes the individual peripheral; anyway, most sociologists downplay the notion that individual cognitions can help in describing or explaining scientific belief change. For the most part, general psychology focuses on the decontextualized individual, neglecting the social context. Thus, its relevance is limited. Historians share the social psychologists interest in understanding individual behavior in social contexts. But, unlike social psychologists, historians do not attempt to give an account of behavior in the present; they rely heavily on archival data bases whose particular biases are usually unknown; and they do not care to use the full array of methodological tools at the social psychologists disposal. Social psychology would also seem to have an advantage over the philosophy of science, which has traditionally sought to influence science studies through its normative rather than its descriptive contributions, and when it has been more descriptive, the philosophy of science has relied on historical case studies that are subject to the same limitations as research in the history of science.

Lest I be charged with naive intellectual imperialism, I want to be clear about what I am and am not arguing. I do not want to suggest that social psychology is the *only* discipline needed in science studies. Even at its best, social psychology would not contribute much to analyses of how science-relevant institutions change over time or to other types of historical change. Metascience needs to be multidisciplinary, even though it is unrealistic to expect (or desire) that all the necessary intellectual perspectives will be present in the same individuals. Instead, my argument is that I would expect social psychology to play a special role in science studies because both social psychology and science studies seek to understand how individuals and their multilevel social contexts are related. For instance, a major need in science studies is to explain how changes in collective belief about scientific propositions occur, and social psychologists have long been concerned with theories of belief and behavioral change that should be relevant to this particular metascientific need.

However, despite its promise for science studies, social psychology has not played a major role in advancing metascience. I try to explain why this is the case and also suggest that the situation is not likely to change in the future unless some trends in academic social psychology are reversed. These trends include studying the individual in isolation from groups and institutions; examining only the perceived social situation and not both the perceived and actual situation together or the actual situation alone; studying a small subset of setting attributes in isolation as opposed to conceiving settings in a complex multivariate way; studying individual cognition as opposed to examining affect, cognition, and behavior as a joint system; the trend towards examining short-term reactions to short-term stimuli rather than examining complex, powerful, and long-lasting stimuli in individual lives; and, finally, examining the unidirectional flow of influence from settings to individuals as opposed to identifying reciprocal causal influences between settings and individuals. My hypothesis is that science studies would benefit more at this time from a social psychology that was more holistic than seems to be the case in the field today.

Elements of a Descriptive Theory of Science

The Individual Scientist

Science is done by individual scientists, and no theory of science is possible without scientists at or near its core. Who becomes a scientist is presumably a complex product of genetic, socialization, and educational factors whose particular importance may vary from one scholarly field to another. To understand who becomes a "successful" scientist (however defined) requires knowledge of all the above and more, including knowledge about the interfaces between hard work, creativity in problem generation and solution, luck in being in the right place at the right time, acumen in recognizing serendipitous research outcomes, being interpersonally skillful enough to make and keep relationships with the major gatekeepers to relevant scholarly communities, and a host of other factors.

Much of the research on individual scientists in psychology deals with problem-solving skills and particularly with a small set of heuristics that humans rely on heavily when processing information and that limit the quality of the solutions generated to scientific puzzles. It is not yet clear whether individual scientists are as prone to these cognitive biases as are other humans, especially when scientists are compared to persons of comparable intelligence but different formal training. However, various practices have been institutionalized in science in order to reduce the influence of such cognitive biases. The multiple anonymous review of knowledge claims is probably foremost among the many mechanisms of public criticism that flourish. However, there can be no guarantee that such practices overcome all the limitations of human information processing, and many historical case studies suggest that there have been periods when all the scientists in a particular time and place shared the same false suppositions about knowledge and so could not overcome their shared cognitive limitations. Indeed, future scholars may come to characterize our own era in a particular science as a *folie à tous* in which some phlogiston-like concept reigned as central in the dominant theory, its flaws undetected by the processes of social criticism designed to protect against individual limits in information processing (and self-interest). The psychologist's current conception of individual scientists suggests that they cannot be demonstrated to have the critical cognitive capacities that normative theories of science demand of them.

Recent research on individual scientists by historians of science has tended to emphasize motivational rather than cognitive factors. Particular interest has been shown in the ambition and competitiveness of scientists and in their willingness to cut corners and neglect others' contributions in furtherance of their own fame and fortune. Historians have also highlighted the pride and obstinacy of scientists who, when faced with evidence apparently disconfirming their pet theory, nonetheless persisted in refuting or ignoring the evidence, presumably because so much of their identity and interest depended on their theory's validity. Finally, historical research has documented many scientists' inability to resist political pressures to orient their work toward acceptable topics or to slant the interpretations they explicitly offer for their results. They thereby create knowledge that they privately know does not serve the truth to which they are supposedly dedicated. At issue here is not just the equivocality of what constitutes the "truth"; it is also the willingness to tilt knowledge claims for personal advantage. The historians' account debunks popular images of scientists as disinterested, concerned only about knowledge growth and tenaciously clinging to scholarly norms that transcend personal interest. Historians see scientists as susceptible to multiple pressures, of which the scientific norms of the day are but one.

This thumbnail sketch does not exhaust the historians' account of science. Also worth noting are the countless biographies of famous scientists that seek to explain their special achievements, often in individual terms relating to how their parents raised them, how they were educated, how they came to work on a particular problem, how they collaborated with colleagues, and so on. This concern is shared by those (relatively few) American sociologists of science who are willing to postulate that some pieces of scientific work are better than others. Such sociologists examine the life of great scientists (e.g., those awarded Nobel Prizes) and explore how their achievements were related to their class background, intellectual training, cognitive style, interpersonal style, and ability to enter and move in scientific networks. The emphasis is on identifying what accounts for special achievements, perhaps in the hope that such research will suggest how less elite scientists might do better work.

These are just a few of the ways in which empirical research suggests that individual scientists influence the process and outcomes of science. I could add many more features from other social science disciplines. Developmentalists would seek to explain the career development of scientists, beginning with their earliest school years and then following them through training and their early, mature, and later careers, probably stratifying by field in order to explain why the rapid growth trajectory of most mathematicians differs so much from that of, say, most literary critics. Economists would presumably want to identify the utilities that individual scientists attach to the major aspects of their professional life—developing new theories, extending current theories, describing aspects of nature, administering science, teaching science, and so forth, probably paying particular attention to the trade-offs that result because most of these goals are pursued simultaneously and, anyway, scientists need to fit their professional preferences into their total life as, for example, spouse, parent, child, friend, and hobbyist.

I could go on and on, suggesting issues from different social science disciplines. But the list would be tedious, would lack any obvious way of prioritizing among the many features offered, and would be tangential to the major point: that science cannot be described without recourse to the thoughts, behaviors, and feelings of individual scientists. As the field most deeply committed to the empirical study of individual behavior of all kinds, psychology has clearly a special relevance for describing individual scientists.

Social Contexts of Individual Scientific Behavior

Very little, if any, science is done by hermits; most is done within a complex set of social relationships that has to be mapped if there is to be a comprehensive account of scientific behavior or performance. There are many ways of characterizing the social context of science. Most sociologists of science prefer to emphasize the contingent (especially the local) over the universal and national, the accidental over the planned, informal networks over individuals, and group power considerations over truth-seeking ones. However, the organizing system I present is somewhat different, although partially overlapping. It is based on distinguishing between formal and informal scientific groups and institutions that operate at the local, national and international levels. My hope is to convince readers of the importance of locating individual scientists within an ever-changing, multilevel, formal and informal system of relationships that all have to be understood if a comprehensive description of science is to be achieved.

At the local level, scientists are surrounded by colleagues who work with and for them in different capacities—as laboratory technicians, graduate assistants, postdoctoral researchers, and colleagues from the same or different departments or research centers. There is presumably variation in recruitment and relationships at this local level that influence scientific production. Campbell (Chapter 2, this volume) has briefly mentioned some of these in his discussion of the differences between the researchers in psychology who (loosely) worked with Tolman at Berkeley and those who worked (much more closely) with Spence at Iowa. Campbell's social–psychological account begins with personality differences between Tolman and Spence that were associated with each establishing a unique local research climate. Spence was more authoritarian and created a research setting in which graduate students who worked with him underwent initiation rites, had no choice but to work on his theory (adapted from Hull), quickly learned to denigrate competing learning theories, and spent much of their time interacting professionally and socially with the other students working on the theory. In return, they were warmly accepted into the local theory group, faculty and senior staff spent time with them, they were taken care of financially, and they were helped to find jobs in prestigious universities where they were expected to proselytize in favor of the theory.

Tolman was more diffident. He made few attempts to proselytize and build a community locally. Students worked with him only very loosely, being free to work on whatever topics they thought were important. Tolman often seemed unconcerned about presenting his own views and, when he did, did not seem to believe in them as fervently as Spence believed in his. Unlike at Iowa, Tolmans students had no chance to submit to powerful social pressures to conform to a well-defined theory, nor could they enjoy the material rewards and feelings of belonging to an elite that characterized Iowa. Although Spence's learning theory won a short-term national victory, Tolman's more cognitive theory eventually prevailed because it helped explain a broader range of cognitive phenomena. However, Tolman's diffidence and reluctance to proselytize may have set back progress in psychology for many years. Did he fail as a "tribal leader," as Campbell implies?

If this social–psychological account of two differences in scientific leadership style is correct (Deutsch's [1992] description of Lewin as a leader provides yet a third style), concepts at three different levels are needed: the individual, the local, and the national. The individual level involves propositions about Spence's authoritarian personality, his closed-minded belief in his theory, and his skill at forging cohesive communities out of disparate individuals as contrasted with Tolman's more open personality and greater tolerance for alternative views. The local level is also indispensable, for Campbell's account presumes that Spence's students were spurred by their group initiation practices, the subsequent bonding rituals, the local evidence that their common enterprise was important, the pains Spence took to promote careers, and the clarity of the signals about what was required to be a loyal group member—specifying Spence's theory better and denigrating competing theories. The national level is also required in Campbell's account. Using his own and his students' papers, Spence fought for his theory at national conventions; he worked to place his students at prestigious research universities and he helped everyone in his theory group publish in prestigious national journals—for which earlier generations of Spence's Iowa graduates were probably reviewers, as they probably were for grant proposals as well. No wonder the national impression seems to have been created—even among graduates of universities other than Iowa—that Spence's was the best learning theory. In contrast, there were no national conferences or other professional meetings where true believers from Berkeley dominated; and there was no network of scholars throughout the country seeking to promote Tolman's ideas. Initially, Tolman's work lacked a national context of acceptance and excitement.

As Campbell reports it, the Spence–Tolman case study has more of a "bottom-up" than a "top-down" flavor. Influence flowed more from Iowa to the national level than vice versa. However, it is not too difficult to think of cases in which influence flows more in the other direction, with a small group of scholars from all over the nation exercising a disproportionate influence over national outlets for ideas, funds, and students and then using their control to disseminate an agenda to the local level. The current genome project may be such a case. Newspapers and general scientific outlets have suggested that the Nobel laureate James Watson and a few of his colleagues played a major role in developing the political and financial support that this giant project requires. But a full account of the project will undoubtedly involve many other forces at the personal and national levels. For instance, the project probably would not have come to pass in its current mammoth form if many local laboratories had not been mobilized to advocate for it in the realization that they would benefit locally from this central planning (and funding) initiative. After all, many local laboratories are required to pull off this massive gene-mapping enterprise. The speculation, then, is that factors at the individual, national, *and local* levels are required for a full description of how the genome project came to be. Just as there can be no decontextualized individual, so there can be no decontextualized scientist. Science studies require us to locate individual scientists in social situations in which local and national factors usually play significant roles.

The importance of social context is evident in other examples. The ultimate in faculty collaboration at the local level is when a university department—or a research department in a private firm—seeks to create a name for itself and goes about developing a strategic plan. Often such plans require putting more resources into a limited number of research programs in the hope that one or more of them will pay off and create a national center of excellence. To this end, a mathematics department might decide to specialize in topographical algebra, an oncology department in breast cancer, or a psychology department in applied psychology. Modifying local priorities in this way can radically alter an individual scientist's life, causing unwanted disruptions for some but generating new synergies for others, increasing their capacities and presumably leading to scientific products that are more than the sum of the individuals involved.

The rationale for multidisciplinary research centers is couched in similar language. The central assumptions are that discourse among people with similar academic training constrains creativity, and that novel ideas are more likely to emerge from the juxtaposition of different disciplinary perspectives. The value of such multidisciplinary research cannot be fully judged yet, but the emergence of cognitive science—created out of computer science, linguistics, and psychology—and neuroscience—created out of physiology, neurology, anatomy, biochemistry and psychology—suggest the possibility of large intellectual returns from multidisciplinary collaboration. Thus, current beliefs about multidisciplinary research centers and departmental strategic plans reflect the assumption that changes in formal institutions at the local level can transform the nature of the knowledge produced.

The same is also possible with larger local units, such as entire universities or research headquarters in multinational corporations. For instance, Carnegie–Mellon University has a distinctly different atmosphere from most other major research universities, being less discipline centered in its structure for teaching and research and more integrated around a single intellectual theme: technology and its human uses. Working there may well point a scientist's career in a direction different from the direction it would have taken had the scientist begun his/her research career elsewhere. Of course, self-selection might explain some of the differences between the faculty chosen by Carnegie–Mellon and the faculty choosing it. However, our concern here is with differences in local norms about structures for research and with differences in the kinds of research activities most valued locally. Each of these can influence the behavior of scientists, both in general and in interaction with the different kinds of selection that draw different scientists to different institutions.

We should not assume, though, that causal influence flows in only one direction—from the local environment to the individual scientist. Scientists also influence institutions in more than the obvious sense that institutions are composed of people. Certain individuals contribute more than others to creating local norms, as the Spence and Tolman examples illustrate for their local theory groups. The same is also true for larger organizations. A history of Carnegie–Mellon would assign major responsibility for its unique climate to such local organizational theorists (and practitioners) as Cyert, March, Newell, and Simon. A history of Stanford University might invoke the names of Sterling and Terman, whose influence on the campus research climate persists two decades after their own campus labors ceased. We forget at our peril the high likelihood of reciprocal influences from institutions to individuals and from individuals (and very small groups of like-minded individuals) to institutions.

To varying degrees, individual scientists are also embedded in formal institutions at the national level. They tend to belong to professional associations that represent the political interests of disciplines and research fields, that produce prestigious research journals which confer legitimacy on individual scientists publishing in them, that promote annual meetings where papers are read and networking takes place, and that give awards for outstanding scientific achievements as well as for maintaining the association itself. Also among the various national institutions of relevance to science are government agencies and foundations that process grant applications and award funds, prestigious councils like the National Academy of Science, and perhaps even major media outlets that write about science.

Full accounts of national scientific organizations would not be possible without examining the individual actors embedded within them. For instance, the American Psychological Association (APA) recently split into two—with the APA continuing to represent mostly mental health practitioners and the new American Psychological Society representing mostly researchers. No history of this split would be complete without an extensive discussion of a few therapists who believed that the financial and intellectual interests of their colleagues were playing second fiddle to the interests of a smaller group of researchers from prestigious universities who were often disdainful of practitioners and did not want to use the APA to lobby for higher and more dependable sources of fees for services. These individual therapists worked hard, year after year, within the APA to make it more responsive to practitioner needs. Even so, while these activist therapists may be necessary for a full account of the discipline split, they and the interests they promoted cannot provide a sufficient explanation. For instance, historical factors are also important. Would psychology have split without the steady post-World War II increase in the number of people providing psychological services for a fee? Would the split have ocurred if so many of the new Psychology Ph.D.'s had not wanted to go into academe or could not find jobs there? We should never forget the historical, political, and social context in which those individuals are embedded who most press for institutional change. Causal influences in science studies have to be reciprocal: Institutions change individuals and individuals, together and in small groups, change institutions.

At a less formal national level, there are "invisible colleges" of scholars who frequently but informally communicate with each other and enjoy a disproportionate influence in setting research agendas, mediating access to publication and grants, and otherwise rewarding some individual scientists over others. Some of these invisible colleges represent splinter groups that disagreed with many of the premises of their parent disciplines. Such groups are important to the life of science and provide a potential source of renewal, sometimes leading to the creation of new cross-disciplinary fields such as biophysics, neuroscience, or adolescent psychiatry. Although it is not totally clear how informal national institutions influence individual scientists' behavior, or how individual scientists and small groups of scientists in turn mold these institutions, it is clear that informal groups of scholars from different parts of the country constitute a crucial part of the context that provides opportunities and constraints for scientific workers and thereby influences scientific products.

It would be a mistake to treat the formal and informal contexts of science as static. Like individual scientists, the contexts change with time and sometimes come to reflect cultural preoccupations that have at best only weak links to science itself. This is probably true, for instance, of current concerns with sexism, racism, and ageism as they impinge on scientific training, research priorities, and allocating professional awards. It is probably also true of concerns that are less obviously person centered, as with the environmental impact concerns that creep even deeper into research agendas in biology, chemistry, and physics. Science exists in a cultural context to which it contributes and by which it is affected. Indeed, anthropologists of science might well tell us that most cultures have some conception of what is and is not societally acceptable as science.

This brings us then to the international level in which scientists are becoming increasingly more embedded. Science looks quite different from a third-world country than from the United States. The material resources available for research are likely to be less, as are the behaviors expected of ordinary scientists and scholars of special distinction. For instance, in many developing countries travel to scientific meetings abroad is much esteemed for many reasons, not all of them strictly scientific and some having to do with matters as mundane as access to hard currencies. And those chosen to travel are not always the best in terms of scientific achievements. Political acceptability to the government sometimes plays a role. In many third-world contexts, a struggle has taken place to define a national or regional scientific identity by pushing against the dominant paradigms and theories emanating from developed countries that relegate most third-world scientists to imitator rather than initiator status. Intellectual resentment against Western hegemony in science is surely widespread, especially in the social sciences where cultural assumptions are so salient that accepting a scientific hegemony necessarily entails accepting an especially unwanted and painfully evident cultural hegemony.

Even within the more developed nations, international organizations, meetings, and exchanges are becoming more important and are likely to become even more so as the American hegemony in science decreases and as the international training of scientists increases. We assume that international contacts influence the behavior of individual scientists and that individual scientists play important roles in establishing who shall enter the networks of international scholars invited to meetings outside their own country. No scientist is an island, and to portray science comprehensively requires attention to individual scientists *and* to the formal and informal contexts at the local, national, and international levels in which their work and life are embedded—a life that is affected by the contexts and that helps codetermine these contexts.

Summary

No descriptive theory of science can be comprehensive unless it does the following:

1. Incorporates attributes of both individual scientists and the contexts in which they operate (it is multilevel);
2. Deals with many different kinds of context at different levels;
3. Presumes that change in both individuals and contexts is a constant fact of scientific life; and
4. Assumes that individual scientists influence scientific institutions locally, nationally, and internationally and are in turn influenced by these institutions (it is transactional).

Also, scientific contexts can be profitably understood as the following:

1. The formal institutions in which scientists work—such as laboratories, faculty departments, universities, professional organizations (at the local, national, and international levels), and public and private entities that provide research funds;
2. The informal settings in which scientists also work, such as local scientific work groups and national or international "invisible colleges";
3. The vast array of social, political, economic, and cultural factors that singly and jointly influence scientists performance, such as the racial and ethnic background of entrants to a science, international events that alter national funding priorities, and cultural factors that create different attitudes to science in some nations or ethnic groups as compared to others;
4. The socially situated developmental changes that occur as individual scientists' careers evolve, often reflecting, for example, where they were born, how they were raised, where they were trained, and what their first job was; and
5. The historical era in which science takes place, for eras differ in, say, who is recruited to science, which topics are favored for study, and how science-promoting *institutions* evolve.

I now ask how well theories and findings in past and current social psychology can help science studies because they are responsive to the above assumptions. But some warnings are in order. I am not a historian, and my account of social psychology is surely biased in unknown respects. Moreover, to characterize the field, I had to paint with broad strokes, and there are surely exceptions to most of the generalizations I make. To reduce these difficulties, I tried to characterize social psychologys development by contrasting the 1954, 1968, and 1985 *Handbooks of Social Psychology*, since these were deliberately written to describe the central issues and orientations in the field. But the *Handbooks* do not cover the last decade; there is some tendency in them to discuss a topic merely because it was discussed in the prior *Handbook* rather than because its importance has remained constant or increased; and no individual chapter writers are so free of idiosyncrasies that they can fully represent how all their disciplinary colleagues look at an issue. Nonetheless, I hope that my account of the history and current state of social psychology will persuade readers to take seriously one hypothesis—that despite the overlap in agenda between social psychology and science studies, and despite the useful insights that social psychology has already provided into science studies (for documentation in this volume, see Campbell, Chapter 2, and Shadish, Fuller, & Gorman, Chapter 1), science studies cannot rely on social psychology to provide many of its key new concepts and theories. For this to happen, social psychology would have to change its dominant metatheory, but since there is no motivation for this, metascientists will have to forage widely within the social sciences, knowing that no single disciplinary key will open all the doors necessary for a comprehensive description of science.

Intellectual Contours of Social Psychology

In Which Senses Do Social Psychologists Study Individuals in Social Contexts?

It is unrealistic to expect social psychology to have clear definitional boundaries that demarcate it from other social sciences. Few would dispute, however, that its subject matter tilts toward understanding the individual in social context. Indeed, Allport (1954) emphasizes this in his general introduction to the field in the first chapter of the first *Handbook of Social Psychology*. Moreover, Lewin's influential work on a metatheory for social psychology had as its starting point the notion that behavior is a joint function of the individual and the settings in which he/she might be found (see Deutsch, 1992). Acknowledgment of this core mandate can also be found in nearly all current introductory textbooks to the field, although understandings of it may vary. At some very general level, then, academic social psychology has always concerned itself with individuals in social contexts and still continues to do so.

But there have been important historical changes in how social psychology understands the relationship between individuals and social contexts. Prior to World War II, most social psychological research took place in stimulus-rich social settings in the "real world." After that war, the environment was still clearly conceptualized as the actual world outside the human head, and Lewin and his students became famous for their manipulations of selected and theoretically significant contextual attributes in laboratory settings. But contexts eventually came to be understood, not in terms of their "actual" characteristics—to use Allports phraseology—but in terms of their "perceived" or "implied" attributes. Allport treated implied and perceived environments as coequal in importance with actual environments. Later, however, the perceived environment became dominant for most social psychologists, and in justifying a cognitive information-processing approach to intergroup relations in his chapter for the 1985 *Handbook*, Stephan wrote: "It is the individual's perception of social reality and his processing of this information that determines individual behavior" (p. 599).

There is not a lot of evidence at present to judge whether it is more productive to emphasize perceived environments over actual ones. In one study, Jessor (1981) showed that perceived neighborhood characteristics correlate more highly with reported behavior than do actual neighborhood characteristics. This is not surprising since most theories of individual behavior predict that proximal variables correlate more highly with behavior than do distal ones. Early social psychologists, and the sociologically trained social psychologists of today, would not agree with the current emphasis on perceived settings. Their work stresses structural attributes of the external world (such as neighborhood poverty levels) and the types of social relationships that occur within worlds (such as levels of neighborhood social cohesion). They worry whether individuals accurately encode situational attributes and whether attributes that are out of individual awareness might be just as important as those that are in it. They also know that perceptions of the environment do not arise out of a void. Perceptions are often complex products of attributes of the objective environment. Given this, why bother making a rigorous distinction between actual and perceived environments, claiming that one is more important than another for a particular discipline or problem? Would we not be better served by theories that show how actual and perceived situations are related to each other and subsequently influence behavior? Such theory is not now available in social psychology, and the dominance of the perceived environment over the actual or over the actual related to the perceived has been well nigh total since the mid 1970s.

Although social psychologists claim that their work deals with the individual in social contexts, it is worth noting that there has never been a social–psychological theory of contexts. Such a theory would presumably conceptualize, justify, and measure the most important aspects of settings and would also specify the ways in which these attributes are interrelated. (Many attempts have been made to construct such theories of individuals, indicating the primacy of the individual when compared to the setting.) Probably the best known theory group within social psychology that was interested in contexts was at the University of Kansas, and its members are known for an intensive longitudinal study of a single town (Barker & Gump, 1964; Barker & Schoggen, 1973; Barker & Wright, 1955). But this interest—and the descriptive, multivariate, and intensive research style that went with it—never caught on in social psychology. This interest was seen as too atheoretical in its origins and results, and I suspect that contexts were seen as too complicated and disorganized to lend themselves to productive study without a prior theory to organize priorities. Perhaps as a consequence, the Kansas researchers turned from social psychology to environmental psychology.

Personality theorists and developmentalists have taken more seriously the task of dealing with situations, particularly in Europe where the work of Magnusson (e.g., 1981) stands out. Magnusson attempted to put the concept of interactions between individuals and contexts back into the core of psychology so that henceforth there would be no more rhetoric about which was more important than the other, so that psychologists would come to see that behavior cannot be abstracted from settings and sociologists would realize that settings cannot be abstracted from individuals. While this approach was greeted with enthusiasm in some areas of personality research and in most of developmental psychology, it fell on deaf ears in social psychology. It is not clear why, but the cool reception may have had to do with the difficulties of conceptualizing situations in a general theoretical way that lends itself easily to measurement, as well as to the absence within the dominant cognitive orientation of exciting hypotheses that absolutely necessitate complex understanding of situations. Social psychology is still without a viable theory or even language of situations, and there is little felt need for such a tool.

However, since contexts are so central to social psychology, they are studied. The intellectual style for doing this has been to use a given substantive theory to specify a restricted number of contextual attributes (usually one) that have to be varied if the theory is to be tested. Thus, in the early days of formal social psychology, attention turned to such issues as how two different types of architectural arrangements influence social interaction patterns, how the complexity of a task influences leadership effectiveness, and whether communicators of high or low credibility are more persuasive. This analytic–reductionist strategy leads to examining just a few contextual variables, and it is not always clear that the attributes selected for manipulation are important in terms of their prevalence or size of impact. Attributes seem to be selected from the armchair because they are thought to be important for a particular domain, such as persuasion or friendship.

If we took a domain such as science and an outcome such as belief revision in science, it would be easy to arrive at a large number of contextual aspects that are presumably important. Indeed, I did so in the first section of this chapter. The difficulty is knowing which attributes are so important that they should be singled out for manipulation and, perhaps especially, knowing how these factors might combine to influence belief revision. In manipulating just a few contextual variables, social psychologists run the risk of neglecting theory about the possibly complex ways in which individual contextual factors interrelate. Persuasion is one of the most researched topics in all social psychology, and contingency theories about contextual variables are most developed there (McGuire, 1985). But we still have no good conception of how the many factors that have been explored to date interact to influence persuasion. Isolating individual contextual variables, or a small number of them amenable to experimentation, does not bode well as a complete strategy for understanding social contexts.

The absence of a general theory of social contexts and the preference for explicating and manipulating just a few domain-specific contextual attributes are also associated with a preference for studying social situations in the laboratory rather than in the field. Immediately after World War II, much social psychological research took place in the field. But Lewin—and especially his students—saw little need for field research as long as theoretical hypotheses about one or two contextual factors were tested, varying these factors by means of simplified analogues to the complex social settings to which generalization was ultimately sought. Hence, to study severity of initiation, the researchers felt no pressure to go to the Marines or to college fraternities. While there are initiation procedures there, they are undoubtedly confounded with other causal factors that need not operate in the researcher-controlled laboratory. So, small groups were set up in the laboratory composed of people who did not know each other and might never meet again, and individuals were randomly assigned to these groups in which they experienced either severe or lenient initiations. However, this movement of context into the laboratory not only abstracted individuals from the rest of their life but also abstracted the few contextual variables examined from all of the other contextual attributes to which they are regularly (but not necessarily inevitably) related. Brunswik's (1956) theory of ecological representativeness in the selection of experimental stimulus materials implies a critique of the strategy social psychologists adopted. Brunswik argues that correlated setting attributes should be brought into laboratory analogues together with the setting attribute of theoretical interest. But Brunswik was not taken seriously in social psychology, and the rule of single variables became dominant.

Would social psychology be any richer if a more comprehensive and holistic conception of contexts were adopted and theorists tried to generate general theories of context or to immerse themselves in a particular class of contexts so that they had some grounded reasons for selecting some attributes as more important than others? There is no way to be sure, but unless social psychologists try these approaches there is also no way to know whether they are now committed to a blind alley. So, my critique is not that social psychology has evolved from an early promising perspective on context to a less productive one that emphasizes the perceived situation over the actual, and that abstracts from obviously multivariate environments only a few attributes for manipulation in controlled laboratory settings. Rather, my critique is that in the person–situation nexus, social psychologists take the individual more seriously than the setting; they have developed no language for describing contexts like the one they have for describing persons; and no intellectual effort seems to be devoted to puzzling through what social situations mean and what methodologies are appropriate for assessing these meanings. Has there been premature closure in social psychology about how contexts should be understood? While this question is, I believe, important in its own right, an answer can only emerge from the task sketched but not undertaken: showing that important knowledge is lost unless a different conceptualization of social situation emerges than that which has been dominant in social psychology since about 1950.

Do Social Psychologists Study a Multiplicity of Contexts?

Earlier I maintained that a descriptive theory of science should involve the individual embedded in a multiplicity of formal and informal contexts at varying distances from the individual. In the aggregate, social psychologists appear to do this. Their work involves family and friendship contexts, school and work situations, neighborhoods, churches, political institutions, and the mass media. Indeed, the three *Handbooks of Social Psychology* contain many chapters whose titles explicitly refer to these formal or informal institutions. (However, the sections are usually in the

Handbook volume devoted to *applied* social psychology, but that is another issue.) So, concern with many types of contexts at many levels removed from individual behavior is part and parcel of how social psychology sees itself and has always seen itself.

Once again, however, historical shifts have taken place that raise questions as to whether the current state of social psychology is as receptive to multilevel theory as it once was. Psychology, with its individual emphasis, has clearly won out over sociology, with its institutional and group emphasis—as the academic discipline the vast majority of social psychologists look to for inspiration. In his introduction to the first *Handbook*, Allport (1954) noted that, until the early 1950s, social psychology was characterized by approximately equal numbers of textbooks with a predominantly psychological and a predominantly sociological flavor, with only one text striking an internal balance between the two (Newcomb, 1950). In his introduction to the second *Handbook*, Allport (1968) remarked that by then, psychologically oriented texts seemed to be about twice as frequent as sociologically oriented ones. The third edition of the *Handbook* (Lindzey & Aronson, 1985) made no reference to the encroaching psychologization of social psychology, but my impressions are that a psychological version of social psychology holds even stronger sway today than in 1968, and that social psychology has become increasingly peripheral within sociology.

This individual psychologization of social psychology is associated with a decrease in research on institutions and small groups. Chapter titles in the 1954 and 1968 *Handbooks* speak to comparable levels of interest in contexts. But the 1985 *Handbook* has strikingly fewer chapters on groups and institutions. Moscovici (1985) places this into a wider context when he claims:

> Until the 1960's, influence phenomena were studied in this context (viz. the context of small groups and the role of these groups in industry, education and psychotherapy)—until the time, that is, when psychology lost interest in the study of groups both large and small and turned its attention to the study of "interpersonal relations." (p. 351)

Some other commentators have noted the waning of interest in group phenomena in academic (but not applied) social psychology (e.g., Steiner, 1974, 1986; Bargel, Gold, & Lewin, 1992).

The individual psychologization of social psychology is also reflected in how groups are studied when they are. Stephan (1985) introduces his *Handbook* chapter with the following statement: "Intergroup relations from the social psychological perspective consists of the systematic study of relations between individuals as they are affected by group membership." In this particular formulation, the study of relationships between groups is reduced to the study of group influences on interpersonal relationships, reflecting the move from groups to interpersonal relationships to which Moscovici drew sad attention. Stephan's formulation of the problem underplays interest in how different groups relate to each other and what the factors are, both within and between groups, that determine the nature and consequences of group relations. This is, of course, a central concern to science studies because scientists tend to organize themselves into theory groups, constructing their relationships within groups so as to build and maintain tribal loyalty while acknowledging the importance of fostering internal dissent and even revolution. Likewise, although tribal leaders often preach that others' theories are to be respected, there is a tension here because they know that refuting others positions will have more salutary consequences for both ingroup cohesiveness and access to additional resources.

Stephans chapter aims to provide a cognitive interpretation of intergroup relations understood as interpersonal relations. Thus, the three main chapter headings read as follows: "A Cognitive Information-Processing Approach to Group Relations"; "Intergroup Cognitions and Behavior"; and "Changing Intergroup Cognitions and Behavior." We see here an individualist conception of social psychology that emphasizes cognitions as opposed to emotions or behavior or the interface between the cognitive, affective, and behavioral. It may seem preposterous at first to use theories of individual cognitions about groups to explain relations between groups but, as already noted, Stephan (1985) claims: "It is the individual's perception of social reality and the processing of this information that influences individual behavior" (p. 599).

Can intergroup behavior be explained almost exclusively in terms of individual cognitive behavior, so that no recourse is needed to the history of a particular group, to its current internal relations, to its relations with other groups, or to its relations with larger institutions in society at the local, national, and international levels? Put into the context of science studies, can the competition between a Spence group and a Tolman group be explained solely by how each group processes information and thinks about the other group? After all, the competition between theories was so oblique at the time, and the Tolman group was so loosely organized? I doubt very much whether Campbells multilayered account of the Spence–Tolman competition could be easily captured by the constructs in Stephans theory. Yet if it is to be a general theory of intergroup relations, should it not do just that? Stephan does not construct an argument to defend his assumption about perceptions of other groups, even though it is central for justifying the study of intergroup relations in terms of individual cognitions about groups. Gone is all sense of *inter*group relations—of two groups being in a relationship to each other and to the rest of society in all of its complexity.

I do not mean to single out Stephan. I use him only as an illustration of the way in which social psychology has evolved from a concern with individuals embedded in a multilayered context provided by families, schools, workplaces, formal and informal networks, neighborhoods, the mass media, and political institutions and has moved toward conceptualizing individuals as processors of information about the social environment. Individuals are presumed to be influenced by these perceptions, thus preserving the traditional individual–context link. But the link is transformed, with perceptions of context looming larger than the contexts themselves, and with the context being reduced to one or two manipulable attributes and no longer seen as a multilayered reality from which many forces impinge upon individuals. Some of these forces are even beyond articulation by those affected, as with the economic–structural changes that in the last decade have dramatically altered job chances in manufacturing for inner-city residents.

The low salience of research on groups in current social psychology is particularly troublesome for metascientists. Groups are of special importance to metascientists because science is done locally by groups, groups bring renewal into science as they coalesce around a new theory or split off from parent disciplines, and groups compete with each other for dominance in socially establishing the "truth." Since social psychology is about individuals in social contexts, and so many of our most immediate social contexts are group ones, we should not be surprised if researchers on science examine current social psychology and ask in puzzlement: Why the low salience of research on groups? I am not sure how social psychologists would answer, but I suspect it would be along two lines. The first is that, despite the efforts of Stephan (1985) and Bar-Tal (1990), groups do not have cognitions like individuals. What is primarily interesting about groups and cognition is the extent to which beliefs are shared, but that is a problem for researchers interested in the social sharing of knowledge rather than in the individual knowledge that interests academic social psychologists in the 1990s. The second response would be that past research on groups has cast doubt on their entitativity. For instance, it seems that groups do not identify or solve problems better than the best individual within them (Kelley & Thibaut, 1968), making it unnecessary to invoke groups as part of the explanation for problem solving. Groups induce conformity, but so do individuals, and we still lack an analysis of how or why conformity is different when groups are the initiating agents rather than individuals (McGuire, 1985). Groups provide social support, but so do individuals, and we still lack an analysis of how processes and outcomes are different when a group opposed to an individual provides the support. The most convincing evidence for the importance of studying groups is that no convincing explanations for observed phenomena can be made in individualistic terms. A relational explanation is the only one plausible.

It may seem odd to outsiders that social psychology has no convincing theories of group entitativity, of phenomena whose only explanation is relational, of effects that are more than the sum of the individual parts. But that seems to characterize the current state of the art. As long as this remains the state of the art, social psychologists will not be much interested in groups, and since they are not interested, the state of the art is not likely to change. The problem of group uniqueness needs resurrecting in social psychology.

But the small groups in which scientists are so importantly found, both locally and in national invisible colleges, are only part of the multilevel reality with which science studies have to deal. Also needed is an emphasis on formal institutions. More is at issue here than theories of self-selection and election into institutions that help describe or explain who joins scientific organizations. More is at issue than theories about the conditions under which individuals actively invest themselves in institutions, theories that might help metascientists better understand who is active in science politics and why. More is at issue than the description of the patterns of institutional involvement that lead to private belief changes, which would help metascientists better understand scientific belief revision. We also need theories to identify processes that lead to the creation of new institutions (in science's case, often new disciplines or new multidisciplinary fields), for such institutionalized subspecialization characterizes modern science and often arises from within existing institutions. Academic social psychology is particularly weak in documenting how institutions change with time, and it is unrealistic to expect the field to illuminate how scientific institutions change as new blood, new theories, and new political environments emerge. After all, institutions and temporal processes are far removed from the concern with the individuals, ahistorical theories, and laboratory research practices that characterize the field.

Do Social Psychologists Assume Constant Change in Individuals and Contexts?

Interest in individual continuity and change has deep historical roots in social psychology. The emphasis was originally on how complex, long-lasting events change individuals, and perhaps the best known study is Newcombs (1943) research on young women who attended Bennington College in the 1930s. The institution was then very liberal, but in terms of their home background and initial attitudes the women were quite conservative. The women have now been followed up twice, the last time in the 1980s when the women were in their 70s.

While some interest remains in such studies in the abstract—McGuire (1985) mentions some in the last *Handbook*—the emphasis has recently faded. More precisely delineated and theory-relevant interventions are preferred, and follow-up measurement is likely to be in terms of weeks rather than years. Moreover, the multivariate description of lives as they change in response to age, birth cohort, and general historical factors has never excited social psychologists, who have left such research to developmentalists within psychology and sociology. Short-term stimuli and response patterns are the order of the day, as is an aggressive ahistoricism that further undermines interest in studying how lives change and what causes them to change in particular historical circumstances. Hence, it is difficult to conclude that current social psychology bothers with any assumptions about individuals and their susceptibility to change in different ages, periods, and economic or social circumstances.

Science studies cannot look to social psychology for leads in understanding how scientific institutions evolve. Chapter titles in the first two *Handbooks of Social Psychology* attest that institutions have been a concern to social psychologists, especially work sites, schools, and settings providing psychological services. But rarely has the interest been in taking the organization as the unit of analysis and seeking to describe or explain how it changes over time in response to external and internal dynamics. Rather, social psychologists have been more interested in organizations as the settings in which individual behavior takes place and in which individual and group performance are affected. But even this kind of interest has waned over the last two decades, presumably because even this amount of focus on institutions renders the individual less salient. Moreover, the analysis of institutions does not lend itself well to the controlled, laboratory-based methods in vogue since the 1950s so that, as early as the first *Handbook*, institution-based research was mostly discussed under the heading "applied research." This implies a certain second-class status in a field that aspires to develop universalist theories that will not be bound in any way by person or setting (or even time) factors. Metascientists should not expect much of a harvest if they look to social psychology to help understand how scientific organizations operate and influence the behavior of individual scientists.

They might expect a greater payoff, however, if they look to organizational psychology, itself basically a historical offshoot of social psychology. But to look there would be an admission of how much social psychology has redrawn its boundaries to exclude most of the few researchers trained in social psychology who have maintained an intellectual interest in groups and/or institutions. The same point can be made about scholars with interests in how individuals change over time in response to other types of external pressures. Such scholars are now likely to be found in developmental psychology and not in social psychology, where they might now not feel comfortable. In a similar vein, the social psychologists from Kansas who were interested in local community found that their work was better received in the newly emerging field of ecological psychology than in social psychology. So, that is where they went. All the sciences are now moving toward ever finer internal subspecialization, and the narrowing of boundary lines around formal academic social psychology may be merely another reflection of this scientific trend. But the boundary lines that are being drawn systematically increase the chances of excluding from social psychology those researchers who seek to study social behavior in social contexts and who assume that social contexts are multivariate, ever changing, and beyond reduction to perceived contexts or to analogues that can be manipulated in laboratory experiments lasting about one hour.

This is not to claim that all analogues have been unproductive for knowledge in general and for science studies in particular. For example, the analogue experiment by Jacobs and Campbell (1961) on the transgenerational perpetuation of arbitrary traditions has proven to be useful for understanding the conditions under which truth reemerges in science after it has been called into doubt. But I do claim that most analogue experiments do not capture the complexity of behavior that occurs in multivariate, multilevel, and constantly changing social contexts. I have little confidence in the relevance of their results for the actual world of science.

How Is Causal Reciprocation Represented in Social Psychology?

Much more emphasis has been placed in social psychology on the situation influencing the individual than vice versa. This is especially true in research on social conformity, attitude change, media influences, social cognition, and group and individual performance. Social psychologists have rarely examined the circumstances associated with individuals changing organizations, although a literature on this exists in other fields—especially social history, political science, and social anthropology where interest in exceptional individuals and elites is strong. From the perspective of a descriptive theory of science, academic social psychology seems to fall particularly short in its analyses of how institutions change and how individuals influence the institutions in which they are embedded.

Social psychologists in Europe (especially Moscovici, 1985) have pioneered the study of how small groups of minorities sometimes influence majority opinion In his chapter in the 1985 *Handbook*, Moscovici acknowledges that, as a European, he dislikes the hegemony of American social psychology, with its current paucity of group research and its earlier construal of social influence in terms of majorities inducing individuals to conform to group norms. He sees the latter emphasis as culture bound, reflecting either American fears about limits to individualism or the manipulative desire to learn how outsiders can be induced to assimilate to mainstream values in a nation needing to socialize new immigrants and foster mass consumption. As Rosenwein (Chapter 8, this volume) and others point out, Moscovici's reorientation of conformity studies is fruitful for science studies because mass belief revision (i.e., theory change) often comes about when a small group of committed scientists persuades the rest of the field to accept a new knowledge claim. The point here is that there was no emphasis before Moscovici on how minorities influence larger and traditionally more dominant groups, though there had been research in the Asch tradition on how larger groups influence individuals and small minorities. In this regard, it is perhaps worth noting that a minority is often the next step after an individual with a strong viewpoint has persuaded a few others to join him/her. It is not much of a step from reversing the causal link between large and small groups to reversing the link between large groups and individuals.

Conclusion

Social psychologists broadly agree that the field's major task is to understand how individuals feel, think, and act in different social contexts. Such understanding would clearly further our knowledge of scientific behavior since scientists are individuals located in many formal and informal networks that influence them and that they influence.

However, social psychology has not made much of a contribution to science studies. I suspect that this is because in social psychology, the individual seems more dominant than the context, and no descriptive theory of contexts is currently available in social psychology or is even felt to be necessary. Social psychologists are content to stress the perceived situation over the actual and to select single attributes of contexts for study instead of the more theoretically disorganized contexts found in the world outside the research laboratory. As long as these orientations remain unchallenged, social psychologists are not likely to feel any need to revise their current perspective on contexts. Such a need would be most keenly felt if data convincingly shattered their key assumption that perceived contexts correlate more highly with behavior than do actual environments, or if data showed that perceived environments do not add to prediction or explanation when compared to actual environments. Yet this is a silly way of presenting the issue, for perceived environments are presumably often influenced in complex ways by actual environments. Hence, full understanding of the importance of contexts will require knowledge of both the actual and the perceived world, particularly as they operate *together* to influence emotions, cognitions, and behavior.

We assume that everything about science is dynamic rather than static. Thus, science interweaves individual scientists with the many formal and informal contexts at different levels that help form them and in which they work. A full description of science would therefore require influences from institutions to small groups and individual scientists and also from individuals to small groups and institutions. Unless such assumptions are mirrored in the metatheory of social psychology, the striking correspondence between the central tasks of social psychology and of science studies will remain more apparent than real.

Unfortunately, my analysis suggests that these assumptions are not clearly mirrored in the current metatheory of social psychology, although they were somewhat better mirrored there when the field was institutionalized shortly after World War II. This is not to claim that science studies have learned nothing from social psychology. Shadish et al. (Chapter 1, this volume) and Campbell (Chapter 2, this volume) have highlighted most of the achievements of social psychology. However, these achievements do not strike me as being as great as the close correspondence between the two fields superordinate purposes would suggest that they should be. But for the two fields' metatheories to draw closer together and reap mutual benefits entails making the "social" more central in social psychology and decreasing the current emphasis on applying information-processing theories from cognitive psychology to features of the social world. The history of social psychology is replete with abrupt transitions in the dominant intellectual problem of the day, so it is not inconceivable to imagine the current assumptions of the field being replaced. If they were replaced by assumptions that explicitly embed individuals in time and social contexts, social psychology, would be brought closer to the vision promulgated by its most famous theorists who wrote immediately after World War II. But do not expect to see such changes any day soon. Few social psychologists care about their theories illuminating science studies. They care more about how their colleagues receive these theories. They are not likely to take seriously a critique of the ahistorical, socially decontextualized, and mostly cognitive assumptions of academic social psychology unless the critique comes from respected sources within their field and provides them with an alternative metatheory that is practical enough to guide their daily research practice. Such conditions do not now exist. But maybe, just maybe, some day, who knows, no, not really, but . . .

Acknowledgment

I would like to acknowledge the feedback offered by Richard Jessor, Will Shadish, Jr., and Arthur Stinchcombe. Of course, none of them is responsible for any of the lapses in fact, judgment, or taste that remain.

References

Allport, G. W. (1954). The historical background of modern social psychology. In G. Lindzey (Ed.), *Handbook of social psychology* (pp. 3–56). Reading, MA: Addison-Wesley.

Allport, G. W. (1968). The historical background of modern social psychology. In G. Lindzey & E. Aronson (Eds.), *Handbook of social psychology* (2nd ed.) (Vol. 1, pp. 1–80). Reading, MA: Addison-Wesley.

Bargel, D., Gold, M., & Lewin, M. (1992). Introduction: The heritage of Kurt Lewis. *Journal of Social Issues, 48*(2), 3–13.

Barker, R. G., & Gump, P. V. (1964). *Big school, small school: High school and student behavior.* Stanford, CA: Stanford University Press.

Barker R. G., & Schoggen, P. (1973). *Qualities of community life.* San Francisco: Jossey-Bass.

Barker, R. G., & Wright, H. F. (1955). *Midwest and its Children.* Evanston, IL: Row Peterson.

Bar–Tal, D. (1990). *Group beliefs.* New York: Springer-Verlag.

Brunswik, E. (1956). *Perception and the representative design of psychological experiments.* Berkeley: University of California Press.

Deutsch, M. (1992). Kurt Lewin: The tough-minded and tender-hearted scientist. *Journal of Social Issues, 48(2),* 31–43.

Jacobs, R. C., & Campbell, D. T. (1961). The perpetuation of an arbitrary tradition through several generations of a laboratory microculture. *Journal of Abnormal and Social Psychology, 62,* 649–658.

Jessor, R. (1981). The perceived environment in psychological theory and research. In D. Magnusson (Ed.), *Toward a psychology of situations: An interactional perspective* (pp. 297–317). Hillsdale, NJ: Erlbaum.

Kelley, H. H., & Thibaut, J. W. (1968). Group problem solving. In G. Lindzey & E. Aronson (Eds.), *Handbook of social psychology* (2nd ed.) (Vol. 4, pp. 1–101). Reading, MA: Addison-Wesley.

Lindzey, G., & Aronson, E. (1985). *Handbook of social psychology* (3rd ed.). Hillsdale, NJ: Erlbaum.

Magnusson, D. (1981). *Toward a psychology of situations*. Hillsdale, NJ: Erlbaum
McGuire, W. J. (1985). Attitudes and attitude change. In G. Lindzey & E. Aronson (Eds.), *Handbook of social psychology* (3rd ed.) (Vol. 2, pp. 233–346). Hillsdale, NJ: Erlbaum.
Moscovici, S. (1985) Social influence and conformity. In G. Lindzey & E. Aronson (Eds.), *Handbook of Social Psychology* (3rd ed.) (Vol. 2, pp. 347–412). Hillsdale, NJ: Erlbaum.
Newcomb, T. N. (1943). *Personality and social change*. New York: Dryden.
Newcomb, T. N. (1950). *Social psychology*. New York: Dryden.
Steiner, I. (1974). Whatever happened to the group in social psychology? *Journal of Experimental Social Psychology, 19*, 251–289.
Steiner, I. (1986). Paradigm: Groups. In L. Berkowitz (Ed.), *Advances in experimental social psychology* (Vol. 19, pp. 251–286). New York: Academic Press.
Stephan, W. G. (1985). Intergroup relations. In G. Lindzey & E. Aronson (Eds.), *Handbook of social psychology* (3rd ed.) (Vol. 2, pp. 599–658). Hillsdale, NJ: Erlbaum.

CHAPTER 16

Editors' Epilogue: Some Reflections

WILLIAM R. SHADISH
STEVE FULLER

The preceding three comments raise provocative and thougtful criticisms of SPS. In this chapter, we would like to respond to and build on these comments.

Tweney's Social Psychology of Science

Tweney (Chapter 14) clearly likes this book and treats it kindly in his generous discussion. Yet Tweney's reading of this volume leaves him with the impression that two assumptions underlie almost everything said: (1) that one can best understand processes by identifying causes and distinguishing them from effects, and (2) that universal laws might be forthcoming in doing so. Any time a respected colleague reads something so radically different from what one reads oneself (or thought one wrote), it is worth exploring the discrepancy. In this case, to our dismay, the discrepancy is severe indeed. The two assumptions are, to the best of our knowledge, never stated anywhere in the book. And while many things can be read as implicit in a work, we see Tweney's two assumptions as inconsistent with the assumptions we think might be implicit in a social psychology of science (SPS). Hopefully, then, this exploration will reveal more agreement with Tweney than not about what the proper assumptions might be.

Tweney's First Assumption

Tweney's first assumption really has two parts. The first part is that SPS (as presented in this book) implicitly claims to have identified the "best approach to understanding processes." (Tweney, Chapter 14, this volume). Social psychologists may be somewhat naive about science studies, but they are not so naive as to think that they have a monopoly on identifying the best approach to doing science studies. Studying causes and effects is one approach to understanding processes, but it is clearly neither the only nor the best approach to doing so. This is not to kowtow to other approaches, but simply to acknowledge the fallibility of all methods in science. Elsewhere, in fact, we have specifically criticized those who maintain that randomized experiments are so well founded as to be beyond the claim that all methods are fallible and let us down in ways we cannot always know well (Shadish, 1989). No method is uniformly best. It is that simple, and nothing we have written should be read as implying that we think otherwise.

The second part of Tweney's first assumption is that one can make a distinction between causes and effects and then try to identify causes that produce particular effects. What is objectionable to Tweney seems not to be the use of causal-sounding language but an implicit assumption about "separability" of cause and effect. Without separability, according to Tweney, causal-sounding language makes no sense: "It only makes sense to speak of one event as the cause of another event when and only when the two events are independent (as with the oft-used billiard ball example)." Tweney follows with a discussion of this concept of separability and how it might apply to science. Although he never clearly defines separability—not a minor omission given the centrality of the concept to his criticism—his brief discussion is reasonably compelling on the notion that very few things associated with science are truly separable.

What is dismaying is that Tweney seems to read SPS as disagreeing, as claiming that this still undefined concept of separability is easy to establish and defend. So we are straightforward in responding that we agree entirely with Tweney. Strict separability is difficult to defend, and one cannot base an experimental SPS on it. But this does not lead us to give up an experimental SPS because separability and experimentation are not as intimately tied as Tweney's first assumption presumes. In particular, a careful reading of those who have written about experimentation (e.g., Cook & Campbell, 1979, 1986) suggests that the fundamental assumption underlying randomized experiments is not separability but manipulability. That is, one must be able to exert some degree of control over the putative cause in order to conduct an experiment where it serves as the independent variable. A variable that is not currently manipulable, such as gender or ethnicity, cannot serve as the independent variable in a randomized experiment. Of course, some things not obviously manipulable today may prove to be manipulable tomorrow, as our increased ability to manipulate genetics suggests.

In fact, manipulability belies separability in both a strong and a weak sense. The strong sense is that the putative cause must not be completely separated from the rest of the world or else we would have no way to manipulate it, nor could it possibly affect an outcome from which it is completely separate. In this sense the whole notion of separability is inconsistent with experimentation. The weak sense is that we commonly accept our ability to manipulate things, events, or processes that we know full well are not separable from other things, events, or processes—we throw circuit breakers that we know to be connected intimately with the electrical circuits and all the things that those circuits power in our house (the circuits having the power to throw circuit breaker themselves during overloads), we read books that seduce and engross us in a way that belies our separation from them, and we consciously try to make a good impression on others by pointing to the connections we want noticed and hiding those we want forgotten. This is apparent even in Tweney's example of three Martians trying to determine how a car works. One of Tweney's Martians rips the wires off the car and finds that it does not run anymore. The only assumption needed here is manipulability, which all of us seem to agree is a warranted assumption. Are the wires separable? Who knows? But the engine did stop. The only thing that an experimental SPS claims on this basis is that we may sometimes learn something from the manipulation, however tentative, contextual, fallible, and difficult to interpret that learning may be. Nor is anyone claiming that what has been learned is that ripping out the wires accounts for 100% of the variance in automobile performance, or that we have identified the one and only cause of that performance. Nothing we see in an experimental SPS requires Tweney's separability assumption.

Tweney's Second Assumption

Tweney then goes on to suggest that SPS also assumes the existence of universal laws. He specifically contrasts this with a more benign assumption of empirical generalization. But this criticism is profoundly misplaced, at least if one is to judge from the writings of the very authors Tweney cites as proponents of misplaced emphasis on experimentation (Campbell & Stanley, 1963; Cook & Campbell, 1979). Experiments tell us only about the probabilisitic effects of some intervention in a particular historical and situational context (Campbell, 1986); the validity of generalizing any causal relationship nearly always depends on further empirical study. Indeed, not only are such generalizations never universal, we are usually hard pressed to find evidence for more than modest generalization at best.

Tweney suggests that all this makes a difference for how one deals with exceptions, expecting them under the empirical generalization rationale but explaining them away under the universal laws rationale. However, these two supposed opposites are not really incompatible. Experimenters have long had a rationale for expecting unexplained variability under so-called random effects models. So experiments are in no way inconsistent with the empirical generalization rationale. Similarly, no researcher has a prior way of knowing whether unexpected variability can be explained or not; barring more radical ontological assumptions than most scientists make, one cannot rule out the possibility of explanations until one has searched for them. To simply say, "Oh, I expected that," in regard to some observed exception will hardly count as good science no matter how one conceptualizes the matter of generalization.

Tweney's Alternative Approaches

It is good to see, then, that up to this point we are in full agreement with Tweney's astute observations about what assumptions might underly SPS. But we part ways modestly with Tweney's final suggestion that an alternative methodology is preferable to that outlined in this book. The parting is only modest because we welcome Tweney's alternative approach as part of the repertoire, only stopping short of elevating it to the preferred status he seems to want it to have over all other methodologies.

Specifically, Tweney says that the experiment takes one (or occasionally a few) independent variable(s) out of the rich, interconnected context in which they normally exist, losing the "dynamic interrelationships of the variables construed as dimensions of an interlocked system." While the criticism is somewhat overstated in that Tweney presents only the crudest versions of experimentation, nonetheless it is fundamentally correct. A hallmark of all experimentation is some effort to "isolate" independent variables, always keeping in mind the virtual impossibility of actually achieving that goal. Tweney's "dynamic modeling" alternative would be far less subject to such a criticism.

But Tweney fails to tell the reader about the trade-offs between the two choices. The experiment loses contextual embeddedness in order to gain fidelity of causal attribution; dynamic modeling gains contextual embeddedness at the cost of losing that fidelity. In dynamic modeling, unless the model is perfectly specified, including all relevant variables and their transformations, each connected in the correct way, researchers have no sure way of knowing if they have reached a correct conclusion about what influences what. Problematically, researchers in many fields can never know in principle whether the model is perfectly specified because their theories are simply not yet strong enough. For reasons we need not elaborate here, experiments do not need to be perfectly specified to achieve this fidelity.

So we prefer to think of experiments and dynamic models as complementary methodologies rather than competing ones. We envision the possibility, for example, that the dynamic modeler may occasionally be interested in seeing whether an intervention into the system will change something. Our claim will simply be that whatever methodology the modeler develops to do so will inevitably look a lot like some version of an experiment, and will inevitably have to cope with all the same logical problems about ruling out alternative explanations that the experimental literature already discusses.

What Epistemology and Ontology Underlie Experimentation?

All this should alert us to the difficulties of trying to intuit what ontologies and epistemologies "really" underlie an experimental approach to science studies. We find it somewhat ironic that Tweney, who throughout most of his response reminds us of the dynamic character of scientific practice, should manifest such a static view of (what is arguably) the core scientific practice, namely, experimentation. To hear Tweney tell it, it is as if all experimentalists, from Galileo to Campbell, have shared the same set of epistemological and ontological assumptions. Of course, these figures have generally assumed that something independent of their actions enabled them to manipulate certain aspects of nature in the course of an experiment. But such an assumption is relatively trivial, saying little of the character of particular scientific enterprises—though we admit the significance of this "minimal realism" in defeating various forms of philosophical skepticism that put up principled roadblocks to the path of empirical inquiry (Bhaskar, 1975). The more interesting point here is the robustness of experimental findings across not merely changing epistemological and ontological assumptions but even more sophisticated experimental designs and subtler understandings of what experimentation is all about (Hacking, 1983). Presumably, a follower of Donald Campbell can redo Galileo's experiments and arrive at results that are understandable as extensions and refinements of Galileo's original experiments. This is not to deny that the Campbellian may see the exact role of experimentation in scientific inquiry somewhat differently from the Galilean. Nor is it to deny that science itself may not offer many incentives for redoing experiments (Collins, 1985). Nevertheless, experimenters must be doing something right if the effects of their experiments can (at least in principle) be reproduced as experimental methodology is improved.

Perhaps the important point in all this is not whether or not one can show that the experiment is "best" or "fatally flawed" by virtue of its supposed consistency with some preferred or reviled assumptions. Rather, the point is to begin to understand how and why such a methodology has remained so trenchantly a part of our scientific practice despite the passing fits and fads of explanatory systems. The experiment is not the Holy Grail, but it must be working reasonably well. After all, the alternative is to think that scientists are stupid folks, unable to understand how self-defeating their own practices are, unable to see the theoretical and empirical barrenness of their work, fundamentally unable to do a good job of science without the guidance of others. Even granting the partial truth of these statements (Fuller, 1992), it seems quite unlikely that they are completely true. We think that this volume demonstrates the potential of an SPS component in science studies to help decide this fundamental issue.

Conclusion

We can agree with many of Tweney's conclusions without agreeing with his premises. SPS and science studies need to use multiple methodologies, and must be particularly careful to include methodologies that respect the contextual embeddedness and richness of scientific practice. The experiment (or meta-analysis) is not necessarily the best way of doing this—even if it is not guilty of all those other questionable assumptions that Tweney notes.

Turner's Interpretivist Challenge to an Experimental SPS

Turner (Chapter 13) is a sociological theorist who has fruitfully challenged the scientistic pretensions of empirical social science. His signature way of offering methodological critique is by juxtaposing what he regards as an atavistic desire for a comprehensively explanatory, causally based science of human behavior with the track record of 20-century attempts at realizing these aspirations, all of which have run afoul of the limitations of statistical and other forms of scientific inference (Turner, 1986). Not surprisingly, then, Turner has seen in SPS's ambitions a likeminded longing for a comprehensive science of science. Like Tweney, he focuses his attack on the SPS's heavy reliance on experimental findings and results.

Turner observes that psychology has been historically a successful science in areas in which it has provided more rigorous accounts of phenomena that already appear in people's folk psychologies. However, it is not clear that a psychology of science offers the prospect of many such areas: Creativity may be one, Turner admits, since that is something that people already psychologize about. However, the concepts that seem to be central to a specifically social psychology of science—such as consensus formation—may not fit the bill so neatly, especially if social psychological concepts are used to explain things that scientists think are caused by the nonhuman events they interact with (i.e., external reality). Thus, the performance of experiments on student subjects working on tasks that are, only in the loosest sense, "analogues" to scientific ones would seem merely to beg the question that Turner wishes to raise. Just because the student subjects work on tasks that resemble scientific ones in certain specifiable respects, it hardly follows that scientists would accept the results of such experiments as telling them anything interesting about how their psychologies work. Moreover, the psychologist who conducts these experiments bears the additional burden of having to explain to scientists why their folk psychological accounts will not explain their behavior just as well as the psychologist's appeal to social psychological concepts.

In effect, Turner is offering a Hobson's choice to social psychologists of science: *Either* the psychologist is going to have to explain the discrepancy with the scientist's account as well as why his/her account of the scientist's psychology is better than the scientist's own, *or* the psychologist is simply going to have to admit that he/she really wants scientists to be doing something other than what they have been doing, which then leaves psychologists with the even more daunting assignment of explaining why scientists should listen to their advice. Given these equally unappetizing alternatives, Turner makes his plea for SPS to scale down its ambitions and expectations.

In response, we grant Turner that a developed SPS could turn the explanation of scientists' behavior into a site of contested authority. After all, the methods of science, especially experimentation, seem to carry their own authority independent of the people who practice those methods. Indeed, this point is often taken to be the mark of science's "objectivity." In that case, we should not be surprised if it becomes possible for one group of people—say, social psychologists of science—to show that another group—say, practicing scientists—does not behave quite in accordance with the methods that both groups officially avow. There is no reason why the accountability of scientists should be any less contestable than the accountability of politicians. As Thomas Kuhn originally observed, and sociologists of science have since confirmed, the professional training of scientists does not make them especially able or inclined to reflect on the methodological and metascientific issues that underpin their practices. They become explicitly focused on these matters only when their paradigm is in a state of crisis. Under normal circumstances, scientists just get on with what they were trained to do, namely, to study a particular piece of reality in a particular way.

Admittedly, the contestability of scientific authority is an uncomfortable prospect for all concerned, especially in a culture that has increasingly come to associate scientists with fraud and misconduct. But, fortunately, SPS is far enough from realizing this possibility that we have time to think about the implications of our project for the evaluation and improvement of scientific performance—an interest that probably engages everyone who has contributed to this volume.

In any case, lest Turner worry too much about SPS becoming the "Big Brother" of the scientific enterprise, the discipline of psychology has become increasingly sensitive to the need for experimenters to weigh seriously the subject's self-understanding that typically emerges in the debriefing session after an experiment. The airing of the traditionally suppressed voice of the subject is both good politics and good methodology. Of course, we do not wish to minimize the problems involved in reconciling what often turn out to be the highly divergent perspectives of experimenters and subjects. But there are some precedents in the debriefing literature that may go some way toward alleviating Turner's concerns. Here it is worth recalling that accounting for the subject's self-understanding is a problem not unique to experimental psychology, but is in fact common to any intensive—including historical and ethnographic—study of the human condition.

But, as in the case of Tweney, we detect a curious asymmetry in Turner's appeal to the "historicity" of psychological inquiry. Turner portrays SPS as the Johnny-come-lately to metascience, struggling desperately to find a niche that has not already been occupied by either scientists' own folk accounts or the metascientific accounts offered by more established fields of history, philosophy, and sociology of science. However, this is to view history as a concatenation of traditions which, once started, continue indefinitely and perhaps even inertially. Consequently, new perspectives are up against what are taken to be centuries of cultural entrenchment. The view is rather akin to the one that Alasdair MacIntyre offers to capture the history of ethics in the West. But we beg to differ with Turner over the long-term entrenchedness of historical, philosophical, sociological, or even folk accounts of science. The character of the scientific enterprise—what people take to be special about science—has changed repeatedly since the so-called scientific revolution. Very often this has come from the methods or theories of one field being imported into another, which implies that disciplinary traditions may not be quite as self-contained and self-sustaining as Turner imagines. Here it is important to remind ourselves that science, too, is a creature of sociohistorical contingency. So, as long as we keep our historical possibilities open and our research agenda alive, we should not be discouraged from thinking that SPS will influence both the understanding and the conduct of scientific inquiry.

Cook's Critique: Our Reach Exceeds Our Grasp

Cook's (Chapter 15) is in some ways the gentlest of the commentaries, praising the idea of SPS as it does and outlining a descriptive theory of science that SPS might aspire to develop. Yet, at the same time, the second half of that commentary is trenchant and penetrating in its critique of the ability of modern academic social psychology as actualized today to help practitioners of SPS to reach that aspiration. The fundamental critique is that today's social psychology, at least among psychologists, is far more a psychology of the individual, and far less one of the social context, than the equal emphasis on both that the appellation of social psychology suggests. This is not a critique that we would even try to refute, mainly because it is correct. We pointed out in the sections on cognitive psychology and social cognition in Chapter 1 that the cognitive revolution has been pervasive in psychology generally, sweeping into social psychology in conquering hordes. Might does not bring right in this arena, but it does bring dominance by any reasonable measure of the quantity of scholarly output. When Cook points out that social psychology has tended more and more toward the study of cognitions about social phenomena, we see immediately that this leads social psychologists to have an impoverished view of all things social. We could not outline the reasons why better than Cook already has. But if we accept that critique, what does it do to the SPS enterprise we have tried so hard to construct in this volume? The answer, we think, lies in putting Cook's critique into a wider context so as to appreciate more fully both its insightfulness and its limits.

First, remember that Cook evaluated SPS against the content of an idealized descriptive theory of science that would, in essence, do all things well. SPS fails against this standard because the parent discipline of social psychology fails against it. Setting the standard so high is fair game; in fact, doing so is probably essential in order to see clearly where the field needs to go. But it would also be fair to pose a minimum performance standard, that a good theory of science would have, as a minimum, both a cognitive and a social component, acknowledging and investigating the importance of both. SPS clears this first hurdle with room to spare; SPS has a substantial enough social component to be a credible pretender to the task Cook outlines in his descriptive theory.

Second, it would also be fair to evaluate SPS not only against some high or low absolute standard but also against the competition, especially against other sociological and psychological approaches to science studies. Here, if Cook's idealized descriptive theory is desirable, SPS does quite well at least in aspiration, for it aspires to bring together both the social and the psychological as true partners. Cognitive psychology of science mostly ignores the social, and most sociological approaches treat many individual psychological factors as either epiphenomenal or of minor consequence at best. Of course, we cannot know for sure that Cook's idealized descriptive theory of science is the right path to pursue, but if it is, it is hard to judge SPS as worse-aimed than its alternatives.

Third, nothing we have written should be interpreted as implying that SPS aspires to be a complete science of science. An idealized descriptive theory of science will have to be social psychological, but not only social psychological. It will also be economic, political, and anthropological and have descriptive, normative and metatheoretical dimensions, and more. Psychology of science is likely to be just one contributor to this larger metascientific literature, always incomplete from the larger perspective.

Fourth, as Cook points out, the balance between social and psychological has been cyclical throughout the history of the field. If things are too cognitive among today's social psychologists, we can be fairly sure that the pendulum will swing back toward the social as time passes. Some limit to this swinging arc is likely, of course. Psychologists will probably never be as social as sociologists, and vice versa, and the day that fact changes will probably be the day that the disciplinary structure of both fields begins to break down (perhaps a welcomed development, but not one we should hold our breath anticipating). But if the pendulum will indeed swing somewhat in the future, SPS can take heart that it has set an agenda prepared to take advantage of the swing.

In the meantime, Cook's critique should serve as fair warning to all of us interested in SPS. On rare occasions, social psychologists of science may extend theorizing in the parent discipline by bringing more social emphasis to bear, but we should not be sanguine that this will occur often. More realistically, SPS will be limited by what the parent discipline has to offer. Hence, SPS will necessarily be vulnerable to cogent attacks from those who claim it is not social enough. They are right. It is not. But it is aspiring to be.

References

Bhaskar, R. (1975). *A realist theory of science*. Leeds, England: Leeds University Press.

Campbell, D. T. (1986). Relabeling internal and external validity for applied social scientists. In W. M. K. Trochim (Ed.), *Advances in quasi-experimental design and analysis* (pp. 67–77). San Francisco: Jossey-Bass.

Campbell, D. T., & Stanley, J. C. (1963). *Experimental and Quasi-Experimental Designs for Research*. Chicago: Rand-McNally.

Collins, H. M. (1985). *Changing order: Replication and induction in scientific practice*. Newbury Park, CA: Sage.

Cook, T. D., & Campbell, D. T. (1979). *Quasi-experimentation: Design and analysis issues for field settings*. Chicago: Rand-McNally.

Cook, T. D., & Campbell, D. T. (1986). The causal assumptions of quasi-experimental practice. *Synthese, 68*, 141–180.

Fuller, S. (1992). Social epistemology and the research agenda of science studies. In A. Pickering (Ed.), *Science as practice and culture* (pp. 390–428). Chicago: University of Chicago Press.

Hacking, I. (1983). *Representing and Intervening*. Cambridge, England: Cambridge University Press.

Shadish, W. R. (1989). Critical multiplism: A research strategy and its attendant tactics. In L. Sechrest, H. Freeman, & A. Mulley (Eds.), *Health services research methodology: A focus on AIDS* (pp. 5–28) (DHHS Publication No. PHS 89–3439). Rockville, MD: National Center for Health Services Research and Health Care Technology Assessment.

Turner, S. (1986). *The search for a methodology of social science*. Dordrecht, The Netherlands: Reidel.

PART IV

For Further Reading

CHAPTER 17

A Guide to the Philosophy and Sociology of Science for Social Psychology of Science

STEVE FULLER

Epistemology and Philosophy of Science

In contemporary analytical philosophy, epistemology and the philosophy of science are two distinct fields practiced, for the most part, by two distinct groups of people. In broadest terms, epistemologists want to know when an individual's belief ought to count as knowledge, whereas philosophers of science want to know when a theory ought to be accepted by a community of researchers. Interestingly, epistemologists tend to treat common sense and science as making equally valid (or invalid, if one is a skeptic) claims to knowledge, while philosophers of science presume that science is superior to common sense and maybe even good enough to defeat the skeptic. As a result, epistemology often appears to be a static enterprise, one devoted to elaborating timeless criteria for knowledge, in contrast to the more dynamically oriented philosophy of science, which is full of arguments about one theory's replacing another. It is worth noting that this dynamism applies just as much to the positivist preoccupation with reductionism (Nagel, 1960/1987, remains the *summa* of this inquiry) as to the post-Kuhnian focus on paradigm shifts and conceptual change. A quick refresher course on the classics in recent philosophy of science—Popper, Kuhn, Lakatos, Feyerabend, Shapere, Laudan—is *Scientific Revolutions* (Hacking, 1981) and the first half of *Representing and Intervening* (Hacking, 1983). Of special interest to psychologists is "A Function for Thought Experiments" (Kuhn, 1981), in which Kuhn explicitly tries to explain paradigm shifts in Piagetian–Gestaltist terms.

With a few notable exceptions (e.g., Hacking among philosophers of science and Richard Rorty [1979] among epistemologists), epistemologists and philosophers of science are united in a common normative interest—in what people *ought* to believe—which has classically been treated as an a priori question, one that must ultimately be defended on the basis of conceptual arguments alone, no matter what the actual empirical record of science might suggest. The move to "naturalize" epistemology and the philosophy of science involves injecting facts about our actual epistemic situation into these normative considerations. In practice, this move tends not to be as radical as it seems. (Kornblith [1985] captures much of the scene.) As a matter of fact, naturalism was strongest early in the 20th century (especially among such pragmatists as Peirce, Dewey, and Mead), when philosophers had no qualms about evaluating science on the basis of its social consequences. But given the social disruptions for which science has been held accountable in the second half of the century, it is perhaps not surprising that philosophers now draw a sharp line between a scientific theory and its applications. Consequently, the more modest naturalist of today will either use an a priori conception of knowledge to determine which historical or psychological facts are relevant to epistemology (e.g., Goldman, 1986) or he/she will use such facts to determine the range of possible norms that warrant further conceptual investigation (e.g., Laudan, 1987). In both cases, naturalism only cuts down the possibility space in which the philosopher thinks about norms for knowledge; it does not dictate any specific solutions.

A more radical form of naturalism argues that epistemology and the philosophy of science are continuous with science itself. In effect, these two philosophical disciplines are really part of the "science of science." The first move in this direction is to deny any hard distinction between "empirical" and "conceptual" issues. On this view, which is mostly closely associated with Quine (1951/1953), a conceptual issue, such as a logical truth or a fundamental scientific principle, is simply one that we would not give up unless we first saw a wide variety of empirical cases in which it did not work. Quine (1969) himself coined "naturalized epistemology" in an essay by the same title. He argues that the positivist distinction between contexts of "discovery" and "justification" can be naturalistically translated into questions of, respectively, perceptual psychology and sociology of knowledge. The former is studied by looking at people's behavioral dispositions to link words to situations, whereas the latter is studied by looking at the communication patterns by which such dispositions are stabilized for a community.

Naturalized epistemology in this more robust sense has come a long way since Quine, although it is still a decidedly minority voice in philosophy. Because Quine was strongly influenced by his Harvard colleague Skinner, he never considered *mechanisms* of belief fixation and change. This side of the naturalistic project has been most thoroughly explored by Campbell's (1988) evolutionary inspired "selective retention" model, which tends to assign the tasks of studying discovery (i.e., "variation") and justification (i.e., "selection") to psychology and sociology, respectively. Toulmin (1972) and Hull (1988) provide an interesting contrast in how the evolutionary model can be deployed. Although Toulmin occasionally alludes to the social and psychological character of science, his major arguments for the evolutionary model are drawn almost entirely from the internal history of science, where personalities and institutions are largely absent. By contrast, Hull portrays the evolution of systematic zoology as an emergent feature of the interpersonal conflicts of scientists he has actually worked with. Although both Toulmin and Hull rely heavily on biological theory to guide their use of the evolutionary model, they take almost nothing from the social sciences literature when it comes to linking the model up to the actual practices of scientists. Instead, they fall back on folk notions of why people do what they do.

Radicalizing the naturalized study of knowledge, Giere (1988) and Fuller (1993) take the social sciences literature seriously, and are quite explicit about their break with the classical tradition. Giere tends to treat the social psychology of science as the aggregation of the individual (specifically cognitive) psychologies of scientists, while Fuller emphasizes the emergent, although still largely cognitive, character of the social. Giere and Fuller are unique in the amount of space they devote to how the naturalist should go about studying science: Giere opts for the historical method, while Fuller makes a special pitch for experiments. Both are concerned with testing normative theories that philosophers have proposed about science (e.g., falsificationism). The most extended historical test has been conducted by Laudan and his associates (1986), as discussed in a special issue of *Synthese*. Various experimental tests are presented in *On Scientific Thinking* (Tweney, Doherty, & Mynatt, 1981), which provides some of the implicit psychology that guided philosophers' original normative proposals.

Social Studies of Science

Although nowadays it is natural to think of the sociology of science as a branch of the sociology of knowledge, this is a development that is due largely to Robert Merton (1977), whose *Sociology of Science* collects his 30 years of essays on the topic, including some historical pieces on the founder of the sociology of knowledge, Karl Mannheim. Since the sociology of knowledge relativizes the justification of knowledge claims to the cultures that sustain them, Mannheim argued that science, because of its universal truth claims, could not itself be an object of sociological study. Otherwise, the sociology of knowledge as a scientific disipline would be reduced to ideology. Against this, Merton argued that even the search for truth requires a specific kind of social order, which he captured in terms of norms that reflected what philosophers have traditionally taken to be the virtues of the scientific method. In this first phase, which lasted well into the 1970s, the sociology of science was done primarily by Americans, who presumed that science's superior epistemological credentials enabled it to play a stabilizing role in modern democracies. Because sociologists at that time were willing to leave the analysis of those credentials to philosophers, they were often said to be uninterested in the "content" of science.

The second phase of the sociology of science took off from the concern that epistemology and the philosophy of science were too important to leave to philosophers, especially given their tendency to see science as a self-directed activity, one insulated from ambient social factors. Bloor (1976) exemplified this trend, which led to the establishment of the "Strong Programme in the Sociology of Knowledge." What made the program strong was its commitment to demonstrating that the very content of scientific knowledge is best explained in terms of the interests of influential groups in the society at large. In fleshing out the details of the Strong Programme, Bloor and his associates at the Edinburgh University Science Studies Unit have tended to rely more on anthropology than on sociology proper for their explanatory models. Bloor (1983) contains many such anthropological appeals in the course of explaining disciplinary boundary formation in early experimental psychology, the management of anomalies to mathematical proofs, and the debates over the definition of community in contemporary social theory. The basic move here is to portray knowledge claims as symbols standing for and manipulated by competing groups in a highly structued agonistic field. The most sustained historical treatment in this vein is *Leviathan and the Air Pump* (Shapin & Schaffer, 1985), which analyzes the social ascendency of experimental knowledge in 17th-century England. An equally sustained Strong Programme treatment of recent science is presented by Pickering (1984).

For all their differences, Merton and Bloor agreed that, in some sense, the science and society of a given period are reflections of one another. Whereas Merton stressed the way in which society is organized to facilitate scientific inquiry, Bloor highlighted the extent of which scientific disputes are social struggles in symbolic disguise. However, almost immediately after the publicatation of Bloor's first book, this reflection thesis was severly challenged from two radically opposing sides. On the one hand, Laudan (1977) argued that Mannheim was right after all: Rational methodology was a sufficient explanation for good science and the sociology of knowledge became relevant only once the epistemic limitations (and hence relative validity) of a particular set of beliefs were revealed. On the other hand, some sociologists started doing ethnographies of science as it was actually being done in laboratories, only to discover a massive disparity between the words and deeds of scientists. The words, as they appeared in journal articles, largely conformed to philosophical canons of rational methodology, but they also represented a highly idealized—if not downright misleading—picture of what took place in the laboratory. The laboratory work itself turned out to be quite chaotic and open ended, even at the level of personal interests and group understanding of the ends of their research. The two major monographs in this vein are by Latour and Woolgar (1979/1986) and Knorr-Cetina (1981). Collins (1985) presents a series of studies that challenge particular method–practice disparities, while Gilbert and Mulkay (1983) offer a guide to the analytical tools used in ethnographic studies of science.

This schismic third phase of the social studies of science has continued unabated, and perhaps has become even more pronounced. Laudan's revival of Mannheim's perspective has recently emboldened some cognitive scientists to argue that the mark of good science is that it can be done on a computer with no social mediation, since the cognitive processes required for such science are purely formal (see "Computer Discovery," 1989; Fuller, DeMey, Shinn, & Woolgar, 1989). But against this move to disembody science, the more radical of the ethnographers have begun to stress the "reflexive" character of sociological inquiry into science, especially the way in which the identities of scientist and ethnographer are jointly constructed in the course of the latter writing about the former. Woolgar (1988b) introduces science studies that culminate in this perspective. *Knowledge and Reflexivity* (Woolgar, 1988a) probes the literary implications of such heightened methodological self-consciousness.

Is there a chance for synthesis, or at least general theories, in the social studies of science? Latour has worked hard to catapult the idea of scientists as Machiavellian "actor–networks" into a general world view, as articulated in the popular *Science in Action* (1987) and the more scholarly *The Pasteurization of France* (1988). Fuller (1988) lays out the terms for a rapprochement between rationalist philosophy of science and a robust sociology of knowledge, which he expects will be played out in matters of science policy. Anthologies have also recently come to the aid of integration. *Science as Practice and Culture* (Pickering, 1992) contains representatives of most of the positions sketched above, as well as feminist critics of science, all of whom have been forced to write about the views of some of the others. This book will probably replace *Science in Context* (Barnes & Edge, 1982), which nevertheless still contains the best social studies of science bibliography. Finally, correlative movements in the social studies of technology are well covered in *The Social Construction of Technological Systems* (Bijker, Hughes, & Pinch, 1987).

References

Barnes, B., & Edge, D. (Eds.). (1982). *Science in context*. Cambridge, MA: MIT Press.

Bijker, W. E., Hughes, T. P., & Pinch, T. J. (Eds.). (1987). *The social construction of technological systems*. Cambridge, MA: MIT Press.

Bloor, D. (1976). *Knowledge and social imagery*. London: Routledge.

Bloor, D. (1983). *Wittgenstein: A social theory of knowledge*. New York: Columbia University Press.

Collins, H. M. (1985). *Changing order: Replication and induction in scientific practice*. Newbury Park, CA: Sage.

Computer discovery and the sociology of scientific knowledge [Special issue] (1989, November). *Social Studies of Science, 19*, 563–696.

Fuller, S. (1988). *Social epistemology*. Bloomington: Indiana University Press.

Fuller, S. (1993). *Philosophy of science and its discontents* (2nd ed.). New York: Guilford Press.

Fuller, S., DeMey, M., Shinn, T., & Woolgar, S. (Eds.). (1989). *The cognitive turn: Sociological and psychological perspectives on science*. Dordrecht, The Netherlands: Kluwer.

Giere, R. (1988). *Explaining science*. Chicago: University of Chicago Press.

Gilbert, N., & Mulkay, M. (1983). *Opening Pandora's box: A sociological analysis of scientists' discourse*. New York: Cambridge University Press.

Goldman, A. (1986). *Epistemology and cognition*. Cambridge, MA: Harvard University Press.

Hacking, I. (Ed.). (1981). *Scientific revolutions*. Oxford, England: Oxford University Press.

Hacking, I. (1983). *Representing and intervening*. Cambridge, England: Cambridge University Press.

Hull, D. (1988). *Science as a process*. Chicago: University of Chicago Press.

Knorr-Cetina, K. (1981). *The manufacture of knowledge: An essay on the constructivist and contextual nature of science*. Oxford, England: Pergamon Press.

Kornblith, H. (Ed.). (1985). *Naturalizing epistemology*. Cambridge, MA: MIT Press.

Kuhn, T. S. (1981). A function for thought experiments. In I. Hacking (Ed.), *Scientific revolutions* (pp. 6–27). Oxford, England: Oxford University Press.

Latour, B. (1987). *Science in action*. Cambridge, MA: Harvard University Press.

Latour, B. (1988). *The pasteurization of France.* Cambridge, MA: Harvard University Press.

Latour, B., & Woolgar, S. (1986). *Laboratory life: The social construction of scientific facts* (2nd ed.). Princeton, NJ: Princeton University Press. (Original work published 1979)

Laudan, L. (1977). *Progress and its problems.* Berkeley: University of California Press.

Laudan, L., Donovan, A., Laudan, R., Barker, P., Brown, H., Ceplin, J., Thagard, P., & Wykstra, S. (1986). *Synthese,* 69:141–223.

Laudan, L. (1987). Progress of rationality? *American Philosophical Quarterly, 24,* 19–31.

Merton, R. (1977). *Sociology of science.* Chicago: University of Chicago Press.

Nagel, E. (1987). *The structure of science.* (2nd ed.). Indianapolis, IN: Hackett. (Original work published 1960)

Overman, E. S. (Ed.). (1988). *Methodology and epistmeology for social science: Selected papers.* Chicago: University of Chicago Press.

Pickering, A. (1984). *Constructing quarks: A sociological history of particle physics.* Chicago: University of Chicago.

Pickering, A. (Ed.). (1992). *Science as practice and culture.* Chicago: University of Chicago Press.

Quine, W. V. (1953). Two dogmas of empiricism. In W. V. Quine, *From a logical point of view* (pp. 26–54). New York: Harper. (Original work published 1951)

Quine, W. V. (1969). *Ontological relativity.* New York: Columbia University Press.

Rorty, R. (1979). *Philosophy and the mirror of nature.* Princeton, NJ: Princeton University Press.

Shapin, S., & Schaffer, S. (1985). *Leviathan and the air pump: Hobbes, Boyle, and the experimental life.* Princeton, NJ: Princeton University Press.

Toulmin, S. (1972). *Human understanding.* Princeton, NJ: Princeton University Press.

Tweney, R. D., Doherty, M. E., & Mynatt, C. R. (Eds.). (1981). *On scientific thinking.* New York: Columbia University Press.

Woolgar, S. (Ed.). (1988a). *Knowledge and reflexivity: New frontiers in the sociology of knowledge.* London: Sage.

Woolgar, S. (1988b). *Science: The very idea.* London: Tavistock.

CHAPTER 18

Social Psychology Readings Relevant to Science: An Annotated Reading List

ARIE W. KRUGLANSKI

This is a highly selective, strongly biased reading list designed to inform about aspects of social psychology relevant to the problem of science. I have attempted to reflect state-of-the-art concerns rather than historical classics in the hope (perhaps somewhat naive) that recency represents progress and that the impactful classics are anyway referred to as points of departure.

Bar-Tal, D., & Kruglanski, A. W. (Eds.). (1988). *The social psychology of knowledge*. Cambridge, England: Cambridge University Press.

An anthology containing a spectrum of recent social psychological contributions to epistemological issues. Most contributors are major figures in the social cognition area, but chapters representative of alternative perspectives are also included, including Gibsonian neorealism (Baron), social constructionism (Gergen), and the history of psychology (Graumann).

Brown, R. (1986). *Social psychology: The second edition*. New York: Free Press.

Very well written, fairly up-to-date, if somewhat selective, coverage of some of the major areas of social psychology. Among the important topics that have been left out are social cognition, group-research, and research methods. On the other hand, the chapter on social psychology of language is excellent. Very readable, enjoyable volume by one of the premier scholars in the field.

Fiske, S. T., & Taylor, S. E. (1991). *Social cognition* (2nd ed.). New York: McGraw-Hill.

The most authoritative text to date on the social cognition approach to social psychology. Features chapters on attribution theory, social inference, person memory, attentional factors in social perception, and attitudes. A useful sourcebook that describes the main ideas and research findings in the social cognition domain. It intended to offer a broad survey rather than provide a critical analysis or a unifying point of view.

Hendrick, C. (1987a). *Group processes*. Newbury Park, CA: Sage.

Hendrick, C. (1987b). *Group processes and intergroup relations*. Newbury Park, CA: Sage.

These two volumes contain a representative, up-to-date, selection of current social psychological research on group processes. Some of the topics covered are minority influence, small-group formation, leadership processes, social loafing, social facilitation, and the interface of cognitive and group phenomena. A good introduction to issues, methods, and research figures currently influential in the groups area.

Higgins, E. T., & Bargh, J. A. (1987). *Social cognition and social perception. Annual Review of Psychology, 38*, 369–425.

Focuses on the main recently active areas of social cognition from the perspective of two of the major contributors to the social cognition "movement." A useful reading for catching up with loci of current research action in social cognition. Highly informative about the major problem areas being presently tackled, and about the substantive and methodological approaches adopted in dealing with those problems.

Janis, I. L. (1982). *Groupthink* (2nd ed.). Boston: Houghton Mifflin.

A somewhat dated, yet still highly interesting account of pressures to uniformity in decision-making groups. At this point, its descriptive value exceeds that of its theoretical insights, which are relatively unsophisticated given recent developments in group theory (in particular, minority influence research), as well as the persuasion and attitudes literature that has seen some important advances in recent years.

Kruglanski, A. W. (1989a). *Lay epistemics and human knowledge: Cognitive and motivational bases*. New York: Plenum Press.

An integration of several major areas of social cognition research (attitudes, attribution, dissonance, social comparison processes) from the standpoint of a unified theory of lay knowledge acquisition. Contains a specific comparison of lay and scientific epistemologies (Chapter 10) and outlines several possible research directions on epistemic phenomena based on the motivational–cognitive approach to knowledge featured in the volume.

Kruglanski, A. W. (1989b). The psychology of being "right": The issue of accuracy in social cognition and perception. *Psychological Bulletin, 106*, 395–405.

A review and a theoretical integration of recent social psychological views of accuracy and error in lay knowledge. Includes a discussion of the Nisbett–Ross–Tversky-Kahneman approach as well as the reactions it generated among social cognition researchers.

Levine, J. M., & Moreland, R. L. (1990). Progress in small group research. *Annual Review of Psychology, 41*, 585–634.

An intelligent and scholarly review of the last 10 years of small-group research. An excellent introduction to issues, approaches, and findings. It turns out that a great deal of the exciting research on groups is carried out these days in industrial and organizational rather than social psychology.

Nisbett, R., & Ross, L. (1980). *Human inference: Strategies and shortcomings of social judgment*. Englewood Cliffs, NJ: Prentice Hall.

Influenced strongly by the work of Amos Tversky and Daniel Kahneman on cognitive heuristics, this strong volume takes the position that human judgment often deviates from "normative" modes of reasoning, represented e.g., via statistics or formal logic. The more fundamental philosophical issue of whether the "normative" modes represent a reasonable or a probable avenue to valid knowledge is not addressed in depth. Intruiguing empirical material highlighting the "irrationality" of lay knowers. Note that this approach has recently evoked a spate of reactions from social cognitive psychologists (for a review, see, e.g. Kruglanski, 1989b).

Paulus, P. B. (Ed.). (1989). *Psychology of group influence* (2nd ed.). Hillsdale, NJ: Erlbaum.

A judiciously selected array of papers representing the hottest current work on the social psychology of groups, including the cognitive approach to group phenomena. Note in particular chapters by Clark and Stephenson, Levine, Miller, and Wicklund. State-of-the-art analyses of small-group research.

Rosenthal, R. (1984). *Meta-analytic procedures for the social and behavioral sciences*. Newbury Park, CA: Sage.

An exceedingly useful compendium of statistical techniques for the assessment of cumulative knowledge in the social and psychological sciences. One way of looking at it is as a correction for the pervasive bias in favor of almost exclusive concern with Type I errors (of falsely accepting the null hypothesis) by the elaboration of cumulative methods that may increase the power of our tests and hence minimize Type II errors (accepting the null when in fact it is false). Of fundamental significance in winnowing the "wheat" from the "chaff" in social science findings, or the signal from the noise. Carries the optimistic implication that often our scientific effects are more substantial (compared to those of other sciences) than we have been willing to assume. I have become a convert.

Zajonc, R. B. (1965). Social facilitation. *Science, 149*, 269–274.

An old classic by one of the most interesting innovators in social psychology. To see how this research has developed, see the review chapters in Hendrick (1987a, 1987b).

CHAPTER 19

Social Psychology and Science Studies

WILLIAM LAWLESS

Social psychology is an important research area for science studies practitioners involved in understanding the causes, effects, and limits of social processes. This section offers a review of the literature relevant to several interesting questions on the social context of scientific decision making. First, one can identify two different traditions of social psychology, sociological social psychology (SSP), practiced by sociologists, and psychological social psychology (PSP), practiced by psychologists. The *Handbook of Social Psychology* (1985) gives a good summary of research in both areas. Current technical reviews of specific topics in both fields, as well as topics in general psychology, can be found in *Psychological Bulletin*; the reviews in *American Psychologist* are especially valuable for a general audience.

The sociological field is less scientistic and more open-minded on what counts as an acceptable practice. It prioritizes realistic insights over empirical reliability in research that includes, for example, symbolic interactionism, ethnomethodology, organizations and power, and life-stage studies. A good overview of the SSP field is provided by Rosenberg and Turner (1981).

In contrast, PSP is symbolized by its concentration on the laboratory method. Consequently, the PSP literature (e.g., *Journal of Personality and Social Psychology*) is more technical and less accessible to outsiders. Despite its more technical approach, some aspects of PSP can be a useful tools in the study of science and in the public dissemination of science studies knowledge. House (1981) discusses the different research areas of PSP, including their rigorous methodologies (cf. Campbell, 1989). PSP areas of interest to the science studies practitioners are:

1. Attitude formation and attitude change through communication, persuasion, and cognitive processes;
2. Group dynamics, such as conformity, power structures, group identification, and minority influences (e.g., divergent thinking, deviance, and alliance formation); and
3. Group decision-making processes, such as consensus making, diffusion of responsibility, norms, risky shift-polarization theory, and groupthink.

These topics are briefly described below.

Sociological Social Psychology

1. Can a scientific theory be explained as a bidirectional product of social construction and social structure mediated by reality?

One area that is closely aligned with science studies is the symbolic interactionist study of the social construction of reality (Stryker, 1981). The core question is how dynamic interactions between individuals lead them to define their situations (Thomas & Thomas, 1928). One problem with symbolic interactionism is that its causal account relies entirely on cognitive processes at the expense of emotional factors. Besides its overly cognitive interpretation of behavior, the theory's claims and predictions are difficult to test or modify. However, despite these caveats, symbolic interactionism provides a useful account of the link between social constructions, social structures, individual behavior, and norms. It can be employed to study scientists' behavior in the laboratory and their interaction with their peers and with policymakers.

Psychological Social Psychology

1. What are the effects and limits of social influences on cognitive processes? What are the personality, cognitive, social, and power differences between experts and nonexperts? How does framing of the situation affect choices made? Is framing an error of thinking?

Scientific beliefs can be treated as attitudes ranging from strong commitments to heuristic schemes, reflecting central or peripheral cognitive process (Petty & Cacioppo, 1981) and subject to change. PSP researchers are interested in questions such as the strength with which different attitudes are held, what it takes to change an attitude, and the computational and inferential processes that individuals employ in making decisions. Examples of the latter include attributions, errors, and biases in thinking (e.g., Ross, Amabile, & Steinmetz, 1977; Ross, Lepper, Strack, & Steinmetz, 1977; Tversky & Kahneman, 1981) and judgment. Expert judgement is a particularly important subject (e.g., Dawes, Faust, & Meehl, 1989, on clinicians [for letters and articles in response to Dawes–Faust–Meehl position, see Faust, Meehl, & Dawes, 1990; Faust & Zisken, 1988; Matarazzo, 1990]; Jussim, 1986, on scientists). A good background review of attitudes and cognitive processes, including a review of communication, persuasion, cognitive dissonance, self-perception, and reinforcement processes, is available from Fiske and Taylor (1990; see also Aronson, 1988).

2. How does the group affect individual thinking processes? What causes private and public discourse to diverge? How do reciprocity and other exchange processes figure in the practice of science and the distribution of power and scientific resources?

Group dynamics is one of the most technical areas in PSP. Although less accessible than others in PSP, this area is one of the more applicable for science studies. A fairly accessible overview of group dynamics is Forsyth's (1990) book, which covers areas such as conformity (original papers by Asch, 1955; Sherif, 1935; reviewed and integrated by Deutsch & Gerard, 1955), obedience (Milgram, 1974), and group identification and racism, specifically ingroup–outgroup effects (Tajfel, 1978). Forsyth also reviews minority influence on group dynamics, for example, divergent thinking (Nemeth, Mayseless, Sherman, & Brown, 1990), deviance (Festinger, 1954; Schachter, 1951), and alliance formation (Caplow, 1956). A popular and insightful review of social influence on cognitive and group processes is provided by Cialdini (1988).

3. What are the limits of the self-fulfilling prophecy and experimenter bias? How do minority scientists affect the majority's decision-making process? How can minority influence be maximized? How can innovation and stability be described?

Another area that should have an important impact on science studies is group decision-making processes, especially those that explain consensus making (indirectly covered by Nemeth, 1986; Tindal, Davis, Vollrath, Nagao, & Hinsz, 1990), diffusion of responsibility (Latane & Darley, 1970), norms (for an example of norm processes also relevant to feminist issues, see Mori, Chaiken, & Pliner, 1987; Ickes, Stinson, Bissonnette, & Garcia, 1990), risky shift-polarization theory (Isenberg, 1986; Myers & Lamm, 1976), and groupthink (Janis, 1982).

4. If transformations lead to data interpretations, how are different interpretations reconciled? If interpretations based on the same data can be qualitatively different between individuals, how is this related to the "master problem" and to relativism? What might intersubjectivity contribute to understanding the master problem?

Most of the research cited above is controversial and does not yet reflect closure. To illustrate this aspect of PSP, a more in-depth example of current research from group dynamics that also touches on research in personality, interdependence, and exchange processes is provided. This particular problem has been described by Allport (1968) as the *master problem* in social psychology. In his terms, it is the difference between "common" (shared by a community) and "person" (restricted to the individual) traits. This has also been described by Durkheim (1951) as the difference between the social and individual mind and by Bachelard (1984) as the discontinuity between common and scientific knowledge. The master problem has the status of an "unsolved" problem (Insko et al., 1987). It is especially relevant to the science studies discussions of social construction, knowledge production through consensus, and influences of power relationships.

One possible reason why researchers have failed to make much headway into this problem is that they have used fixed or static explanatory parameters. Static factors include personality and a fixed picture of the environment (Simon, 1990). Both these factors could be treated as dynamic in themselves (e.g., Freud's theory of the internal dynamics of an invariant postoedipal personality and Skinner's theory of a dynamic environment that invariantly selects behavior from passive individuals). However, recent studies such as a personal account of the social construction of depression (Coyne & Downey, 1991) have cast doubts on the validity of a noninteractionist account of human behavior based on static causes. Further, static factors have been unable to satisfactorily account for the master problem, the social construction process, or the relationship between intersubjective phenomena and reality. It is suspected that all these phenomena are causally related.

On the other hand, because of the complexity inherent in dynamic process, few testable theories of interpersonal dynamics exist (Snyder & Ickes, 1985). Lewin's (1951) interdependence theory, which was one of the first accounts of interpersonal dynamics, is premised on a social construction process for individuals and for groups. Lewin theorized that behavior was a function of personality that subjectively transformed the objective world into perceptions mediated by dynamic interdependence processes. This served as a starting point for exchange theory, which later developed into interpersonal theory (Kelley, 1979), but even in its mature form, it remained static (see Kelley & Thibaut, 1978). Despite this fixity, Kelley managed to lay out a blueprint for integrating interpersonal theory and Lewin's dynamic interdependence theory that has been tested and confirmed (Lawless, 1992). Kelley's idea was that social and individual preferences guide the transformation of perceptions of the environment and produce reliable attributions that are intersubjectively dependent on or egotistically independent of attributions made by others.

References

Allport, G. W. (1968). *The person in psychology. Selected essays*. Boston: Beacon Press.

Aronson, E. (1988). *The social animal*. (5th ed.). New York: Freeman.

Asch, S. E. (1955). Opinions and social pressure. *Scientific American, 193*, 31–35.

Bachelard, G. (1984). *The new scientific spirit* (A. Goldhammer, Trans.). Boston: Beacon Press. (Original work published 1934)

Campbell, D. T. (1989). Fragments of the fragile history of psychological epistemology and theory of science. In B. Gholson, W. R. Shadish, Jr., R. A. Neimeyer, & A. C. Houts (Eds.), *Psychology of science: Contributions to metascience* (pp. 21–46). Cambridge, England: Cambridge University Press.

Caplow, T. (1956). Coalition formation. *American Sociological Review, 21*, 489–493.

Cialdini, R. B. (1988). *Influence: Science and practice* (2nd ed.). Glenview, IL: Scott Foresman.

Coyne, J. C., & Downey, G. (1991). Social factors and psychopathology: Stress, social support, and coping processes. *Annual Review of Psychology, 42*, 401–425.

Dawes, R. M., Faust, D., & Meehl, P.E. (1989). Clinical versus actuarial judgment. *Science, 243*, 1668–1673.

Deutsch, M. & Gerard, H. (1955). A study of normative and informational social influence on individual judgment. *Journal of Abnormal and Social Psychology, 51*, 629–636.

Durkheim, E. (1951). *Suicide. A study in sociology* (J. A. Spaulding & G. Simpson, Trans.). New York: Free Press.

Faust, D., Meehl, P. E., & Dawes, R. M. (1990). Clinical and actuarial judgment. [Letter to editor in response to Kleinmuntz]. *Science, 247*, 651–652.

Faust, D., & Zizken, J. (1988). Psychiatiric diagnosis. [Letter to editor]. *Science, 242*, 651–652.

Festinger, L. (1954). A theory of social comparison processes. *Human Relations, 7*, 117–140.

Fiske, S. T., & Taylor, S. E. (1990). *Social cognition* (2nd ed.). Reading, MA: Addison-Wesley.

Forsyth, D. R. (1990). *Group dynamics* (2nd ed.). Pacific Grove, CA: Brooks/Cole.

House, J. S. (1981). Social structure and personality. In M. Rosenberg & R. H. Turner (Eds.), *Social psychology, sociological perspectives* (pp. 525–561). New York: Basic Books.

Ickes, W., Stinson, L., Bissonnette, V., & Garcia, S. (1990). Naturalistic social cognition: Empathic accuracy in mixed-sex dyads. *Journal of Personality and Social Psychology, 59*, 730–742.

Insko, C. A., Pinkley, R. L., Hoyle, R. H., Dalton, B., Hong, G., Slim, R. M., Landry, P., Holton, B., Ruffin, P. F., & Thibaut, J. (1987). Individual versus group discontinuity: The role of intergroup contract. *Journal of Experimental Social Psychology, 23*, 250–267.

Isenberg, D. J. (1986). Group polarization: A critical review and meta-analysis. *Journal of Personality and Social Psychology, 50*(6), 1141–1151.

Janis, I. L. (1982). *Groupthink* (2nd ed.). Boston: Houghton Mifflin.

Jussim, L. (1986). Self-fulfilling prophecies: A theoretical and integrative review. *Psychological Review, 93*(4), 429–445.

Kelley, H. H.(1979). *Personal relationships: Their structures and processes*. Hillsdale, NJ: Lawrence Erlbaum.

Kelley, H. H., & Thibaut, J. (1978). *Interpersonal relations: A theory of interdependence.* New York: Wiley.

Latane, B., & Darley, J. M. (1970). *The unresponsive bystander: Why doesn't he help?* New York: Appleton-Century-Crofts.

Lawless, W .F. (1991). *Social support from a Lewinian perspective*. Unpublished Dissertation, Virginia Tech, Blacksburg, VA.

Lawless, W. F. (1992). *Lewinian interdependence, dynamics, and tension systems. An application to social support and game theory.* A dissertation presented to the faculty at Virginia Tech, Blacksburg, VA.

Lewin, K. (1944/1951). *Field theory and social science.* New York: Harper.

Lindzey, G., & Aronson, E. (Eds.). (1985). *Handbook of social psychology* (3rd ed.). Hillsdale, NJ: Erlbaum.

Matarazzo, J. D. (1990). Psychological assessment versus psychological testing. Validation from Binet to the school, clinic, and courtroom. *American Psychologist, 45*, 999–1017.

Milgram, S. (1974). *Obedience to Authority*. NY: Harper & Row.

Mori, D., Chaiken, S., & Pliner, P. (1987). "Eating lightly" and the self-presentation of femininity. *Journal of Personality and Social Psychology, 53*, 693–702.

Myers, D. G., & Lamm, H. (1976). The group polarization phenomena. *Psychological Bulletin, 83*, 602–627.

Nemeth, C. (1986). Differential contributions of majority and minority influence. *Psychological Review, 93*, 1–10.

Nemeth, C., Mayseless, O., Sherman, J., & Brown, Y. (1990). Exposure to dissent and recall of information. *Journal of Personality and Social Psychology, 58*(3), 429–437.

Petty, R. E., & Cacioppo, J. T. (1981). The effects of involvement on responses to argument quality and quantity: Central and peripheral routes to persuasion. *Journal of Personality and Social Psychology, 46*, 69–81.

Rosenberg, M., & Turner, R. H. (Eds.). (1981). *Social Psychology. Sociological Psychology, 46*, 69–81.

Ross, L. D., Amabile, T. M., & Steinmetz, J. L. (1977). Social roles, social control, and biases in social-perception processes. *Journal of Personality and Social Psychology, 35*, 485–494.

Ross, L. D., Lepper, M. R., Strack, F., & Steinmetz, J. L. (1977). Social explanation and social expectations: Effects of reevaluation and hypothesis explanations on subjective likelihood. *Journal of Personality and Social Psychology, 35*, 817–829.

Schachter, S. (1951). Deviation, rejection and communication. *Journal of Personality and Social Psychology, 46*, 190–208.

Sherif, M. (1935). Experiments in norm formation. *Archives of Psychology, 27* (187), 1–60.

Simon, H. A. (1990). Invariants of human behavior. *Annual Review of Psychology, 41*, 1–19.

Snyder, M., & Ickes, W. (1985). Personality and social behavior. In G. Lindzey & E. Aronson (Eds.), *Handbook of social psychology* (3rd ed.) (pp. 883–948). Hillsdale, NJ: Erlbaum.

Stryker, S. (1981). Symbolic interactionism: Themes and variations. In M. Rosenberg & R. H. Turner (Eds.), *Social psychology: Sociological perspectives* (pp. 3–29). New York: Basic Books.

Tajfel, H. (1978). The social psychology of intergroup relations. *Annual Review of Psychology, 33*, 1–40.

Thomas, W. I., & Thomas, D. S. (1928). *The child in America*. New York: Knopf.

Tindal, R. S., Davis, J. H., Vollrath, D. A., Nagao, D. H., & Hinsz, V. B. (1990). Asymmetrical social influence in freely interacting groups: A test of three models. *Journal of Personality and Social Psychology, 58*(3), 438–449.

Tversky A., & Kahneman, D. (1981). The framing of decisions and the psychology of choice. *Science, 211*, 453–458.

Index

D

E

F

Q

R

T

U

V

W